Natural Degradation: Management of Polymer Degradation

Natural Degradation: Management of Polymer Degradation

Editors

Alexandre Vetcher
Alexey Iordanskii

MDPI

Basel • Beijing • Wuhan • Barcelona • Belgrade • Novi Sad • Cluj • Manchester

Editors

Alexandre Vetcher
Scientific and educational center "Nanotechnologies" at the Institute of Biochemical Technology and Nanotechnology (IBTN) of Peoples' Friendship University of Russia n.a. P. Lumumba (RUDN)
Moscow
Russia

Alexey Iordanskii
N. N. Semenov Federal Research Center for Chemical Physics Academy of Science Chemical Physics Academy of Science
Moscow
Russia

Editorial Office
MDPI
St. Alban-Anlage 66
4052 Basel, Switzerland

This is a reprint of articles from the Special Issue published online in the open access journal *Polymers* (ISSN 2073-4360) (available at: www.mdpi.com/journal/polymers/special_issues/3FXG0G80XN).

For citation purposes, cite each article independently as indicated on the article page online and as indicated below:

Lastname, A.A.; Lastname, B.B. Article Title. *Journal Name* **Year**, *Volume Number*, Page Range.

ISBN 978-3-7258-0264-7 (Hbk)
ISBN 978-3-7258-0263-0 (PDF)
doi.org/10.3390/books978-3-7258-0263-0

Contents

Preface

The rapid development of polymer science has brought convenience to people's lives. At the same time, more and more waste is produced from the use of polymer products, which causes harm to the environment. In order to protect the living environment, scholars have researched and developed the recycling technology of waste plastics. On the other hand, many contemporary medicines contain polymers as a component, so a detailed study of the decay of such polymers, at least to their exit from the body, is required.

Alexandre Vetcher and Alexey Iordanskii
Editors

polymers

Article

Biorenewable Oxypropylated Pentane-1,2,5-triol as a Source for Incorporation in Rigid Polyurethane Foams

Georgy Grancharov [1,*], Mariya-Desislava Atanasova [1], Radostina Kalinova [1], Pencho Tuleshkov [1], Petar D. Petrov [1], Maya K. Marinova [2], Martin A. Ravutsov [2] and Svilen P. Simeonov [2,3]

1 Institute of Polymers, Bulgarian Academy of Sciences, Acad. G. Bontchev Str. bl. 103A, 1113 Sofia, Bulgaria; m.atanasova@polymer.bas.bg (M.-D.A.); kalinova@polymer.bas.bg (R.K.); pen.tul@polymer.bas.bg (P.T.); ppetrov@polymer.bas.bg (P.D.P.)

2 Institute of Organic Chemistry with Centre of Phytochemistry, Bulgarian Academy of Sciences, Acad. G. Bontchev Str. bl. 9, 1113 Sofia, Bulgaria; maya.marinova@orgchm.bas.bg (M.K.M.); martin.ravutsov@orgchm.bas.bg (M.A.R.); svilen.simeonov@orgchm.bas.bg (S.P.S.)

3 Research Institute for Medicines (iMed.ULisboa), Faculty of Pharmacy, Universidade de Lisboa, Av. Prof. Gama Pinto, 1649-003 Lisbon, Portugal

* Correspondence: granchar@polymer.bas.bg

Abstract: In this study, as a product from the efficient Achmatowicz rearrangement and mild subsequent hydrogenation–reduction reactions of biorenewable C5 alcohols derived from lignocellulose, pentane-1,2,5-triol was successfully used after oxypropylation in the preparation of rigid polyurethane foams—one of the most important classes of polymeric materials. Despite the broad range of applications, the production of polyurethanes is still highly dependent on petrochemical materials considering the need of renewable raw materials and new process technologies for the production of polyol or isocyanate components as a key point for the sustainable development of polyurethane foams. The synthesized oxypropylated pentane-1,2,5-triol was analyzed using proton NMR spectroscopy, hydroxyl number, and viscosity, whereas the newly obtained foams incorporated with up to 30% biorenewable polyol were characterized using compressive stress, thermogravimetry, dynamic mechanical analysis, and scanning electron microscopy. The modified rigid polyurethanes showed better compressive strength (>400.0 kPa), a comparable thermal degradation range at 325–450 °C, and similar morphological properties to those of commercial polyurethane formulations.

Keywords: pentane-1,2,5-triol; polyurethanes; Achmatowicz rearrangement; biorenewable C5 alcohols; oxypropylation

Citation: Grancharov, G.; Atanasova, M.-D.; Kalinova, R.; Tuleshkov, P.; Petrov, P.D.; Marinova, M.K.; Ravutsov, M.A.; Simeonov, S.P. Biorenewable Oxypropylated Pentane-1,2,5-triol as a Source for Incorporation in Rigid Polyurethane Foams. *Polymers* **2023**, *15*, 4148. https://doi.org/10.3390/polym15204148

Academic Editors: Alexey L. Iordanskii and Alexandre Vetcher

Received: 30 August 2023
Revised: 10 October 2023
Accepted: 16 October 2023
Published: 19 October 2023

1. Introduction

Nowadays, polyurethanes are a key class of polymeric materials and, among them, polyurethane foams (PUFs) correspond to 67% of the total polyurethane consumption [1,2]. PUFs show better thermal insulation properties than other commercially available insulating materials such as mineral wool or expanded polystyrene, making them the first choice for a wide range of applications such as automotive, electronics, furnishing, footwear, packaging, or construction materials [3–6]. Efforts of the polyurethane industry are currently focused on the replacement of petro-based feedstocks with bio-based ones [7–12].

The first attempts for a green and sustainable PUFs were the inclusion of polyols derived from natural sources, such as vegetable oils [13,14], agricultural wastes [15,16], and lignocellulosic biomass [17–19]. Different vegetable oils, such as castor oil [20,21], soybean oil [13,22], palm oil [23,24], rapeseed oil [25,26], tung oil [15,27], mustard seed oil [28,29], or canola oil [30,31], were used as precursors for the preparation of bio-polyols and the corresponding environmentally friendly bio-based PUFs [32,33]. Several studies have reported the preparation of bio-polyols or addition of post-agricultural products from corn stalks [34,35], cotton stalks [36], and wheat stalks [37–39] to develop PUF formulations. Among the bioresources utilized for the production of bio-based polyurethane foams, elastomers, coatings, and adhesives, lignin plays a key role, being the most abundant product on the Earth and a major component of lignocellulose [40–42]. Usually, for the preparation of the desired PUFs, lignin was used as an unmodified additive [43,44], liquefied by hydroxyl-containing compounds [45,46], or converted into bio-polyol via the oxypropylation process [47–49]. The latter case comprises the use of a suitable catalyst, temperature, and pressure during the course of the oxypropylation [50]. Valuable polyurethane structures were obtained on the basis of bio-renewable polyols such as self-healing and antibacterial materials, as well as controlled-release fertilizers [51]. In order to obtain or improve key parameters of bio-based PUFs, some inorganic filling materials such as nano-silica, aluminum silicate, and ferroso-ferric oxide nanoparticles were added into polyurethane formulations to enhance their physical–mechanical [52], dielectric [53], and magnetic properties [54], respectively.

The other major lignocellulose component, hemicellulose, possesses an abundance of xylans and is mainly composed by pentoses, which are used in the production of furfural. Furfural is defined as an easily accessible, cheap, and important bio-based compound, finding application in the synthesis of different chemicals, but its main part during production is usually hydrogenated to the C5 alcohol form—furfuryl alcohol [55,56]. The C5 alcohols are important materials for the industry assuming their application in polyurethane, polyester, polyether, and fuel additive production [57], and they were synthesized via hydrolysis or hydrogenolysis of the furan ring in very harsh conditions [58–60]. The synthesis of pentane-1,2,5-triol (125PTO) is not well studied, since Simeonov et al. reported a high-yielding preparation of 125PTO from furfuryl alcohol [61,62]. Here, the catalytically challenging reactions have been avoided using the Achmatowicz rearrangement product and easily accessible catalysts in mild conditions. Initially, furfuryl alcohol was converted into 6-hydroxy-(2H)-pyran-3(6H)-one (Achmatowicz intermediate) via the Achmatowicz rearrangement reaction in the presence of titanium silicalite, and next, 125PTO was obtained after Pd/C hydrogenation and $NaBH_4$ reduction of the Achmatowicz intermediate. Furthermore, a flow chemistry synthesis based on Ru/C-catalyzed hydrogenation has also been reported [61]. Alternatively, 125PTO can be prepared via the gas-phase hydrogenation reaction catalyzed by modified Ni and/or Pt mesoporous silica catalysts. The conversion and selectivity of the newly prepared catalysts reached approximately 100% based on the intermediate product [62].

In the present study, the preparation of the oxypropylated biorenewable pentane-1,2,5-triol obtained via Achmatowicz rearrangement and batch reduction reactions was investigated. The successful inclusion of that bio-based 125PTO polyol into new polyurethane formulations at quantities up to 30%, and its subsequent influence on the physical–mechanical, thermal, and morphological properties of modified rigid PUFs was also reported.

2. Experiment

2.1. Materials

Catalyst titanium silicalite was synthesized as described previously with some modifications [61]. Rigid PUF insulating system "Elastopor" (BASF) was used for PUF preparation. It includes polyol A component containing polyether polyol, catalyst, surfactant, foam agent, and isocyanate B component consisting of polymeric diphenylmethane 4, 4′-diisocyanate (pMDI—31 wt% of isocyanate groups). Furfuryl alcohol, sodium borohydride ($NaBH_4$), propylene oxide (PrO), pyridine, acetic anhydride, potassium hydroxide, potassium carbonate, magnesium sulfate, hydrochloric acid, acetonitrile, ethyl acetate, hexane, methylene chloride, ethanol, methanol, and isopropanol (Sigma Aldrich–Merck, Darmstadt, Germany) were used as received.

2.2. Instruments and Methods

^{1}H NMR spectra were recorded on a Bruker Avance II apparatus operating at 600 MHz. For the measurements, 12 mg of sample were dissolved in 0.7 mL deuterated solvents (DMSO-d_6, and $CDCl_3$). The viscosity of samples was measured on a Rheometer RheoStress 600 (Thermo Haake, Waltham, MA, USA) using 500 µL drops at 25 °C. The compressive strength and the compressive modulus of elasticity were tested at a 5 mm/min displacement rate of the compression plate on a "TIRA TEST 2300" (Schalkau, Germany) testing machine using ISO 844 [63]. The maximum force inducing a 10% relative strain was determined by decreasing the foam height in relation to the initial height according to the direction of foam growth on 5 × 5 × 5 cm cubes. Thermogravimetric analyses were performed on a TGA4000 (PerkinElmer, Shelton, CT, USA) combined with a gas chromatograph and mass-selective detector. The equipment is supplied with PYRIS6 software (PerkinElmer, Shelton, CT, USA) measuring mass change in inert gas flow (argon) with a controlled-temperature elevation of approximately 10 mg samples. Dynamic Mechanical Analyzer-Q800 (TA Instruments, New Castle, DE, USA) was used for determining the viscoelastic properties of PUFs as a function of temperature on bar samples with dimensions 25 × 5 × 5 mm. The foam structure was analyzed using scanning electron microscope (SEM) Philips 515 (Eindhoven, The Netherlands) with secondary electron image detectors (SEI), acceleration voltage 30 kV, magnification 40,000×, and 5 × 5 mm thin sheets. The hydroxyl number (OH no) of samples was determined using an acylation method with acetic anhydride in pyridine as the medium. An excess of acetic anhydride after hydrolysis and the obtained acetic acid was titrated by standard potassium hydroxide solution and phenolphthalein as an indicator. The foaming process was analyzed in accordance with ASTM D7487-13e [64] through implementation of the PUF cup test and measuring PUF's cream, gel, and tack-free times. The apparent density of foams (the ratio of foam weight to its geometrical volume) was determined for cube-shaped samples with a side length of 50 mm in accordance with ISO 845 [65]. The closed cell content and the water absorption of the obtained PUFs were determined in accordance with standard procedures ISO 4590 and ISO 2896, respectively [66,67]. Dimensional stability of the foams was carried out in the thermostating process of samples at temperatures −25 °C and 100 °C in 48 h. The result of this test included a change in linear dimensions of PUFs in accordance with ISO 2796 [68]. The bio-based mass content of the PUFs was calculated as a percentage of the total mass of samples available in accordance with ISO 16620-4 [69].

2.3. Synthesis of 6-Hydroxy-(2H)-pyran-3(6H)-one

Furfuryl alcohol (20.0 g), distilled before experiment, was dissolved in acetonitrile (200 mL) and then the catalyst TS-1 (2.0 g) was slowly added to the solution. It was followed by dropwise addition of aq. H_2O_2 (37%, 30 mL), and the reaction mixture was heated and stirred at 40 °C for 5 h. The reaction progress was monitored through thin layer chromatography (EtOAc/Hexane = 1:1) following the full consumption of furfuryl alcohol. The crude mixture was filtered and the obtained filtrate was evaporated. Next, it was dissolved in methylene chloride and dried over magnesium sulfate to give 6-hydroxy-(2H)-pyran-3(6H)-one (21.6 g, 94% yield), without the need of further purification. The product was obtained as pale-yellow oil that crystallizes at −12 °C.

^{1}H NMR (600 MHz, $CDCl_3$): d = 6.98 (dd, 1H), 6.18 (d, 1H), 5.64 (d, 1H), 4.58 (d, 1H), 4.15 (d, 1H), 3.80 ppm (s, 1H).

2.4. Synthesis of Pentane-1,2,5-triol

A mixture containing 6-hydroxy-(2H)-pyran-3(6H)-one (12.6 g, 0.11 mol), ethanol (280 mL), and catalyst Pd/C (1.3 g, 10 wt%) was allowed to react in H_2 atmosphere at room temperature for 5 h. The obtained mixture was then filtered through a layer of Celite® and the solvent was evaporated to yield 12.0 g of pale-yellow oil. The crude product was dissolved in methanol (200 mL) and $NaBH_4$ (11.7 g, 0.31 mol) was slowly added in portions at 0 °C. The reaction mixture was warmed to RT and stirred for 24 h. After that, a prepared solution of i-PrOH/HCl was slowly added to the reaction suspension up to pH = 2 in order to decompose the boronic complexes. This was followed by addition of potassium carbonate (saturated solution in MeOH) to neutralize the added acid. The mixture was filtered through a paper filter and nylon membrane filter (pore size 0.45 μm, diameter 47 mm), followed by solvent evaporation and preparation of the desired product as oil (12.1 g, 97% yield).

^{1}H NMR (400 MHz, DMSO-d_6): d = 3.44–3.34 (m, 3H), 3.29–3.22 (m, 2H), 1.56–1.47 (m, 1H), 1.47–1.35 ppm (m, 2H), 1.24–1.15 ppm (m, 1H);

2.5. Treatment of Pentane-1,2,5-triol by Propylene Oxide

Equipped with a magnetic stirrer, thermometer, reflux condenser, and dropping funnel, 4.3 g of biorenewable pentane-1,2,5-triol and 0.13 g of potassium hydroxide (~3.0% by weight) were charged into a glass reactor as a catalyst. The reaction system was heated at 120 °C and propylene oxide was charged into the dropping funnel. The temperature increased to 130 °C and, at this point, PrO (~8.0 mL) was added dropwise until the reaction temperature started to decrease, and the addition of propylene oxide was stopped. The reaction mixture was stirred additionally at 130 °C for 4 h and finally was subjected to vacuum drying (10 mm Hg, 110–115 °C) until a constant mass of the oxypropylated polyol (10.2 g).

^{1}H NMR (400 MHz, DMSO-d_6), d = 5.12–4.80 (m, 3.6H), 3.80–3.10 (m, 13.4H), 1.85–1.18 (m, 4H), 1.04–0.92 ppm (m, 10.9H);

2.6. Preparation of Rigid Polyurethane Foams

The rigid PUFs were prepared by mixing polyol A component and isocyanate B component of the rigid PUF insulating system at a weight ratio 1:1.2 (isocyanate index = 110). After intensive stirring for 10 s, the mixture was poured into a mold of dimensions 180/130/130 mm. Additionally, to initial polyurethane formulation (F-1) without biorenewable polyol, three types of rigid PUFs were prepared in this study, incorporating 10 wt% (F-2), 20 wt% (F-3), and 30 wt% (F-4) of biorenewable polyol through partial replacement of standard petrochemical polyol. Obtained PUFs were conditioned for 24 h at normal conditions after removal from the mold and cut to appropriate size for the different characterizations.

3. Results and Discussion

Furfuryl alcohol, the hemicellulose-based product, was chosen as a starting compound for efficient preparation of pentane-1,2,5-triol. First, it was converted via Achmatowicz rearrangement to Achmatowicz intermediate, 6-hydroxy-(2H)-pyran-3(6H)-one, and the latter yielded the desired 125PTO after hydrogenation via mild batch reduction reaction. The synthesis of 6-hydroxy-(2H)-pyran-3(6H)-one via Achmatowicz rearrangement from furfuryl alcohol was achieved in 94% conversion using the titanium silicalite (TS-1) catalyst and 30% hydrogen peroxide system in acetonitrile at 40 °C (Scheme 1). Next, the conversion of the Achmatowicz intermediate to pentane-1,2,5-triol was performed via batch hydrogenation induced by the Pd/C catalyst, and immediate mild $NaBH_4$ reduction at room temperature resulted in a 97% conversion (Scheme 1). Oxypropylation of the thus-prepared pentane-1,2,5-triol with propylene oxide was obtained in the presence of KOH as the catalyst at 130 °C, yielding bio-based polyol. It was found that pentane-1,2,5-triol and propylene oxide reacted at a ratio of 1:3.0 measured gravimetrically and 1:3.6 determined by ^{1}H NMR data, and therefore, all three hydroxyl groups of 125PTO reacted at least to one equivalent of PrO (Scheme 1).

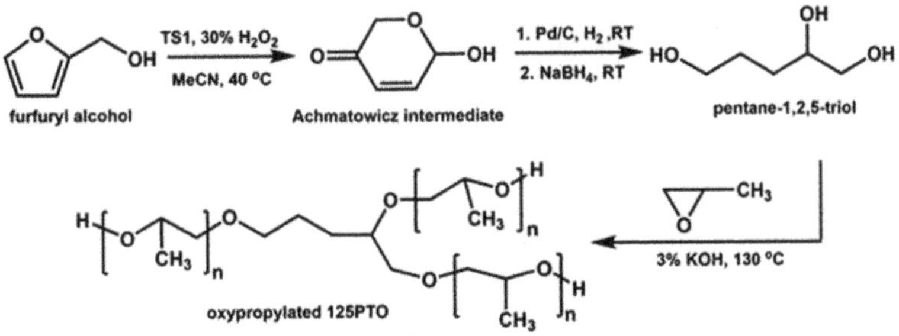

Scheme 1. Synthesis of bio-based polyol from biorenewable hemicellulose product furfuryl alcohol.

The three-step preparation of oxypropylated 125PTO achieved on the basis of biorenewable furfuryl alcohol was followed by ^{1}H NMR spectroscopy. Firstly, hemicellulose derivative C5 furfuryl alcohol was reacted with the titanium silicalite catalyst—a 30% hydrogen peroxide system—in a rearrangement reaction, leading to the Achmatowicz intermediate possessing the ^{1}H NMR spectrum shown in Figure 1a. The high yield conversion of the furfuryl alcohol precursor into the Achmatowicz rearrangement intermediate was confirmed through ^{1}H NMR spectroscopy, revealing sharp signals assigned to protons of the cyclic product containing alkene—6.98 ppm (dd, 1H) and 6.18 ppm (d, 1H); methine—5.64 (d, 1H); methylene—4.58 ppm (d, 1H) and 4.15 ppm (d, 1H); and hydroxyl protons—3.80 ppm (s, 1H). In the second step, an approximately quantitative conversion of the Achmatowicz intermediate to pentane-1,2,5-triol with the ^{1}H NMR spectrum presented in Figure 1b was achieved via combined batch hydrogenation induced by the Pd/C catalyst and immediate mild $NaBH_4$ reduction. ^{1}H NMR spectroscopy data show signals assigned to protons of linear product 125PTO—3.44–3.34 (m, 3H, for -OCH- and -OCH_2-), 3.29–3.22 (m, 2H, for -OCH_2-), 1.56–1.47 (m, 1H, for -CH_2-), 1.47–1.35 ppm (m, 2H, for -CH_2-), and 1.24–1.15 ppm (m, 1H, for -CH_2-). The third step completes the reaction of 125PTO with propylene oxide in the presence of potassium hydroxide as the catalyst. The successful preparation of oxypropylated pentane-1,2,5-triol was also attested via ^{1}H-NMR analysis (Figure 1c) to show signals of protons at 5.12–4.80 (m, 3.6H, for -OCH-), 3.80–3.10 (m, 13.4H, for -OCH- and -OCH_2-), 1.85–1.18 (m, 4H, for -CH_2-), and 1.04–0.92 ppm (m, 10.9H, for -CH_3-).

The properties of the hydroxyl products related to PUF formulations are shown in Table 1.

Table 1. Properties of hydroxyl products related to rigid polyurethane foam system.

Property	Pentane-1,2,5-triol	Oxypropylated 125PTO	Polyol Component
Viscosity (mPa·s)	60	180	430
OH no. (mg KOH/g)	1380 *	410	300

* Theoretically calculated value.

The treatment of pentane-1,2,5-triol by propylene oxide reduces the hydroxyl number to 410 mg KOH/g and increases the viscosity to 180 mPa·s of bio-based polyol. This helps the newly obtained 125PTO polyol to approach properties similar to those of the initial polyol component used for the preparation of rigid PUFs and, in addition, to fulfil the requirements that a given polyol should possess when used in rigid polyurethane foam formulations—a hydroxyl number between 300 and 800, and a viscosity below 300 Pa·s [47,48].

The renewable 125PTO-based polyol was introduced into the rigid PUF compositions in quantities up to 30 wt%, replacing the original system's polyol component. Initially, the oxypropylated product was added to the polyol A component of the polyurethane system and thoroughly mixed. Next, an isocyanate-containing B component of the rigid PUF insulating system was included to such a prepared polyol mixture. After short intensive stirring, the whole mixture was poured into a mold where polyurethane foam started to rise and form (Scheme 2).

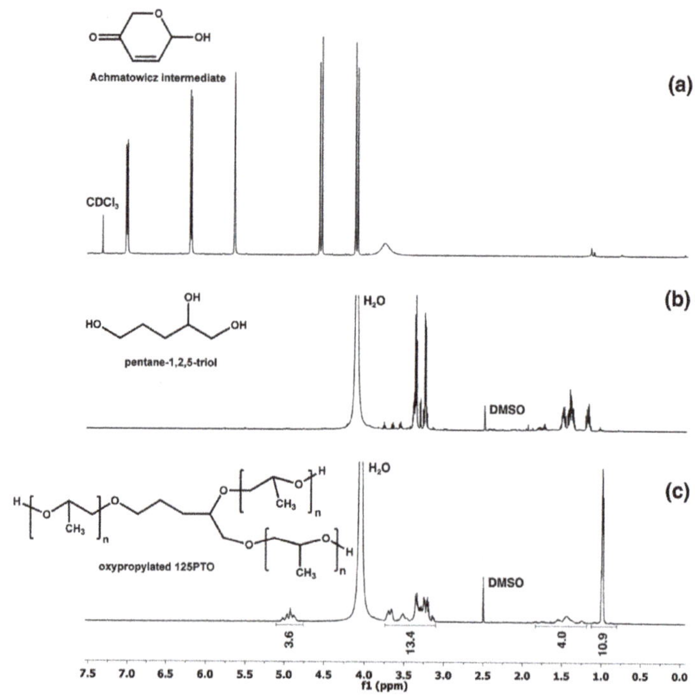

Figure 1. ^{1}H NMR spectrum of the products synthesized during preparation of biorenewable oxypropylated pentane-1,2,5-triol: (**a**) 6-hydroxy-(2H)-pyran-3(6H)-one (Achmatowicz intermediate), (**b**) pentane-1,2,5-triol, and (**c**) oxypropylated pentane-1,2,5-triol.

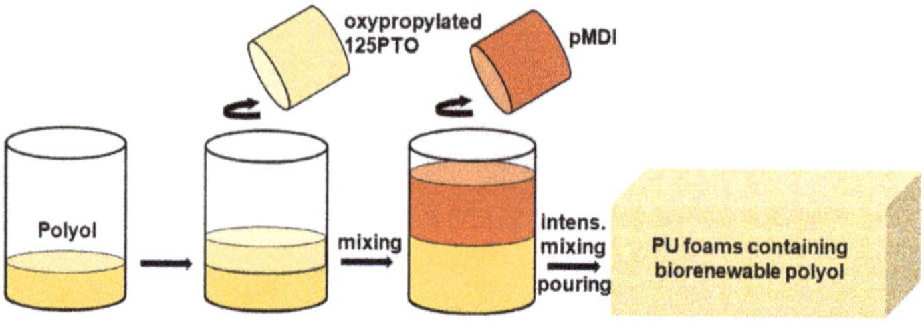

Scheme 2. Preparation of rigid polyurethane foams containing biorenewable oxypropylated pentane-1,2,5-triol.

The investigation of the reactivity of the new insulating PUFs shows that the cream time, gel time, and tack-free time are similar to those of the standard system (Table 2). The density of the new rigid PUFs increases with the amount of added 125PTO polyol, whereas the percentage of the closed cell content decreases with the quantity of incorporated biorenewable polyol (Table 2). The bio-based mass content of rigid PUFs with 10, 20, and 30% of added bio-based polyol is enhanced to 1.95, 3.90, and 5.85%, respectively (Table 2).

Table 2. Characteristic data, reaction times, and properties of obtained rigid polyurethane foams.

Characteristic Data	F-1 (0%)	F-2 (10%)	F-3 (20%)	F-4 (30%)
Polyol (g)	10.0	9.0	8.0	7.0
pMDI (g)	12.0	12.0	12.0	12.0
Oxypropylated 125PTO (g)	0	1.0	2.0	3.0
wt (%) (based on polyol component)	0	10	20	30
wt (%) bio-based content (based on PU foam)	0	1.95	3.90	5.85
Reaction time at 25 °C cup test				
- Cream time (s)	20.2 ± 0.4	19.7 ± 0.4	19.5 ± 0.3	19.7 ± 0.3
- Gel time (s)	133.8 ± 2.5	119.1 ± 2.3	112.8 ± 2.2	116.5 ± 2.1
- Tack-free time (s)	185.6 ± 2.6	203.5 ± 2.8	198.7 ± 2.7	186.8 ± 2.5
Apparent density (kg/m^3)	52.8 ± 1.5	56.0 ± 1.7	59.1 ± 1.8	64.3 ± 2.0
Closed cell content (%)	97.4 ± 1.3	97.2 ± 1.4	93.0 ± 1.4	90.4 ± 1.5

The presence of the oxypropylated bio-based product in the rigid polyurethane compositions causes a favorable effect on their physical–mechanical properties as well. The compressive strength at 10% relative deformation increases from 389.9 kPa for the standard composition (F-1) to 442.4, 409.9, and 474.9 kPa for rigid PUFs with 10, 20, and 30% of added bio-based polyol (F-2, F-3, and F-4), respectively, as shown in Figure 2. A similar relationship can be observed for the modulus of elasticity at compressive strength. It increases from 7.78 MPa for composition F-1 to 8.83, 8.21, and 9.47 MPa for compositions F-2, F-3, and F-4, respectively (Figure 2). The obtained results show that biorenewable 125PTO polyol can be successfully implemented in the preparation of new rigid PUF insulations with improved exploitation characteristics such as higher density and better mechanical properties.

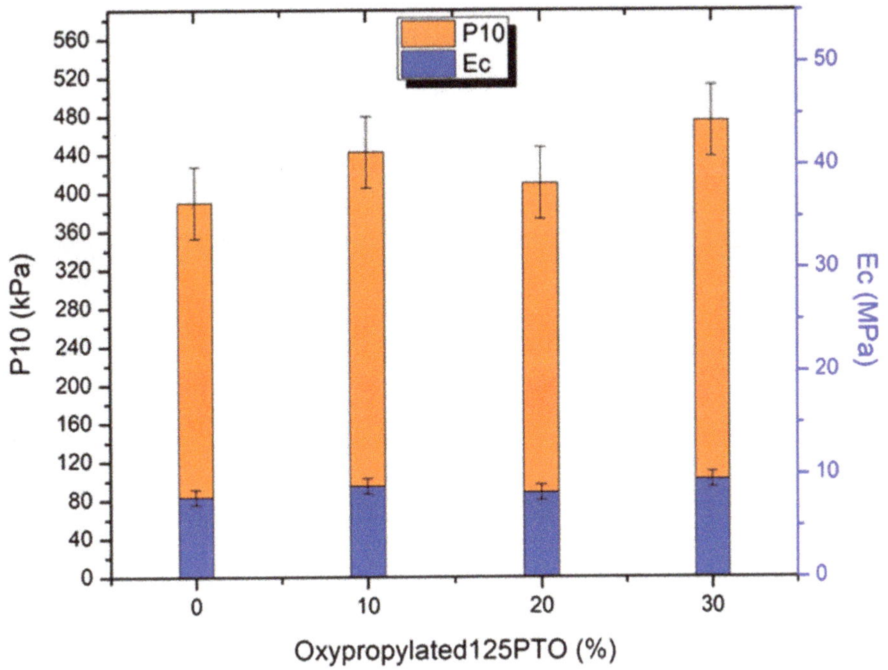

Figure 2. Influence of the content of oxypropylated renewable product on the compressive strength and the modulus of elasticity at compressive strength of polyurethane foams.

An investigation of dimensional stability of rigid PUFs shows that polyurethane foams containing oxypropylated 125PTO (F-2, F-3, and F-4) possess identical and even better dimensional stability to the standard polyurethane composition F-1 determined at −25 °C and 100 °C for 48 h (Table 3).

Table 3. Dimensional stability of rigid polyurethane foams.

Dimensional Stability (%)	F1 (0%)	F2 (10%)	F3 (20%)	F4 (30%)
48 h, −25 °C				
- Length (%)	2.1 ± 0.4	1.8 ± 0.3	1.1 ± 0.3	0.0 ± 0.0
- Width (%)	2.2 ± 0.5	0.0 ± 0.0	0.9 ± 0.2	0.0 ± 0.0
- Thickness (%)	2.9 ± 0.8	0.9 ± 0.2	1.0 ± 0.3	0.0 ± 0.0
48 h, 100 °C				
- Length (%)	−2.1 ± 0.3	−0.9 ± 0.2	0.0 ± 0.0	−0.4 ± 0.1
- Width (%)	2.6 ± 0.5	1.7 ± 0.4	1.8 ± 0.4	0.6 ± 0.2
- Thickness (%)	0.3 ± 0.1	0.7 ± 0.2	−0.7 ± 0.2	0.0 ± 0.0

Thermal degradation properties of PUFs were studied through thermogravimetric analyses (TGAs). The TGA data in Figure 3 depict that polyurethane foams with added bio-based polyol (F-2, F-3, and F-4), and the standard polyurethane foam (F-1) possess no significant difference in their thermal stability. The first weight loss at about 5% can be observed between 175 and 250 °C. It corresponds to the evaporation of low-molecular-weight products such as water, blowing agent, and some monomers. Most of the PUFs' chemical bonds have not begun to break up in this stage. The major decomposition step for all samples indicating weight loss at about 75% starts at 275 °C and finishes at 475 °C. Here, the first stage at 275–375 °C is mainly connected with the initial degradation of urethane, urea, and isocyanurate bonds [70]. In the second degradation stage at 375–475 °C, the decomposition of urethane bonds continues and accomplishes a dissociation of polyol segments [71,72]. The char residue at the end of the degradation process is 15.0% for standard polyurethane composition F-1, whereas it is 18.5, 21.7, and 20.7% for those modified with oxypropylated polyol compositions F-2, F-3, and F-4, respectively.

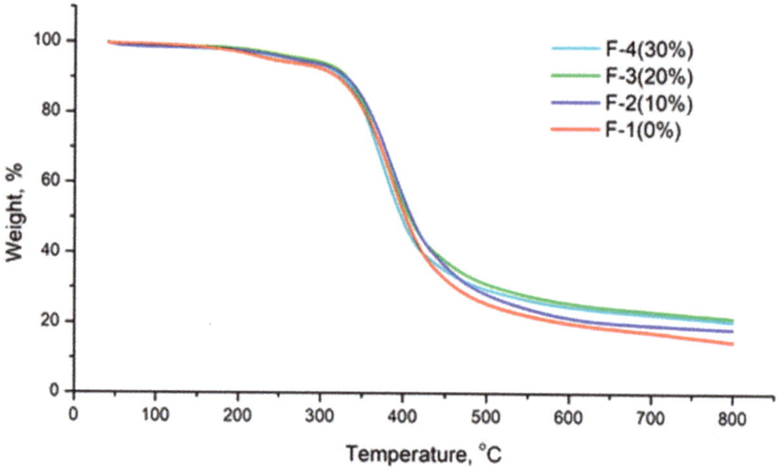

Figure 3. Thermogravimetric analyses of rigid polyurethane foams with added bio-based polyol (F-2, blue line; F-3, green line; and F-4, cyan line) and standard polyol component (F-1, red line).

Further investigation of the thermal properties of PUFs was achieved through dynamic mechanical analysis (DMA), which measures the response of a given material to a cyclic deformation as a function of temperature. Usually, DMA results exhibit three main parameters: (i) the storage modulus, showing the elastic response to the deformation; (ii) the loss modulus, yielding the plastic response to the deformation; and (iii) tan δ, which is the loss-to-storage-modulus ratio, useful for the determination of occurrence of molecular mobility transitions known as the glass transition temperature (Tg). The sample with 10% of biorenewable 125PTO-based polyol shows an average Tg of −57.9 °C and −52.7 °C, whereas the foam with standard polyol exhibits a Tg of −59.9 °C and −67.1 °C, parallel and perpendicular to the growth direction, respectively (Figure 4). The increment in Tg was also observed for the sample with 20% of biorenewable polyol, showing an average Tg of −36.7 °C and −44.8 °C, and for foam with 30% of bio-based polyol, representing an average Tg of −51.0 °C and −35.9 °C, parallel and perpendicular to the growth direction, respectively. Obviously, an increase in the glass transition temperature for modified samples F-2, F-3, and F-4 in both foam growth directions can be seen in the investigated temperature range compared with standard sample F-1. This also indicates a possible increase in branching points, some restricted segmental motions, and a slight decrease in the elastic properties of the modified samples containing renewable polyol [72].

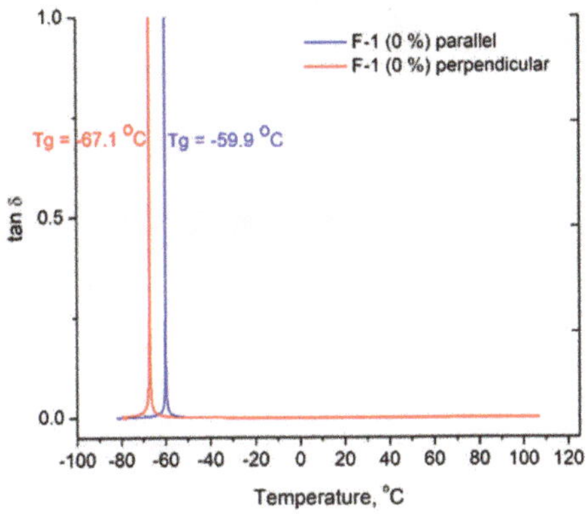

(**a**) F-1 (0% content)

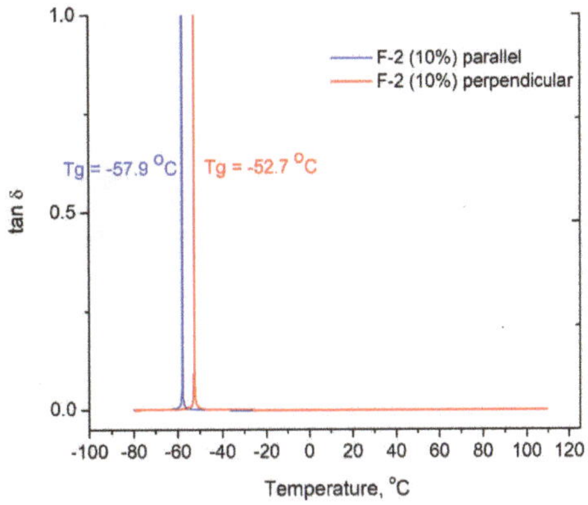

(**b**) F-2 (10% content)

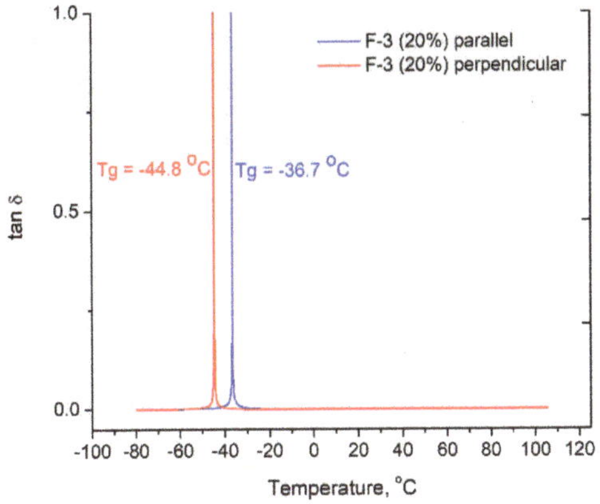

(**c**) F-3 (20% content)

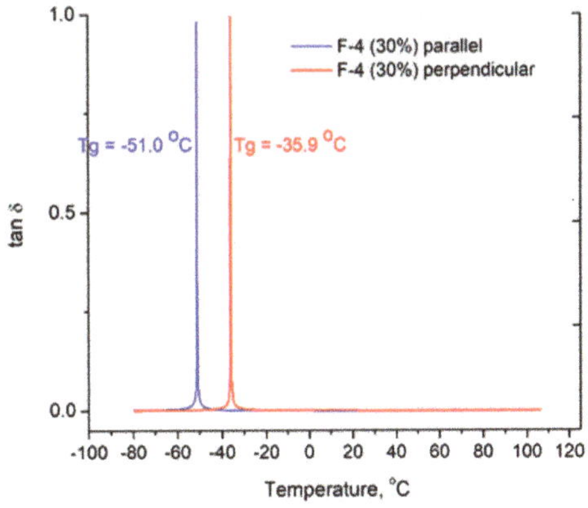

(**d**) F-4 (30% content)

Figure 4. Dynamic mechanical analysis showing tan δ of rigid polyurethane foams with different contents of oxypropylated biorenewable product determined parallel and perpendicular to foam growth direction.

The morphological properties of those modified with 125PTO-based polyol PUFs were analyzed using scanning electron microscopy. PUF samples for SEM observation were cut from the perpendicular orientation to the foam growth direction. It can be seen from Figure 5 that the cell size of standard foam (F-1) and the foam with 10% of added biorenewable polyol (F-2) is smaller than that of the foams with 20 and 30% contents of bio-based polyol (F-3 and F-4), respectively. The cell size of foams with a lower content of 125PTO-based polyol varies in the range of 350 ± 50 μm, whereas the cell size of foams with a higher amount of added oxypropylated product is altered in the range of 480 ± 80 μm. Therefore, the use of oxypropylated 125PTO in quantities up to 30% is beneficial for the enhancement in cell size, i.e., thermal insulation ability of PUFs improves even the slight increase in their apparent foam density.

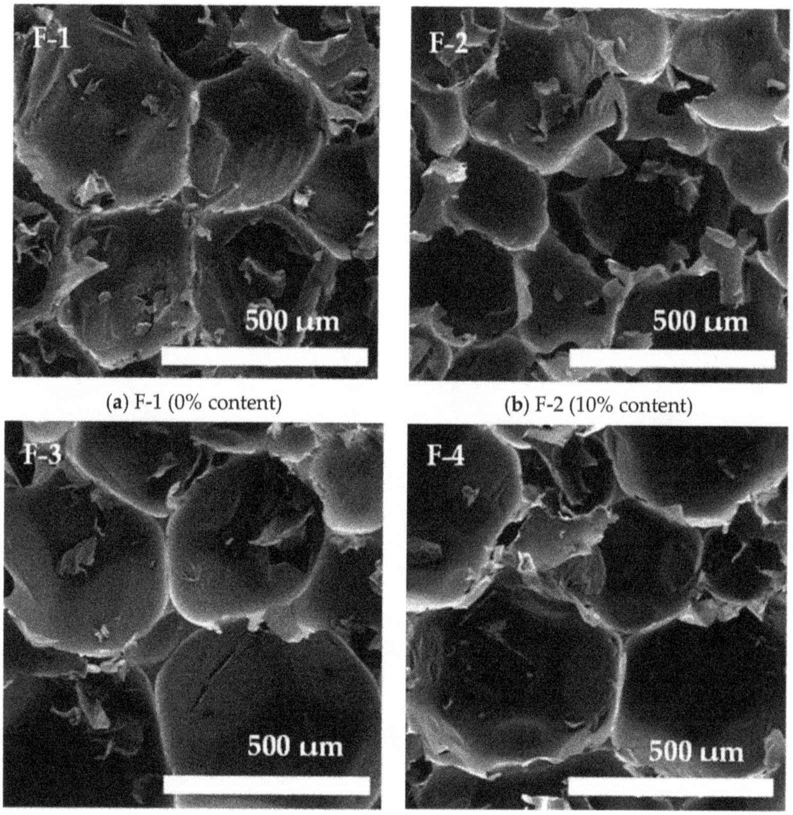

(**a**) F-1 (0% content) (**b**) F-2 (10% content)

(**c**) F-3 (20% content) (**d**) F-4 (30% content)

Figure 5. SEM analysis of rigid polyurethane foams prepared with different contents of oxypropylated bio-based product.

4. Conclusions

Biorenewable oxypropylated pentane-1,2,5-triol was synthesized from furfuryl alcohol via Achmatowicz rearrangement, subsequent batch hydrogenation–reduction reactions, and the terminative oxypropylation process. Bio-based polyol was successively introduced in new rigid polyurethane foam compositions at amounts up to 30% in place of standard polyol. The obtained polyurethane foams containing renewable polyol attained better compressive strength (>400.0 kPa), a comparable thermal degradation range at 325–450 °C, and similar morphological properties to those of the initial petro-based commercial polyurethane formulation. The modified rigid foams can be successfully implemented in the preparation of construction, industrial, and household insulations.

Author Contributions: Conceptualization, G.G., P.T., P.D.P., M.K.M. and M.A.R.; Methodology, G.G., R.K., P.T., P.D.P., M.A.R. and S.P.S.; Validation, G.G., R.K., P.D.P., M.K.M. and S.P.S.; Formal analysis, G.G., M.-D.A., R.K., P.T., P.D.P., M.K.M. and M.A.R.; Investigation, G.G., M.-D.A., R.K., P.T., P.D.P., M.K.M., M.A.R. and S.P.S.; Data curation, M.-D.A., R.K., M.K.M., M.A.R. and S.P.S.; Writing—original draft, G.G.; Writing—review & editing, G.G.; Visualization, M.-D.A.; Supervision, G.G. and S.P.S.; Project administration, S.P.S.; Funding acquisition, S.P.S. All authors have read and agreed to the published version of the manuscript.

Funding: This research was funded by the National Science Fund (Agreement No. КП-06-ОПР01-2/2018).

Institutional Review Board Statement: Not applicable.

Informed Consent Statement: Not applicable.

Data Availability Statement: The data presented in this study are available on request from the corresponding author.

Acknowledgments: This research was supported by the National Science Fund (Agreement No. КП-06-ОПР01-2/2018) for funding. The project leading to this application has received funding from the European Union's Horizon 2020 research and innovation program under grant agreement No. 951996. INFRAMAT project (part of the Bulgarian National Roadmap for Research Infrastructures, supported by the Bulgarian Ministry of Education and Science) provided the research equipment that was used in the investigation.

Conflicts of Interest: The authors declare no conflict of interest.

References

1. Ionescu, M. *Chemistry and Technology of Polyols for Polyurethanes*; Smithers Rapra Publishing: Shrewsbury, UK, 2005.
2. Szycher, M. *Szycher's Handbook of Polyurethanes*, 2nd ed.; CRC Press: New York, NY, USA, 2006.
3. Gama, N.V.; Ferreira, A.; Barros-Timmons, A. Polyurethane foams: Past, present, and future. *Materials* **2018**, *11*, 1841. [CrossRef]
4. Eaves, D. *Handbook of Polymer Foams*; Rapra Technology Ltd.: Shawbury, UK, 2004.
5. Das, S.; Heasman, P.; Ben, T.; Qiu, S. Porous organic materials: Strategic design and structure–function correlation. *Chem. Rev.* **2017**, *117*, 1515–1563. [CrossRef]
6. Ashida, K. *Polyurethane and Related Foams: Chemistry and Technology*; Taylor & Francis Group: Boca Raton, FL, USA, 2007.
7. Simón, D.; Borreguero, A.M.; de Lucas, A.; Rodríguez, J.F. Recycling of polyurethanes from laboratory to industry, a journey towards the sustainability. *Waste Manag.* **2018**, *76*, 147–171. [CrossRef]
8. Furtwengler, P.; Avérous, L. Renewable polyols for advanced polyurethane foams from diverse biomass resources. *Polym. Chem.* **2018**, *9*, 4258–4287. [CrossRef]
9. Chung, H.; Washburn, N.R. Chemistry of lignin-based materials. *Green Mater.* **2013**, *1*, 137–160. [CrossRef]
10. Kazzaz, A.E.; Feizi, Z.H.; Fatehi, P. Grafting strategies for hydroxy groups of lignin for producing materials. *Green Chem.* **2019**, *21*, 5714–5752. [CrossRef]
11. Ma, Y.; Xiao, Y.; Zhao, Y.; Bei, Y.; Hu, L.; Zhou, Y.; Jia, P. Biomass based polyols and biomass based polyurethane materials as a route towards sustainability. *React. Funct. Polym.* **2022**, *175*, 105285. [CrossRef]
12. Kaur, R.; Singh, P.; Tanwar, S.; Varshney, G.; Yadav, S. Assessment of bio-based polyurethanes: Perspective on applications and bio-degradation. *Macromol* **2022**, *2*, 284–314. [CrossRef]
13. Tan, S.; Abraham, T.; Ference, D.; Macosko, C.W. Rigid polyurethane foams from a soybean oil-based polyol. *Polymer* **2011**, *52*, 2840–2846. [CrossRef]
14. Brenes-Granados, D.; Cubero-Sesin, J.M.; Gutiérrez, F.O.; Vega-Baudrit, J.; Gonzalez-Paz, R. Variation of physical properties of rigid polyurethane foams synthesized from renewable sources with different commercial catalysts. *J. Renew. Mat.* **2017**, *5*, 280–289. [CrossRef]
15. Da Silva, V.R.; Mosiewicki, M.A.; Yoshida, M.I.; Da Silva, M.C.; Stefani, P.M.; Marcovich, N.E. Polyurethane foams based on modified tung oil and reinforced with rice husk ash I: Synthesis and physical chemical characterization. *Polym. Test.* **2013**, *32*, 438–445. [CrossRef]
16. Czlonka, S.; Bertino, M.F.; Strzelec, K. Rigid polyurethane foams reinforced with industrial potato protein. *Polym. Test.* **2018**, *68*, 135–145. [CrossRef]
17. Li, Y.; Ren, H.; Ragauskas, A.J. Rigid polyurethane foam/cellulose whisker nanocomposites: Preparation, characterization, and properties. *J. Nanosci. Nanotechnol.* **2011**, *11*, 6904–6911. [CrossRef] [PubMed]
18. Li, Y.; Ragauskas, A.J. Kraft lignin-based rigid polyurethane foam. *J. Wood Chem. Technol.* **2012**, *32*, 210–224. [CrossRef]
19. Saffar, T.; Bouafif, H.; Braghiroli, F.L.; Magdouli, S.; Langlois, A.; Koubaa, A. Production of bio-based polyol from oxypropylated pyrolytic lignin for rigid polyurethane foam application. *Waste Biomass Valorization* **2020**, *11*, 6411–6427. [CrossRef]
20. Tavares, L.B.; Boas, C.V.; Schleder, G.R.; Nacas, A.M.; Rosa, D.S.; Santos, D.J. Bio-based polyurethane prepared from Kraft lignin and modified castor oil. *Express Polym. Lett.* **2016**, *10*, 927–940. [CrossRef]

21. Ionescu, M.; Radojcic, D.; Wan, X.; Shrestha, M.L.; Petrovic, Z.S.; Upshaw, T.A. Highly functional polyols from castor oil for rigid polyurethanes. *Eur. Polym. J.* **2016**, *84*, 736–749. [CrossRef]
22. Miao, S.; Sun, L.; Wang, P.; Liu, R.; Su, Z.; Zhang, S. Soybean oil-based polyurethane networks as candidat biomaterials: Synthesis and biocompatibility. *Eur. J. Lipid Sci. Technol.* **2012**, *114*, 1165–1174. [CrossRef]
23. Adnan, S.; Tuan Noor, M.T.I.; 'Ain, N.H.; Devi, K.P.P.; Mohd, N.S.; Kian, Y.S.; Idris, Z.B.; Campara, I.; Schiffman, C.M.; Pietrzyk, K.; et al. Impact of the hard-segment concentration on highly resilient polyurethane foams based on palm olein polyol. *J. Appl. Polym. Sci.* **2017**, *134*, 45440. [CrossRef]
24. Zhou, X.; Sain, M.M.; Oksman, K. Semi-rigid biopolyurethane foams based on palm-oil polyol and reinforced with cellulose nanocrystals. *Compos. Part A* **2016**, *83*, 56–62. [CrossRef]
25. Kairyte, A.; Vejelis, S. Evaluation of forming mixture composition impact on properties of water blown rigid polyurethane (PUR) foam from rapeseed oil polyol. *Ind. Crops Prod.* **2015**, *66*, 210–215. [CrossRef]
26. Prociak, A.; Kuranska, M.; Cabulis, U.; Kirpluks, M. Rapeseed oil as main component in synthesis of bio-polyurethane-polyisocyanurate porous materials modified with carbon fibers. *Polym. Test.* **2017**, *59*, 478–486. [CrossRef]
27. Da Silva, V.R.; Mosiewicki, M.A.; Yoshida, M.I.; Da Silva, M.C.; Stefani, P.M.; Marcovich, N.E. Polyurethane foams based on modified tung oil and reinforced with rice husk ash II: Mechanical characterization. *Polym. Test.* **2013**, *32*, 665–672. [CrossRef]
28. Borowicz, M.; Paciorek-Sadowska, J.; Lubczak, J.; Czuprynski, B. Biodegradable, flame-retardant, and bio-based rigid polyurethane/polyisocyanurate foams for thermal insulation application. *Polymers* **2019**, *11*, 1816. [CrossRef] [PubMed]
29. Paciorek-Sadowska, J.; Borowicz, M.; Czupryński, B.; Tomaszewska, E.; Liszkowska, J. New bio-polyol based on white mustard seed oil for rigid PUR-PIR foams. *Pol. J. Chem. Technol.* **2018**, *20*, 24–31. [CrossRef]
30. Kong, X.; Liu, G.; Curtis, J.M. Novel polyurethane produced from canola oil based poly(ether ester) polyols: Synthesis, characterization and properties. *Eur. Polym. J.* **2012**, *48*, 2097–2106. [CrossRef]
31. Karimi, M.B.; Khanbabaei, G.; Sadeghi, G.M.M. Unsaturated canola oil-based polyol as effective nucleating agent for polyurethane hard segments. *J. Polym. Res.* **2019**, *26*, 253. [CrossRef]
32. Miao, S.; Wang, P.; Su, Z.; Zhang, S. Vegetable-oil-based polymers as future polymeric biomaterials. *Acta Biomater.* **2014**, *10*, 1692–1704. [CrossRef] [PubMed]
33. Czlonka, S.; Strakowska, A.; Strzelec, K.; Kairyte, A.; Kremensas, A. Bio-based polyurethane composite foams with improved mechanical, thermal, and antibacterial properties. *Materials* **2020**, *13*, 1108. [CrossRef] [PubMed]
34. Yan, Y.; Pang, H.; Yang, X.; Zhang, R.; Liao, B. Preparation and characterization of water-blown polyurethane foams from liquefied cornstalk polyol. *J. Appl. Polym. Sci.* **2008**, *110*, 1099–1111. [CrossRef]
35. Hu, S.; Li, Y. Two-step sequential liquefaction of lignocellulosic biomass by crude glycerol for the production of polyols and polyurethane foams. *Bioresour. Technol.* **2014**, *161*, 410–415. [CrossRef] [PubMed]
36. Wang, Q.; Tuohedi, N. Polyurethane foams and bio-polyols from liquefied cotton stalk agricultural waste. *Sustainability* **2020**, *12*, 4214. [CrossRef]
37. Li, H.; Feng, S.; Yuan, Z.; Wei, Q.; Xu, C. Highly efficient liquefaction of wheat straw for the production of bio-polyols and bio-based polyurethane foams. *Ind. Crops Prod.* **2017**, *109*, 426–433. [CrossRef]
38. Li, H.; Xu, C.; Yuan, Z.; Wei, Q. Synthesis of bio-based polyurethane foams with liquefied wheat straw: Process optimization. *Biomass Bioenergy* **2018**, *111*, 134–140. [CrossRef]
39. Serrano, L.; Rincon, E.; García, A.; Rodríguez, J.; Briones, R. Bio-degradable polyurethane foams produced by liquefied polyol from wheat straw biomass. *Polymers* **2020**, *12*, 2646. [CrossRef]
40. Xu, C.; Ferdosian, F. Conversion of lignin into bio-based chemicals and materials, Ch. 8. Lignin-based polyurethane (PU) resins and foams. In *Green Chemistry and Sustainable Technology*; Springer Nature: Berlin, Germany, 2017; pp. 133–156.
41. Alinejad, M.; Henry, C.; Nikafshar, S.; Gondaliya, A.; Bagheri, S.; Chen, N.; Singh, S.K.; Hodge, D.B.; Nejad, M. Lignin-based polyurethanes: Opportunities for bio-based foams, elastomers, coatings and adhesives. *Polymers* **2019**, *11*, 1202. [CrossRef]
42. Borges da Silva, E.A.; Zabkova, M.; Araujo, J.D.; Cateto, C.A.; Barreiro, M.F.; Belgacem, M.N.; Rodrigues, A.E. An integrated process to produce vanillin and lignin-based polyurethanes from Kraft lignin. *Chem. Eng. Res. Des.* **2009**, *87*, 1276–1292. [CrossRef]
43. Pan, X.; Saddler, J.N. Effect of replacing polyol by organosolv and kraft lignin on the property and structure of rigid polyurethane foam. *Biotechnol. Biofuels* **2013**, *6*, 12. [CrossRef]
44. Huang, X.; De Hoop, C.F.; Xie, J.; Hse, C.Y.; Qi, J.; Hu, T. Characterization of Biobased Polyurethane Foams Employing Lignin Fractionated from Microwave Liquefied Switchgrass. *Int. J. Polym. Sci.* **2017**, *2017*, 4207367. [CrossRef]
45. Bernardini, J.; Cinelli, P.; Anguillesi, I.; Coltelli, M.-B.; Lazzeri, A. Flexible polyurethane foams green production employing lignin or oxypropylated lignin. *Eur. Polym. J.* **2015**, *64*, 147–156. [CrossRef]
46. Muller, L.C.; Marx, S.; Vosloo, H.C.M.J. Polyol preparation by liquefaction of technical lignins in crude glycerol. *Renew. Mater.* **2017**, *5*, 67–80. [CrossRef]
47. Nadji, H.; Bruzzese, C.; Belgacem, M.N.; Benaboura, A.; Gandini, A. Oxypropylation of lignins and preparation of rigid polyurethane foams from the ensuing polyols. *Macromol. Mater. Eng.* **2005**, *290*, 1009–1016. [CrossRef]
48. Cateto, C.A.; Barreiro, M.F.; Rodrigues, A.E.; Belgacem, M.N. Optimization study of lignin oxypropylation in view of the preparation of polyurethane rigid foams. *Ind. Eng. Chem. Res.* **2009**, *48*, 2583–2589. [CrossRef]
49. Cateto, C.A.; Barreiro, M.F.; Ottati, C.; Lopretti, M.; Rodrigues, A.E.; Belgacem, M.N. Lignin-based rigid polyurethane foams with improved biodegradation. *J. Cell. Plast.* **2014**, *50*, 81–95. [CrossRef]

50. Berrima, B.; Mortha, G.; Boufi, S.; El Aloui, E.; Belgacem, M.N. Oxypropylation of soda lignin: Characterization and application in polyurethane foams production. *Cellul. Chem. Technol.* **2016**, *50*, 941–950.
51. Li, H.; Dun, C.; Jariwala, H.; Wang, R.; Cui, P.; Zhang, H.; Dai, Q.; Yang, S.; Zhang, H. Improvement of bio-based polyurethane and its optimal application in controlled release fertilizer. *J. Control. Release* **2022**, *350*, 748–760.
52. Zhang, Q.; Lin, X.; Chen, W.; Zhang, H.; Han, D. Modification of rigid polyurethane foams with the addition of nano-SiO_2 or lignocellulosic biomass. *Polymers* **2020**, *12*, 107. [CrossRef]
53. Baysal, G.; Aydın, H.; Hosgoren, H.; Uzan, S.; Karaer, H. Improvement of synthesis and dielectric properties of polyurethane/Mt-QASs+ (novel synthesis). *J. Polym. Environ.* **2016**, *24*, 139–147. [CrossRef]
54. Moghaddam, S.T.; Naimi-Jamal, M.R. Reinforced magnetic polyurethane rigid (PUR) foam nanocomposites and investigation of thermal, mechanical, and sound absorption properties. *J. Thermoplast. Compos. Mater.* **2019**, *32*, 1224–1241. [CrossRef]
55. Machado, G.; Leon, S.; Santos, F.; Lourega, R.; Dullius, J.; Mollmann, M.E.; Eichler, P. Literature review on furfural production from lignocellulosic biomass. *Nat. Resour.* **2016**, *7*, 115–129. [CrossRef]
56. Nanni, G.; Heredia-Guerrero, J.A.; Paul, U.C.; Dante, S.; Caputo, G.; Canale, C.; Athanassiou, A.; Fragouli, D.; Bayer, I.S. Poly(furfuryl alcohol)-polycaprolactone Blends. *Polymers* **2019**, *11*, 1069. [CrossRef]
57. Sun, D.; Sato, S.; Ueda, W.; Primo, A.; Garcia, H.; Corma, A. Production of C4 and C5 alcohols from biomass-derived materials. *Green Chem.* **2016**, *18*, 2579–2597. [CrossRef]
58. Chen, S.; Wojcieszak, R.; Dumeignil, F.; Marceau, E.; Royer, S. How catalysts and experimental conditions determine the selective hydroconversion of furfural and 5-hydroxymethylfurfural. *Chem. Rev.* **2018**, *118*, 11023–11117. [CrossRef] [PubMed]
59. Nakagawa, Y.; Tamura, M.; Tomishige, K. Catalytic reduction of biomass-derived furanic compounds with hydrogen. *ACS Catal.* **2013**, *3*, 2655–2668. [CrossRef]
60. Nakagawa, Y.; Tamura, M.; Tomishige, K. Catalytic conversions of furfural to pentanediols. *Catal. Surv. Asia* **2015**, *19*, 249–256. [CrossRef]
61. Simeonov, S.P.; Ravutsov, M.A.; Mihovilovic, M.D. Biorefinery via Achmatowicz rearrangement: Synthesis of pentane-1,2,5-triol from furfuryl alcohol. *ChemSusChem* **2019**, *12*, 2748–2754. [CrossRef] [PubMed]
62. Simeonov, S.; Lazarova, H.; Marinova, M.; Popova, M. Achmatowicz rearrangement enables hydrogenolysis-free gas-phase synthesis of pentane-1,2,5-triol from furfuryl alcohol. *Green Chem.* **2019**, *21*, 5657–5664. [CrossRef]
63. *ISO 844:2021*; Rigid Cellular Plastics—Determination of Compression Properties. International Organization for Standardization: Geneva, Switzerland, 2021.
64. *ASTM D7487-13*; Standard Practice for Polyurethane Raw Materials: Polyurethane Foam Cup Test. ASTM International: West Conshohocken, PA, USA, 2016.
65. *ISO 845:2006*; Cellular Plastics and Rubbers—Determination of Apparent Density. International Organization for Standardization: Geneva, Switzerland, 2006.
66. *ISO 4590:2016*; Rigid Cellular Plastics—Determination of the Volume Percentage of Open Cells and of Closed Cells. International Organization for Standardization: Geneva, Switzerland, 2016.
67. *ISO 2896:2001*; Rigid Cellular Plastics—Determination of Water Absorption. International Organization for Standardization: Geneva, Switzerland, 2001.
68. *ISO 2796:1986*; Cellular Plastics, Rigid—Test for Dimensional Stability. International Organization for Standardization: Geneva, Switzerland, 1986.
69. *ISO 16620-4:2016*; Plastics—Biobased Content—Part 4: Determination of Biobased Mass Content. International Organization for Standardization: Geneva, Switzerland, 2016.
70. Jiao, L.; Xiao, H.; Wang, Q.; Sun, J. Thermal degradation characteristics of rigid polyurethane foam and the volatile products analysis with TG-FTIR-MS. *Polym. Degrad. Stab.* **2013**, *98*, 2687–2696. [CrossRef]
71. Septevani, A.A.; Evans, D.A.C.; Chaleat, C.; Martin, D.J.; Annamalai, P. K A systematic study substituting polyether polyol with palm kernel oil based polyester polyol in rigid polyurethane foam. *Ind. Crops Prod.* **2015**, *66*, 16–26. [CrossRef]
72. Gosz, K.; Kowalkowska-Zedler, D.; Haponiuk, J.; Piszczyk, L. Liquefaction of alder wood as the source of renewable and sustainable polyols for preparation of polyurethane resins. *Wood Sci. Technol.* **2020**, *54*, 103–121. [CrossRef]

polymers

MDPI

Review

Recent Advantages on Waste Management in Hydrogen Industry

Alexander V. Shchegolkov [1], Aleksei V. Shchegolkov [2], Natalia V. Zemtsova [3], Yaroslav M. Stanishevskiy [4] and Alexandre A. Vetcher [4,5,*]

1 Institute of Technology of the Department of Technology and Methods of Nanoproducts Manufacturing, Tambov State Technical University, 392000 Tambov, Russia
2 Department of Chemical Technology, Platov South-Russian State Polytechnic University (NPI), 132 Enlightenment Str., 346428 Novocherkassk, Rostov Region, Russia
3 Department "Technique and Technology for Obtaining Nanoproducts", Tambov State Technical University, 106 Sovetskaya Str., 392000 Tambov, Russia
4 Institute of Biochemical Technology and Nanotechnology (IBTN), Peoples' Friendship University of Russia (RUDN), 6 Miklukho-Maklaya Str., 117198 Moscow, Russia
5 Complementary and Integrative Health Clinic of Dr. Shishonin, 5 Yasnogorskaya Str., 117588 Moscow, Russia
* Correspondence: avetcher@gmail.com

Abstract: The turn to hydrogen as an energy source is a fundamentally important task facing the global energetics, aviation and automotive industries. This step would reduce the negative man-made impact on the environment on the one hand, and provide previously inaccessible power modes and increased resources for technical systems, predetermining the development of an absolutely new life cycle for important areas of technology, on the other. The most important aspect in this case is the development of next-generation technologies for hydrogen industry waste management that will definitely reduce the negative impact of technology on the environment. We consider the approaches and methods related to new technologies in the area of hydrogen storage (HS), which requires the use of specialized equipment equipped with efficient and controlled temperature control systems, as well as the involvement of innovative materials that allow HS in solid form. Technologies for controlling hydrogen production and storage systems are of great importance, and can be implemented using neural networks, making it possible to significantly improve all technological stages according to the criteria of energy efficiency reliability, safety, and eco-friendliness. The recent advantages in these directions are also reviewed.

Keywords: hydrogen storage (tank); nanocomposite(s); nanotubes; waste management

Citation: Shchegolkov, A.V.; Shchegolkov, A.V.; Zemtsova, N.V.; Stanishevskiy, Y.M.; Vetcher, A.A. Recent Advantages on Waste Management in Hydrogen Industry. *Polymers* **2022**, *14*, 4992. https://doi.org/10.3390/polym14224992

Academic Editor: Vijay Kumar Thakur

Received: 10 October 2022
Accepted: 15 November 2022
Published: 18 November 2022

1. Introduction

Enhancement of environmental friendliness at all levels of energy resources employment is still a priority task that can be solved by considering hydrogen as an energy source [1]. Hydrogen is well-known as a carbon-free energy source, and its properties have been exhaustively studied by generations of scientists. Its wide application could potentially replace hydrocarbons and, accordingly, emissions of gaseous carbon in a variety of forms.

The application of hydrogen is based on the employment of fuel cells, which are efficient energy converters with significant potential for use in transport and other areas of energy production [2]. Fuel cells have an energy conversion efficiency of about 60–70%, which is significantly higher than for devices using the Carnot cycle. Currently, fuel cells have been demonstrated to be safe and efficient devices that can ensure a fast refueling process and energy efficiency [3]. Cathodic and anodic reaction—implemented on FC with a pronounced anode and cathode—is characterized by the fact that hydrogen is ionized, and its energy is released with the accumulation of electrons on the FC anode's surface. At the same time, oxygen is reduced at the cathode, which indicates anodic oxidation and cathodic reduction [4].

It should be noted that the main material for creating fuel cells is titanium, which corrodes during operation. To reduce the intensity of the corrosion process, a cathodic deposition of tungsten trioxide on the titanium surface can be used [5].

There are several main directions for the use of hydrogen that are widespread at the moment. The most important applications of hydrogen are:

- Chemical industry—synthesis of ammonia, methanol, and hydrocarbons, as well as the recovery of metals from their oxide form [6].
- Nanotechnology—in the process of CVD synthesis for the reduction of metal oxide catalysts in the synthesis of carbon nanotubes (CNTs) [7,8]. The reaction takes place at 600–1000 °C in propane–butane flow.
- Energetics—an energy source for electric and thermal power engineering [9].
- Petrochemistry—oil refining (hydrogenation purification of petroleum products—hydrodesulfurization) [10].
- Transportation—cars running on gaseous and liquid hydrogen [11,12].

In this latter case, a distinction should be made between passenger cars [13] and commercial cargo and passenger transportation means [14,15].

The implementation of hydrogen as the main energy source in various types of buses with FC in their design features the fact that hydrogen is converted into electrical energy, and the by-product is water vapor, which condenses into water in the environment [16]. The approach presented—in which H_2 is generated electrically to split water into O_2 and H_2 or by chemical conversion of methane to H_2 (loop conversion of methane with steam on anti-coking compounds $CeO_2/La_{0.9}Sr_{0.1}Fe_{1-x}Ni_xO_3$) [17]—makes it possible to abandon the use of petroleum products such as motor oils. This will also have a positive and tangible impact on the ecological situation of megacities, as it will eliminate the need to recycle used engine oils. The transition to a hydrogen energy system is likely to be based on H_2, obtained as a result of reforming natural gas or electrolysis.

The most widespread use of hydrogen is primarily in the field of motor transport, which needs environmentally friendly and affordable energy sources that are capable of replacing hydrocarbons. Another option that should be considered is the generation of energy at a thermal power plant [18], and in this case, one practice involves the partial mixing of hydrogen with methane or other gaseous fuels based on hydrocarbons [19].

The transition to hydrogen as the main type of fuel could form and transform the design of new types of internal combustion engine, and in particular, hydrogen rotary Wankel engines could find distribution [20]. Other types of vehicles in which hydrogen can be used include aircraft or air transport [21] and unmanned aerial vehicles [22]. Commercial hydrogen production currently depends mainly on steam natural gas reforming [23] and coal partial oxidation [24]. Clean production using both biomass and solar energy production methods is on its way [25].

Thus, there is a widespread practice of using hydrogen, which has the possibility of serving as a basis for an entire direction of research. For the successful dissemination of these achievements, however, several fundamentally important and significant problems need to be solved, including the safe generation and storage of hydrogen. At the same time, it should be borne in mind that polymer waste can be used as a source of cheap raw materials for producing hydrogen and related high-performance materials. There is also a fundamental possibility when using new control technologies related to artificial intelligence in the process of hydrogen synthesis and storage.

2. Design and Thermodynamics of HS Tank

HS is currently a "bottleneck" for the implementation of the use of hydrogen as renewable energy. A key challenge for the full development of hydrogen-based technologies is the safe, efficient, and economical storage of hydrogen.

Practical materials for HS should have the ability to undergo a reversible hydrogenation/dehydrogenation process, which is determined by the binding energy of hydrogen atoms. The PCT curve is a graph representing the dependence of pressure on composition at different temperatures. Figure 1 demonstrates a graphical interpretation of the Van 't Hoff equation [26], which indicates the dependence of the logarithm of the equilibrium desorption pressure on the reciprocal temperature (T) (Figure 1a), as well as the dependence of the amount of hydrogen accumulation on pressure (Figure 1b).

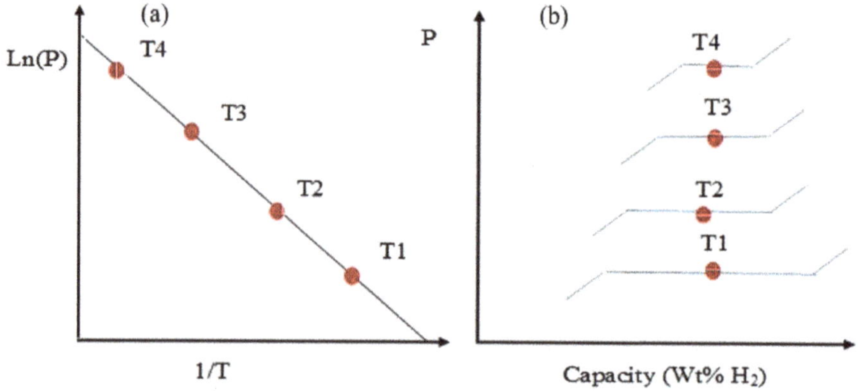

Figure 1. Van 't Hoff diagram of metal hydride (MH) (**a**) and phase diagram (**b**).

It is necessary to consider the temperature regimes that are characteristic of the storage of liquid hydrogen. Transient heat transfer plays a leading role in multilayer insulation (MLI) in combination with a steam-cooled shield (SCS) used in liquid HS tanks. In [27], the profile of the transition temperature and the change in the heat flux of MLI and SCS were predicted and analyzed, which can help optimize the operating parameters for liquid HS.

Considering the technological aspects of HS in porous materials, a new design concept for a portable HS tank was identified [28]. Within the framework of this concept, a storage method is used in which low pressure and cryogenic temperature are realized. To maintain the cryogenic temperature, three-layer insulation was used, allowing for at least 12.5 days without the need for an external cooling circuit to maintain the optimum temperature. The HS tank is portable and can be used in various types of FS electric vehicles (FCEVs). As a tank filler, porous absorbents can be used, which form such storage conditions at which a temperature of 77 K and a pressure below 100 bar is maintained. The presented parameters are significantly lower than the internal pressure of 700 bar in commercial type IV tanks [29], thus improving safety and reducing the risk of explosion.

The safety of the use of HS tanks with TPRD under fire conditions can be improved by using composite materials [30]. HS tank model inputs include thermal parameters of the hydride and tank materials, fire heat flow into the tank, diameter TPRD, and initiation delay time TPRD. Non-stationary heat transfer from the environment through the tank wall and lining to hydrogen leads to decomposition of the composite resin for wrapping and melting of the lining. The lower limit of the diameter of the TPRD hole is sufficient to prevent the tank from bursting in the case of fire.

With respect to the option of storing hydrogen in liquid or solid form, storage in the solid state is preferable. This is due to the improvement in explosion and fire safety, and also provides better volumetric and gravimetric density, which improves the weight and size parameters of hydrogen accumulators. It should be noted that hydrogen in solid-state storage is bound by physicochemical forces [31]. The strength of the interaction between hydrogen and the carrier material varies from weak van der Waals interactions, which are characteristic of the physisorption binding of molecular hydrogen, to the strong chemisorption binding of atomic hydrogen [32]. The storage density of hydrogen can be improved by using hydride-type materials; hydrogen is packed with a HH distance of up to 170 kg/m^3, which is more than twice the density of liquid hydrogen.

When storing hydrogen based on MH, it is necessary to use specialized heat exchangers or thermal control systems due to both endothermic and exothermic reactions taking place during the filling and unloading of MH tanks (Figure 2). In the absence of a heat exchanger or a suitable temperature range, the operation of the MP will have a negative influence, leading to the instability of the supply of hydrogen in FS systems. In [33], the influence of the tank surface temperature on the hydrogen consumption and hydrogen flow characteristics for the HS system MH of an electric vehicle operating on hydrogen FC was studied. Various temperature values were provided with the help of an external heat circulation device and a heat exchanger inside the MH tank. The FC operated in a power range from 200 to 600 W, and was regulated depending on the temperature and flow rate of the pumped reservoir [34].

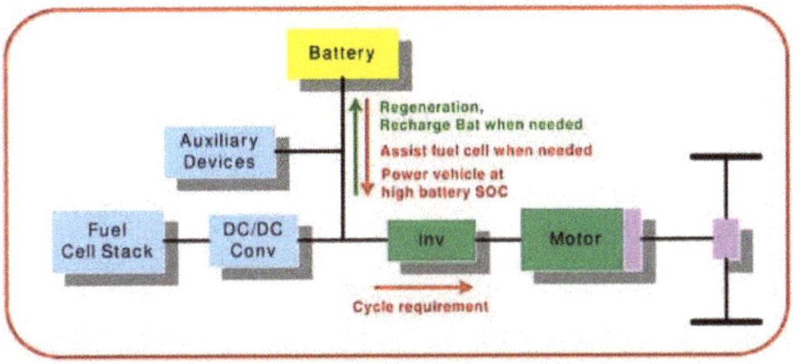

(a)

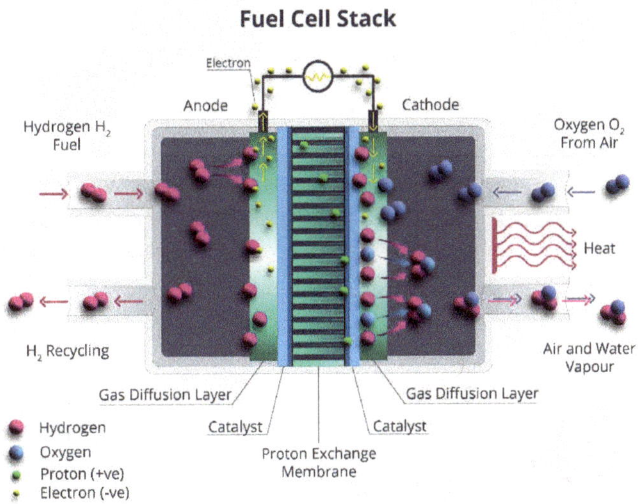

(b)

Figure 2. (**a**) Schematic of FC-based electric vehicle [29]. (**b**) The process of charging the cell with hydrogen.

MH cartridges based on the hydride of $La_{0.75}$ $Ce_{0.25}$ Ni_5 can be used for HS [35]. The low thermal conductivity of MH is a limiting factor with respect to the technological problems of HS. To improve the thermal and physical characteristics of MH, metal foam with a porosity gradient can be used [36].

3. Composite Materials for HS—Thermoset Composites

The safe storage of a reasonable amount of hydrogen is associated with many problems related to the method and materials. Hydrogen accumulator materials can be of different types:

1. Dissociative material in which molecular hydrogen is dissociated into hydrogen atoms occupying internodes;
2. Materials with chemically bonded hydrogen;
3. Materials that adsorb molecular hydrogen, in which molecular hydrogen attaches to the surface due to weak interactions, such as the Van der Waals force or physical sorption.

The ability of certain materials to accumulate hydrogen depends on the structure and type of interaction with hydrogen. There are some new materials for HS. Storage of hydrogen in solid form can be briefly divided into the following categories:

1. MHs;
2. Hydrides based on light metals;
3. Chemical hydrides (complex hydrides);
4. Nanostructured materials (adsorption of molecular hydrogen).

Intermetallic compounds can be used as hydrogen accumulators [37,38]. This is due to the peculiarities of their atomic structure, in which interstices with the optimal binding energy for hydrogen are observed, forming the process of absorption or desorption under conditions close to standard. In this regard, for the storage of hydrogen, the class of compounds of the Laves phase with the formula unit (AB) needs to be taken into consideration [39]. Because they change from one to the other on heating and cooling (usually C14 at high temperatures and C15 at low temperatures), hydrogen absorption–desorption can be thought of as a phase change process. Structure C15 is an fcc structure containing six atoms per unit cell, while structures C14 and C36 are hexagonal structures containing 12 and 24 atoms per unit cell. Figure 3 demonstrates the crystal structures of C14- and C15-type alloys. Ideally, the lattice parameters are closely related in each structure and between structures. However, in real MH alloys with a predominance of C14, the c/a ratio is slightly lower than the theoretical value (223–$\sqrt{}$≅1633) [40]. Three types of position are available for tetrahedral hydrogen filling positions (A2B2, AB3 and B4) for both C14 and C15 structures, as shown in Figure 3. There are no octahedral positions at all in the Laves phases, so further discussion will focus only on tetrahedral positions [41].

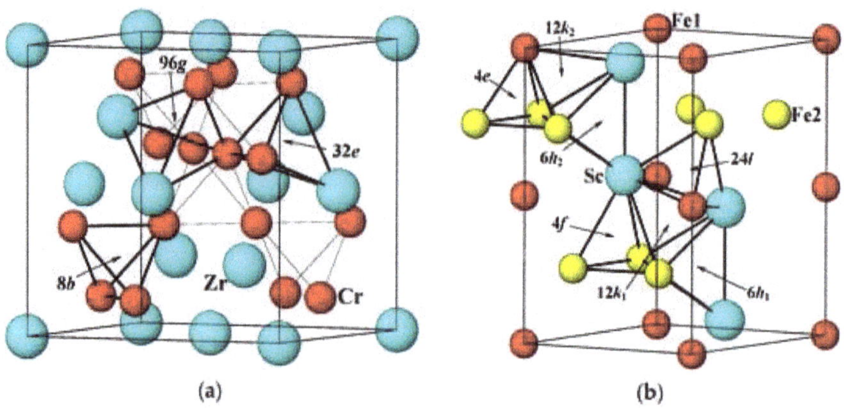

Figure 3. Elementary cells for the structures $ZrCr_2$ (C15) (**a**) and $ScFe_2$ (C14) (**b**). Various tetrahedral hydrogen filling sites (A2B2, AB3 and B4) are indicated by arrows [34].

High-temperature MH, such as MgH_2, is considered one of the most promising HS technologies [42]. However, there are two main bottlenecks, including the low rate of H_2 absorption and the low power of the MH reactor. In this regard, heat removal from the

MG tank plays a crucial role in the HS process. The results show that the charging time is significantly reduced by increasing the number of air channels, as the heat transfer rate is significantly improved. When the initial coolant temperature rises, the charging time increases; however, as the Reynolds number of the coolant increases and the hydrogen inlet pressure increases, the absorption process accelerates. The recommended configuration of the heat exchanger is introduced taking into account both the loading time and production constraints. It is shown that the loading of a new multi-zone hydrogen energy storage using four air channels is approximately 30 min, which provides a more applicable hydrogen fuel system.

Ref [43] presents experimental studies concerning the absorption of H_2 in a solid-state HS device based on $LmNi_{4.91}Sn_{0.15}$ with integrated cooling tubes (ECT). MH with ECT 36 and 60 loaded with 2.75 kg LmNi served as the basis for a hydrogen accumulator that implements various modes of supply pressure (10–35 bar), absorption temperature (20–30 °C) and coolant flow (2.2–30 L/min).

It has been found that at any given absorption temperature, the rate of H_2 absorption and the amount of absorbed H_2 increase when the H_2 supply pressure rises to about 35 bar. Assuming the supply of H_2 at a pressure of 35 bar and an absorption temperature of 30 °C, using oil as a coolant at a flow rate of 3.2 L/min, the maximum absorption of hydrogen is ≈1.2 wt.% for 10 min for 36 ECT and 8 min for 60 ECT. Under absorption conditions with a supply pressure of 25 bar, a water flow rate of 30 L/min and an absorption temperature of 30 °C, the absorption time in the reactor with 60 ECT was reduced to 5 min. Most metals are able to absorb hydrogen reversibly. Undoubtedly, the MG reactor (MR) is the main device used to achieve the desired stability and integrated operation of the HS system.

It has been found that at any given absorption temperature, the rate of hydrogen absorption and the amount of hydrogen absorbed increases as the hydrogen supply pressure rises to about 35 bar. Assuming a hydrogen supply pressure of 35 bar and an absorption temperature of 30 °C, using oil as the heat transfer medium at a flow rate of 3.2 L/min, the maximum absorbed hydrogen is ≈1.18 wt% per 10 min for 36 ECT and 8 min for 60 ECT. Under absorption conditions with a supply pressure of 25 bar, water flow 30 L/min and absorption temperature 30 °C, the absorption time in the reactor from 60 ECT is reduced to 5 min. The majority of metals can reversibly absorb hydrogen. Undoubtedly, the MH reactor (MHR) is the main device used to achieve the stable and integrated operation of HS systems desired.

Furthermore, each of the materials of this class in the form of nanocomposites will be considered to give a reasonable explanation for the improvement in the storage conditions of hydrogen as an energy source.

4. Hydrogen Generation and Storage on the MoS_2-Containing Materials

It appears that MoS_2 possesses great prospects as a cost-effective replacement for Pt for catalysis of the hydrogen evolution reaction (HER) in water, despite its claimed catalytic efficiency being lower than that of Pt. The latter is known to be the best HER catalyst, but it appears to be too expensive for mass production in the hydrogen industry. Monolayer (ML) MoS_2 films were grown using CVD processes. Substrates are able to affect the catalytic activity of MoS_2 in two ways: by forming an interfacial tunnel barrier with MoS_2; and by changing the chemical nature of MoS_2 through charge transfer (proximity doping).

The catalytic characteristics can be further improved such that they are even better than those of Pt by crumpling films on flexible substrates, since the Tafelian slope of the film is significantly reduced in the presence of compression deformation caused by crumpling [44]. MoS_2 can be used for the hydrogen evolution reaction (HER). Thermal effects formed an effective electron transfer in the atomic MS MoS_2 and at the electrolyte–catalyst interface, leading to an increase in the activity of GWR [45].

Vertically grown MoS_2 nanolists supported by conductive carbon nanotubes and reduced graphene oxide (CNT-rGO) on traditional Vietnamese paper (MoS_x/CNT-rGO/VTP) can be used for the electrochemical reaction of hydrogen evolution (HER). The catalyst demonstrates excellent electrocatalytic activity of HER, including a low initial potential of 190 mV, a Taffel slope of 59 mV/°, and excellent stability in an acidic electrolyte solution [46].

The Taffel slope shows the dependence of the catalytic reaction rate on the applied overvoltage. The smaller the Taffel slope, the faster the reaction rate increases when applying overvoltage. Typically, the Tafel slope (n) of an electrochemical reaction is dictated by the rate-determining reaction stage, and can be written as a function of the number of electrons involved (z) and the charge transfer coefficient (α) of the limiting factor. The extraction step is n = 2.3zRT/aF, where R is the ideal gas constant, T is the absolute temperature, F is the Faraday constant [47]. For MoS2 materials, which are usually found in an acidic environment, three reaction stages can be involved:

- Primary discharge stage (Vollmer reaction):

  $H_3O^+ + e^- \rightarrow Hads + H_2O$

- Electrochemical desorption stage (Geyrovsky reaction):

  $Hads + H_3O^+ + e^- \rightarrow H_2 + H_2O$

- Recombination stage (Tafel reaction):

  $Hads + Hads \rightarrow H_2$

The edge regions of MoS_2 are catalytically active in hydrogen evolution reactions (HER). F atoms with high electronegativity are doped into the edge nodes of MoS_2, leading to a fivefold increase in activity compared to the initial edges, which is explained by the more moderate binding energy for hydrogen particles [48].

The photocatalyst of the MoS_2@ MoO_3 heterojunction with the (S)-scheme stage was prepared by partial sulfidation in situ. The excellent design of the MoS_2@ MoO_3 interface nanomaterials provides a high rate of surface reaction of hydrogen evolution. The on-site vulcanization strategy gradually causes corrosion from the outside to the inside. The rate of hydrogen formation is 12,416.8 mmol/h·g [49].

The improvement of MoS_2 properties can be achieved by the formation of a chemical bond between the MoS_2 nanolayer on graphene and vacancies. There is a clear decrease in the metallic state of the MoS_2 nanolayer as electrons are transferred to form a strong contact with the restored graphene substrate. The absence of a metallic state associated with unsaturated atoms in the peripheral regions in turn changes the activity of hydrogen release. The easiest path of evolution is from the marginal regions of the Mo, and the presence of graphene leads to a decrease in the energy barrier from 0.17 to 0.11 eV. The evolution of H_2 from the S-edge is complicated due to an increase in the energy barrier from 0.43 to 0.84 eV [50].

The formation of S-vacancies on the basic plane of MoS_2 by electrochemical desulfurization makes it possible to improve the process of hydrogen generation. The formation of S-vacancies is possible on various 2H- MoS_2 nanostructures. By changing the applied desulfurization potential, it is possible to vary the degree of desulfurization and the resulting hydrogen release activity [51].

A three-dimensional hierarchical hybrid composite of MoS_2, reduced graphene oxide (GO) and CNT demonstrates excellent electrocatalytic activity and stability in the hydrogen evolution reaction, with a low initial potential of only 35 mV, a Taffel slope of ~38 mV/°, and an apparent exchange current density of 74.25 mA/cm^2. The excellent hydrogen release activity is due to the synergistic effect of MoS_2 with its electrocatalytically active edge regions and excellent electrical coupling with the underlying graphene and CNT grid [52].

Hierarchical MoP- MoS_2 electrocatalysts on hollow carbon spheres co-doped with N, P and S (MoP- MoS_2/HCS) demonstrate remarkable HER characteristics. MoP- MoS_2/HCS not only exhibit significant electrocatalytic activity at low overvoltages (71 mV and 125 mV in 1.0 M KOH and 0.5 M H_2SO_4, respectively) at a current density of 10 mA/cm^2, they also exhibit outstanding stability with respect to its process [53,54].

5. Processing of Composite Materials from the Hydrogen Industry

Global industry actively uses various types of plastic, which inevitably leads to the formation of a large amount of plastic waste. More than 60% of used plastic ends up in landfills or is incinerated, which harms the environment and the ecosystem as a whole [55]. Thermal processing by pyrolysis and gasification of plastic waste into fuel and chemical products has been identified as a promising technology for solving the problems of plastic waste [56]. Pyrolysis is a method of thermochemical treatment of plastic waste, which can solve such pollution problems, as well as restore valuable energy and products such as oil and gas. Pyrolysis of solid plastic waste has become important because it offers great advantages in terms of environmental pollution and reducing the carbon footprint of plastic products by minimizing carbon monoxide and carbon dioxide emissions compared to combustion and gasification [57].

In [58], a simple and highly efficient method initiated by microwave plasma discharge for the decomposition of plastics into hydrogen and carbon nanotubes was proposed. Iron-based catalysts applied to activated carbon calcined at 400 °C showed the best catalytic activity due to excellent physicochemical properties. H2 was rapidly released in 25 s, with a hydrogen efficiency of more than 85%.

In [59], pyrolysis and catalytic decomposition of polypropylene were carried out in the technological process for the production of hydrogen and carbon nanomaterials. A series of new Fe/Ni catalysts was prepared, and the effect of the active metal component of the catalyst (Fe, Ni, FeNi) and the synthesis method (sol–gel and impregnation) was studied. The results showed that the production of hydrogen and solid matter occurred in descending order with loading of Fe-Ni, Fe and Ni, while the catalysts prepared by sol–gel were more catalytic than their impregnated counterparts. FeNi (SG) demonstrated optimal activity in the production of 25.14 mmol/g of hydrogen plastic and 360 mg/g of high-quality plastic made of carbon nanomaterials.

In [60], the use of Ni-Fe catalysts was studied for the catalytic pyrolysis of plastic waste to produce hydrogen and CNT, as well as the influence of the composition of the catalyst and carrier materials. The bimetallic Ni-Fe catalyst showed higher catalytic activity in H_2 yield than monometallic Ni or Fe catalysts due to the optimal interaction between the metal and the carrier. The effect of steam supply and catalyst temperature on the yield of CNT (287 mg/g of plastic) and hydrogen (31.8 mmol H_2/g of plastic) is optimal at 800 °C in the presence of a bimetallic Ni-Fe/γ-Al_2O_3 catalyst (Figure 4).

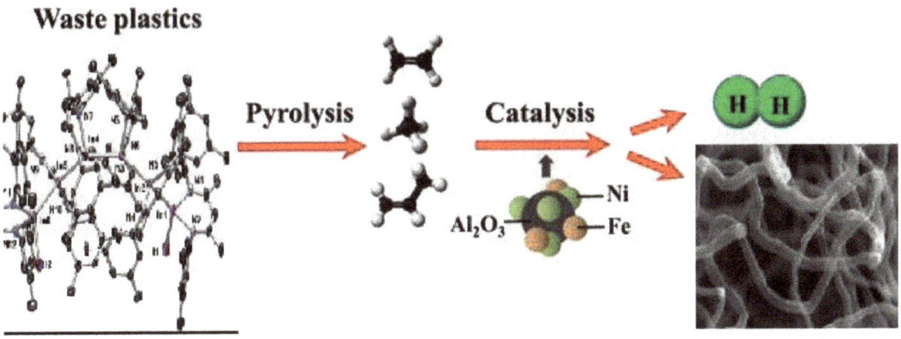

Figure 4. Schematic of pyrolysis of polymer with catalyst and hydrogen release.

For the catalytic pyrolysis of plastic waste, a two-stage fixed-bed reactor (with pyrolysis zone, height 310 mm; bottom: pyrolysis zone, height 310 mm) was reported (Figure 5 [60]). Three series of experiments were conducted to determine the technological parameters for the generation of hydrogen and CNTs: with the use of Fe/γ-Al_2O_3, Fe/α-Al_2O_3, Ni/γ-

Al_2O_3, Ni/α-Al_2O_3, and Ni-Fe/γ-Al_2O_3 catalysts; with ratios of mass of steam to mass of plastic of 0, 0.3, 1, and 2.6; and with catalytic temperatures of 700, 800, and 900 °C.

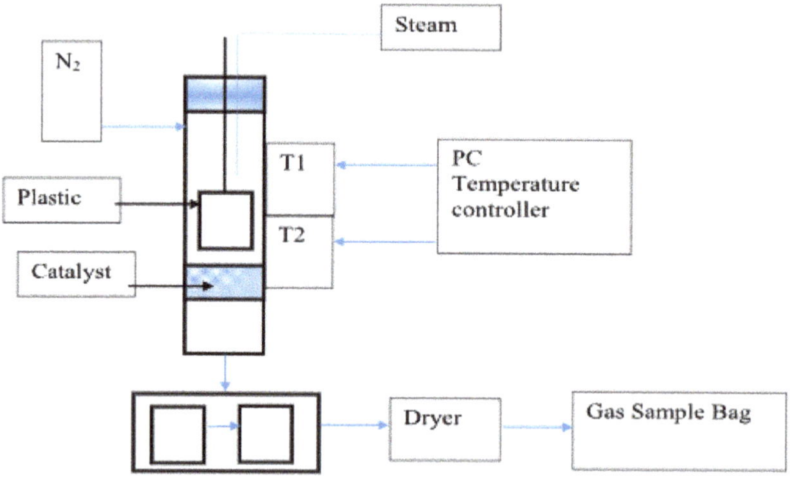

Figure 5. Schematic diagram of the pyrolysis–catalysis process of plastic waste.

The analysis of waste gases and thermodynamic calculations [61] showed that the H_2 emission, through the decomposition of the by-product CH_4, acted as a thermal micronization medium, where Fe_xO_y is gradually restored after the waste is converted into activated carbon (CA). The resulting CA is then effectively involved in the catalytic decomposition of H_2O, leading to the microgeneration of secondary H_2, creating a controllable system. Thus, the fast release of H_2 from the system was eliminated, resulting in improved recovery of Fe_xO_y due to a simplified H_2 microgeneration/regeneration process.

A two-stage catalytic pyrolysis steam reforming process with MSM-41 mesoporous Fe-Ni bimetallic catalysts was used to produce hydrogen-enriched synthesis gas from a simulated mixture of waste plastics (SMWP) [62]. Various weight ratios of Fe:Ni catalyst materials (0:20, 5:15, 10:10, 15:5, 20:0) were investigated to determine the effect on H_2 production. The results showed that the combined presence of Fe and Ni leads to a synergistic increase in the total gas yield and the formation of hydrogen and carbon monoxide. The highest gas yield of 95 wt.%, the highest H_2 yield of 46.1 mmol H_2/g plastic, and the highest CO yield at 31.8 mm/g plastic are characteristic of the Fe/Ni/MCM-41 catalyst (1:1). This catalyst gives a hydrogen yield of 46.7 vol.% and a CO yield of 32.2 vol.% [62].

Ni/SiO_2 and Fe/SiO_2 catalysts with metal particles of different sizes were studied in the production of hydrogen and CNTs during the catalytic processing of polypropylene waste using a two-stage fixed-bed reaction system we reported [63]. The results show that Fe-based catalysts, in particular those with large particle size (~80 nm), gave the highest hydrogen yield (~25.60 mmol H_2/g of plastic) and the highest carbon yield (29 wt. %), as well as the largest proportion of graphitic carbons (according to the analysis of the TPO of the reacted catalyst).

In the process of hydrogen production, a more complex three-component catalyst can be used [64]. The yield of hydrogen increased with an increase in the gasification temperature from 600 to 900 °C for both Ni-Mg-Al and industrial nickel catalysts. The maximum hydrogen production was 52% of the maximum theoretical amount of hydrogen available in polypropylene, which is 22–38 g H_2/100 g of polypropylene obtained with a Ni-Mg-Al catalyst, at a gasification temperature of 800 °C and a water flow rate with an injection speed of 28–46 g/h.

Real waste plastics contain dissimilar materials. In [64], the production of H_2 from pyrolysis–catalytic steam reforming of polyethylene, polystyrene (PS), and polyethylene terephthalate waste plastics was considered. The highest yield of hydrogen (125 mmol/g plastic) was obtained with PS at a catalyst temperature of 900 °C and a steam hourly space velocity of 7.59 g/(h/g catalyst) with 10 wt% Ni/Al_2O_3) catalyst. The high catalyst temperature (900 °C) and the optimized steam rate significantly increase the hydrogen yield. The authors found that Ni/Al_2O_3 has the highest selectivity and catalytic activity for hydrogen yield.

Comparison of energy spendings of various methods is presented in Table 1, from which it can be observed that that with the highest energy consumption is water electrolysis.

Table 1. Comparison of energy efficacy of different processes of hydrogen generation.

Process of Hydrogen Generation	Specific Heat Consumption for Endothermic Reactions, q_x (kJ/kg H_2)	Specific Consumption of Reference Fuel (rf) to Provide Endothermic Reactions, b (kg rf/kgH_2)	Specific rf Consumption for Production of kg H_2, b^* (kg rf/kgH_2)	The Ratio of the Calorific Value of the Total Amount of Fuel to H_2 on kg H_2 Δ (kJ/kJ)
Conversion of Methane with Steam in Reactors with a Fluidized Bed of a Dispersed Catalyst $CH_4 + 2H_2O \rightarrow CO_2 + 4H_2$ (by-product CO_2)	34,987	1.32	4.74	1.12
Carbon gasification of solid fuel with water vapor $C + 2H_2O \rightarrow CO_2 + 2H_2$ (by-product CO_2)	67,958	2.89	6.24	1.37
CH_4 pyrolysis at 1350 °C $CH_4 \rightarrow 2H_2 + C$ (by-product C)	18,922	0.72	7.56	1.815
Water electrolysis $2H_2O \rightarrow H_2 + 0.5O_2 + H_2O$ (by-product O_2)	214,268.4	27.77	27.77	1.77

6. Conclusions

The use of hydrogen includes a variety of directions from chemical technologies to unmanned aerial vehicles. The field of motor transport is a key one for the mass development of water generation and storage technologies.

New technologies in the field of HS are based on the use of specialized equipment with temperature control systems, as well as the use of innovative materials that make it possible to store hydrogen in solid form. For the storage of hydrogen in solid form, Mg- or Li-based MHs in the form of classical composites or having nanostructured morphology show the greatest efficiency. Hydrides of nanometals can be obtained using the "bottom-up" and "top-down" strategies. It should be noted the principal possibility of using carbon materials for HS, namely carbon nanotubes, both single-layer and multi-layer. To explain the mechanisms of the catalytic effect of impurities in the metallic alloy on the properties of CNTs with respect to hydrogen accumulation, a spillover mechanism or hydrogenation of the Cubas type can be used.

MoS_2 and the wide variety of composites based on it exhibit serious prospects in technologies for the production and storage of hydrogen. It is obvious that further technological development will provide novel solutions.

To produce hydrogen, waste polymer products are used—the processing of which is realized on the basis of catalytic pyrolysis. Catalytic pyrolysis also makes it possible to obtain carbon nanotubes that can be used for HS. Control technologies for hydrogen production and storage systems are implemented on the basis of neural networks, making it possible to significantly improve all technological stages according to the criteria of energy efficiency and reliability, as well as safety. An analysis of the energy costs for hydrogen production shows that direct current electrolysis is the most expensive, and thermal decomposition (pyrolysis) is less expensive.

It is worth underlining that the contemporary situation in the energy market does not demonstrate logical tendencies and therefore does not allow us to make clear predictions on the directions of hydrogen industry development in the closest future. However, the long-term prospective requires continued scientific research in this direction [65–67].

Author Contributions: Conceptualization, A.V.S. (Alexander V. Shchegolkov) and A.A.V.; Methodology, A.V.S. (Alexander V. Shchegolkov), A.V.S. (Aleksei V. Shchegolkov); Software, A.V.S. (Alexander V. Shchegolkov), A.V.S. (Aleksei V. Shchegolkov); Validation, A.V.S. (Alexander V. Shchegolkov), A.V.S. (Aleksei V. Shchegolkov); Formal analysis, A.V.S. (Alexander V. Shchegolkov), A.V.S. (Aleksei V. Shchegolkov); Investigation, A.V.S. (Alexander V. Shchegolkov), A.V.S. (Aleksei V. Shchegolkov); Resources, A.V.S. (Alexander V. Shchegolkov), A.V.S. (Aleksei V. Shchegolkov); Data curation, A.V.S. (Alexander V. Shchegolkov), A.V.S. (Aleksei V. Shchegolkov); Writing—original draft preparation, A.V.S. (Alexander V. Shchegolkov), A.V.S. (Aleksei V. Shchegolkov), A.A.V., Y.M.S., N.V.Z.; Writing—review and editing, A.A.V.; Visualization, A.V.S. (Alexander V. Shchegolkov), A.V.S. (Aleksei V. Shchegolkov), A.A.V.; Supervision, Y.M.S. and A.A.V.; Project administration, A.V.S. (Alexander V. Shchegolkov), A.V.S. (Aleksei V. Shchegolkov); Funding acquisition, A.A.V., A.V.S. (Alexander V. Shchegolkov), A.V.S. (Aleksei V. Shchegolkov). All authors have read and agreed to the published version of the manuscript.

Funding: This paper was supported by the RUDN University Strategic Academic Leadership Program (recipient A.V.)

Data Availability Statement: The data presented in this study are available on request from the corresponding author.

Acknowledgments: Alexandre A. Vetcher (IBTN (RUDN)) gratefully acknowledges that this paper has been supported by the RUDN University Strategic Academic Leadership Program (for A.A.V.).

Conflicts of Interest: The authors declare no conflict of interest. The funders had no role in the design of the study; in the collection, analysis, or interpretation of data; in the writing of the manuscript, or in the decision to publish the results.

References

1. Yang, Y.; Tong, L.; Yin, S.; Liu, Y.; Wang, L.; Qiu, Y.; Ding, Y. Status and challenges of applications and industry chain technologies of hydrogen in the context of carbon neutrality. *J. Clean. Prod.* **2022**, *376*, 134347. [CrossRef]
2. Nimir, W.; Al-Othman, A.; Tawalbeh, M.; Al Makky, A.; Ali, A.; Karimi-Maleh, H.; Karimi, F.; Karaman, C. Approaches towards the development of heteropolyacid-based high temperature membranes for PEM fuel cells. *Int. J. Hydrogen Energy* 2021, *in press*. [CrossRef]
3. Modarres, M.; Kaminskiy, M.; Krivtsov, V.V. *Reliability Engineering and Risk Analysis: A Practical Guide*, 3rd ed.; CRC Press: Boca Raton, FL, USA, 2017.
4. Olabi, A.G.; Abdelkareem, M.A.; Wilberforce, T.; Sayed, E.T. Application of graphene in energy storage device—A review. *Renew. Sustain. Energy Rev.* **2021**, *135*, 110026. [CrossRef]
5. Shchegolkov, A.V.; Lipkin, M.S.; Shchegolkov, A.V. Preparation of WO_3 Films on Titanium and Graphite Foil for Fuel Cell and Supercapacitor Applications by Electrochemical (Cathodic) Deposition Method. *Russ. J. Gen. Chem.* **2022**, *92*, 1161–1167. [CrossRef]
6. Hu, Y.; Xue, N.; Zhang, Y.; Hu, P. An eco-friendly vanadium precipitation method through solution-phase hydrogen reduction with nickel catalysis. *J. Taiwan Inst. Chem. Eng.* **2022**, *134*, 104334. [CrossRef]
7. Machado, G.; Coelho, C. Vertically-aligned carbon nanotube at low pressure by cold-wall thermal CVD using a two-phase deposition step. *Carbon Trends* **2021**, *5*, 100087. [CrossRef]
8. Chan, K.; Maznam, N.; Hazan, M.; Ahmad, R.; Sa'Ari, A.; Azman, N.; Mamat, M.; Rahman, M.; Tanemura, M.; Yaakob, Y. Multi-walled carbon nanotubes growth by chemical vapour deposition: Effect of precursor flowing path and catalyst size. *Carbon Trends* **2022**, *6*, 100142. [CrossRef]
9. Rejeb, O.; Alirahmi, S.M.; Assareh, E.; Assad, M.E.H.; Jemni, A.; Bettayeb, M.; Ghenai, C. Innovative integrated solar powered polygeneration system for green Hydrogen, Oxygen, electricity and heat production. *Energy Convers. Manag.* **2022**, *269*, 116073. [CrossRef]
10. Speight, J.G. Effects in Refining. In *High Acid Crudes*; Gulf Professional Publishing: Houston, TX, USA, 2014; Chapter 4; pp. 77–109.
11. Jafari, H.; Safarzadeh, S.; Azad-Farsani, E. Effects of governmental policies on energy-efficiency improvement of hydrogen fuel cell cars: A game-theoretic approach. *Energy* **2022**, *254*, 124394. [CrossRef]
12. Ku, A.Y.; Reddi, K.; Elgowainy, A.; McRobie, J.; Li, J. Liquid pump-enabled hydrogen refueling system for medium and heavy duty fuel cell vehicles: Station design and technoeconomic assessment. *Int. J. Hydrogen Energy* **2022**, *47*, 25486–25498. [CrossRef]

13. Boretti, A. Comparison of fuel economies of high efficiency diesel and hydrogen engines powering a compact car with a flywheel based kinetic energy recovery systems. *Int. J. Hydrogen Energy* **2010**, *35*, 8417–8424. [CrossRef]
14. Coleman, D.; Kopp, M.; Wagner, T.; Scheppat, B. The value chain of green hydrogen for fuel cell buses—A case study for the Rhine-Main area in Germany. *Int. J. Hydrogen Energy* **2020**, *45*, 5122–5133. [CrossRef]
15. Charters, D. A comparison of energy vectors in powering hybrid buses. *Renew. Energy Focus* **2016**, *17*, 73–74. [CrossRef]
16. Emonts, B.; Schiebahn, S.; Görner, K.; Lindenberger, D.; Markewitz, P.; Merten, F.; Stolten, D. Re-energizing energy supply: Electrolytically-produced hydrogen as a flexible energy storage medium and fuel for road transport. *J. Power Sources* **2017**, *342*, 320–326. [CrossRef]
17. Zhao, K.; Fang, X.; Cui, C.; Kang, S.; Zheng, A.; Zhao, Z. Co-production of syngas and H_2 from chemical looping steam reforming of methane over anti-coking $CeO_2/La_{0.9}Sr_{0.1}Fe_{1-x}Ni_xO_3$ composite oxides. *Fuel* **2022**, *317*, 123455. [CrossRef]
18. Peláez-Peláez, S.; Colmenar-Santos, A.; Pérez-Molina, C.; Rosales, A.-E.; Rosales-Asensio, E. Techno-economic analysis of a heat and power combination system based on hybrid photovoltaic-fuel cell systems using hydrogen as an energy vector. *Energy* **2021**, *224*, 120110. [CrossRef]
19. Karimkashi, S.; Kahila, H.; Kaario, O.; Larmi, M.; Vuorinen, V. Numerical study on tri-fuel combustion: Ignition properties of hydrogen-enriched methane-diesel and methanol-diesel mixtures. *Int. J. Hydrogen Energy* **2020**, *45*, 4946–4962. [CrossRef]
20. Wang, H.; Ji, C.; Shi, C.; Yang, J.; Ge, Y.; Wang, S.; Chang, K.; Meng, H.; Wang, X. Parametric modeling and optimization of the intake and exhaust phases of a hydrogen Wankel rotary engine using parallel computing optimization platform. *Fuel* **2022**, *324*, 124381. [CrossRef]
21. Gomez, A.; Smith, H. Liquid hydrogen fuel tanks for commercial aviation: Structural sizing and stress analysis. *Aerosp. Sci. Technol.* **2019**, *95*, 105438. [CrossRef]
22. Shi, C.; Zhang, Z.; Ji, C.; Li, X.; Di, L.; Wu, Z. Potential improvement in combustion and pollutant emissions of a hydrogen-enriched rotary engine by using novel recess configuration. *Chemosphere* **2022**, *299*, 134491. [CrossRef]
23. Collodi, G. Hydrogen Production via Steam Reforming with CO_2 Capture. *Chem. Eng. Trans.* **2010**, *19*, 37–42.
24. Santhanam, K.S.; Press, R.J.; Miri, M.J.; Bailey, A.V.; Takacs, G.A. *Introduction to Hydrogen Technology*; John Wiley & Sons: Hoboken, NJ, USA, 2017.
25. Takeda, S.; Nam, H.; Chapman, A. Low-carbon energy transition with the sun and forest: Solar-driven hydrogen production from biomass. *Int. J. Hydrogen Energy* **2022**, *47*, 24651–24668. [CrossRef]
26. Lakhlifi, A.; Dahoo, P.R.; Picaud, S.; Mousis, O. A simple van't Hoff law for calculating Langmuir constants in clathrate hydrates. *Chem. Phys.* **2015**, *448*, 53–60. [CrossRef]
27. Jiang, W.; Sun, P.; Li, P.; Zuo, Z.; Huang, Y. Transient thermal behavior of multi-layer insulation coupled with vapor cooled shield used for liquid hydrogen storage tank. *Energy* **2021**, *231*, 120859. [CrossRef]
28. Nguyen, D.H.; Kim, J.H.; Vo, T.T.N.; Kim, N.; Ahn, H.S. Design of portable hydrogen tank using adsorption material as storage media: An alternative to Type IV compressed tank. *Appl. Energy* **2022**, *310*, 118552. [CrossRef]
29. Su, Y.; Lv, H.; Zhou, W.; Zhang, C. Review of the Hydrogen Permeability of the Liner Material of Type IV On-Board Hydrogen Storage Tank. *World Electr. Veh. J.* **2021**, *12*, 130. [CrossRef]
30. Molkov, V.; Dadashzadeh, M.; Kashkarov, S.; Makarov, D. Performance of hydrogen storage tank with TPRD in an engulfing fire. *Int. J. Hydrogen Energy* **2021**, *46*, 36581–36597. [CrossRef]
31. Ahamed, M.I.; Shakeel, N.; Anwar, N.; Khan, A.; Asiri, A.M.; Dzudzevic-Cancar, H. 4—Graphene-based nanocomposite for hydrogen storage application. In *Micro and Nano Technologies, Nanomaterials for Hydrogen Storage Applications*; Sen, F., Khan, A., Asiri, A.M., Eds.; Elsevier: Amsterdam, The Netherlands, 2021; pp. 57–78, ISBN 9780128194768. [CrossRef]
32. Zhang, L.; Ren, D.; Ding, W. High hydrogen storage ability of a decorated g-C_3N_4 monolayer decorated with both Mg and Li: A density functional theory (DFT) study. *Int. J. Hydrogen Energy* **2022**, *47*, 28548–28555. [CrossRef]
33. Özdoğan, E.; Hüner, B.; Süzen, Y.O.; Eşiyok, T.; Uzgören, I.N.; Kıstı, M.; Uysal, S.; Selçuklu, S.B.; Demir, N.; Kaya, M.F. Effects of tank heating on hydrogen release from metal hydride system in VoltaFCEV Fuel Cell Electric Vehicle. *Int. J. Hydrogen Energy*, 2022; *in press*. [CrossRef]
34. Manoharan, Y.; Hosseini, S.E.; Butler, B.; Alzhahrani, H.; Senior, B.T.F.; Ashuri, T.; Krohn, J. Hydrogen Fuel Cell Vehicles; Current Status and Future Prospect. *Appl. Sci.* **2019**, *9*, 2296. [CrossRef]
35. Malleswararao, K.; Aswin, N.; Kumar, P.; Dutta, P.; Murthy, S.S. Experiments on a novel metal hydride cartridge for hydrogen storage and low temperature thermal storage. *Int. J. Hydrogen Energy* **2022**, *47*, 16144–16155. [CrossRef]
36. Bai, X.-S.; Yang, W.-W.; Yang, Y.-J.; Zhang, K.-R.; Yang, F.-S. Multi-variable optimization of metal hydride hydrogen storage reactor with gradient porosity metal foam and evaluation of comprehensive performance. *Int. J. Hydrogen Energy* **2022**, *47*, 35340–35351. [CrossRef]
37. Banerjee, S.; Mukhopadhyay, P. Interstitial Ordering. In *Pergamon Materials Series*; Elsevier: Pergamon, Turkey, 2007; Chapter 8, Volume 12, pp. 717–781.
38. Matysina, Z.A.; Gavrylyuk, N.A.; Kartel, M.T.; Veziroglu, A.; Veziroglu, T.N.; Pomytkin, A.P.; Schur, D.V.; Ramazanov, T.S.; Gabdullin, M.T.; Zolotarenko, A.D.; et al. Hydrogen sorption properties of new magnesium intermetallic compounds with $MgSnCu_4$ type structure. *Int. J. Hydrogen Energy* **2021**, *46*, 25520–25532. [CrossRef]
39. Chang, S.; Young, K.-H.; Ouchi, T.; Meng, T.; Nei, J.; Wu, X. Studies on Incorporation of Mg in Zr-Based AB_2 Metal Hydride Alloys. *Batteries* **2016**, *2*, 11. [CrossRef]

40. Young, K.-H.; Nei, J.; Wan, C.; Denys, R.V.; Yartys, V.A. Comparison of C14- and C15-Predomiated AB_2 Metal Hydride Alloys for Electrochemical Applications. *Batteries* **2017**, *3*, 22. [CrossRef]
41. Eisapour, A.H.; Eisapour, M.; Talebizadehsardari, P.; Walker, G.S. An innovative multi-zone configuration to enhance the charging process of magnesium based metal hydride hydrogen storage tank. *J. Energy Storage* **2021**, *36*, 102443. [CrossRef]
42. Anbarasu, S.; Muthukumar, P.; Mishra, S.C. Tests on $LmNi_{4.91}Sn_{0.15}$ based solid state hydrogen storage device with embedded cooling tubes—Part A: Absorption process. *Int. J. Hydrogen Energy* **2014**, *39*, 3342–3351. [CrossRef]
43. Kunwar, B.; Cheng, H.N.; Chandrashekaran, S.R.; Sharma, B.K. Plastics to fuel: A review. *Renew. Sustain. Energy Rev.* **2016**, *54*, 421–428. [CrossRef]
44. Li, G.; Chen, Z.; Li, Y.; Zhang, D.; Yang, W.; Liu, Y.; Cao, L. Engineering Substrate Interaction to Improve Hydrogen Evolution Catalysis of Monolayer MoS_2 Films beyond Pt. *ACS Nano* **2020**, *14*, 1707–1714. [CrossRef]
45. Qu, J.; Li, Y.; Li, F.; Li, T.; Wang, X.; Yin, Y.; Ma, L.; Schmidt, O.G.; Zhu, F. Direct Thermal Enhancement of Hydrogen Evolution Reaction of On-Chip Monolayer MoS_2. *ACS Nano* **2022**, *16*, 2921–2927. [CrossRef]
46. Tekalgne, M.A.; Van Nguyen, K.; Nguyen, D.L.T.; Nguyen, V.-H.; Nguyen, T.P.; Vo, D.-V.N.; Trinh, Q.T.; Hasani, A.; Do, H.H.; Lee, T.H.; et al. Hierarchical molybdenum disulfide on carbon nanotube-reduced graphene oxide composite paper as efficient catalysts for hydrogen evolution reaction. *J. Alloys Compd.* **2020**, *823*, 153897. [CrossRef]
47. Li, Y.; Wang, H.; Xie, L.; Liang, Y.; Hong, G.; Dai, H. MoS_2 Nanoparticles Grown on Graphene: An Advanced Catalyst for the Hydrogen Evolution Reaction. *J. Am. Chem. Soc.* **2011**, *133*, 7296–7299. [CrossRef]
48. Zhang, R.; Zhang, M.; Yang, H.; Li, G.; Xing, S.; Li, M.; Xu, Y.; Zhang, Q.; Hu, S.; Liao, H.; et al. Creating Fluorine-Doped MoS_2 Edge Electrodes with Enhanced Hydrogen Evolution Activity. *Small Methods* **2021**, *5*, 2100612. [CrossRef]
49. Zhang, L.; Jin, Z.; Tsubaki, N. Activating and optimizing the MoS_2@MoO_3 S-scheme heterojunction catalyst through interface engineering to form a sulfur-rich surface for photocatalyst hydrogen evolution. *Chem. Eng. J.* **2022**, *438*, 135238. [CrossRef]
50. Liao, T.; Sun, Z.; Sun, C.; Dou, S.X.; Searles, D.J. Electronic Coupling and Catalytic Effect on H_2 Evolution of MoS_2/Graphene Nanocatalyst. *Sci. Rep.* **2014**, *4*, 6256. [CrossRef]
51. Tsai, C.; Li, H.; Park, S.; Park, J.; Han, H.S.; Nørskov, J.K.; Zheng, X.; Abild-Pedersen, F. Electrochemical generation of sulfur vacancies in the basal plane of MoS_2 for hydrogen evolution. *Nat. Commun.* **2017**, *8*, 15113. [CrossRef]
52. Khan, M.; Yousaf, A.B.; Chen, M.; Wei, C.; Wu, X.; Huang, N.; Qi, Z.; Li, L. Molybdenum sulfide/graphene-carbon nanotube nanocomposite material for electrocatalytic applications in hydrogen evolution reactions. *Nano Res.* **2016**, *9*, 837–848. [CrossRef]
53. Wang, X.; Dai, J.; Xie, H.; Yang, C.; He, L.; Wu, T.; Liu, X.; Xu, Y.; Yuan, C.; Dai, L. In-situ construction of ultrathin MoP-MoS_2 heterostructure on N, P and S co-doped hollow carbon spheres as nanoreactor for efficient hydrogen evolution. *Chem. Eng. J.* **2022**, *438*, 135544. [CrossRef]
54. Chen, J.; Li, S.; Tao, Z. Novel hydrogen storage properties of MoS_2 nanotubes. *J. Alloys Compd.* **2003**, *356–357*, 413–417. [CrossRef]
55. Passamonti, F.J.; Sedran, U. Recycling of waste plastics into fuels. LDPE conversion in FCC. *Appl. Catal. B Environ.* **2012**, *125*, 499–506. [CrossRef]
56. Al-Salem, S.M.; Antelava, A.; Constantinou, A.; Manos, G.; Dutta, A. A review on thermal and catalytic pyrolysis of plastic solid waste (PSW). *J. Environ. Manag.* **2017**, *197*, 177–198. [CrossRef]
57. Zhang, P.; Liang, C.; Wu, M.; Chen, X.; Liu, D.; Ma, J. High-efficient microwave plasma discharging initiated conversion of waste plastics into hydrogen and carbon nanotubes. *Energy Convers. Manag.* **2022**, *268*, 116017. [CrossRef]
58. Jagodzińska, K.; Jönsson, P.G.; Yang, W. Pyrolysis and in-line catalytic decomposition of excavated landfill waste to produce carbon nanotubes and hydrogen over Fe- and Ni-based catalysts—Investigation of the catalyst type and process temperature. *Chem. Eng. J.* **2022**, *446*, 136808. [CrossRef]
59. Yao, D.; Zhang, Y.; Williams, P.T.; Yang, H.; Chen, H. Co-production of hydrogen and carbon nanotubes from real-world waste plastics: Influence of catalyst composition and operational parameters. *Appl. Catal. B Environ.* **2018**, *221*, 584–597. [CrossRef]
60. Assefi, M.; Mofarah, S.S.; Maroufi, S.; Nekouei, R.K.; Wang, W.; Kert, E.; Sahajwalla, V. Regeneration of hydrogen through thermal micronisation of end-of-life polymers for sustainable reduction of iron oxide. *Fuel Process. Technol.* **2022**, *226*, 107038. [CrossRef]
61. Zhang, Y.; Huang, J.; Williams, P.T. Fe–Ni–MCM-41 Catalysts for Hydrogen-Rich Syngas Production from Waste Plastics by Pyrolysis–Catalytic Steam Reforming. *Energy Fuels* **2017**, *31*, 8497–8504. [CrossRef]
62. Liu, X.; Zhang, Y.; Nahil, M.A.; Williams, P.T.; Wu, C. Development of Ni- and Fe-based catalysts with different metal particle sizes for the production of carbon nanotubes and hydrogen from thermo-chemical conversion of waste plastics. *J. Anal. Appl. Pyrolysis* **2017**, *125*, 32–39. [CrossRef]
63. Wu, C.; Williams, P.T. Hydrogen from waste plastics by way of pyrolysis–gasification. *Proc. Inst. Civ. Eng.-Waste Resour. Manag.* **2014**, *167*, 35–46. [CrossRef]
64. Aminu, I.; Nahil, M.A.; Williams, P.T. High-yield hydrogen from thermal processing of waste plastics. *Proc. Inst. Civ. Eng.-Waste Resour. Manag.* **2022**, *175*, 3–13. [CrossRef]
65. Barthelemy, H.; Weber, M.; Barbier, F. Hydrogen storage: Recent improvements and industrial perspectives. *Int. J. Hydrogen Energy* **2017**, *42*, 7254–7262. [CrossRef]
66. Yusaf, T.; Fernandes, L.; Abu Talib, A.R.; Altarazi, Y.S.; Alrefae, W.; Kadirgama, K.; Ramasamy, D.; Jayasuriya, A.; Brown, G.; Mamat, R.; et al. Sustainable Aviation—Hydrogen Is the Future. *Sustainability* **2022**, *14*, 548. [CrossRef]
67. Razi, F.; Dincer, I. Challenges, opportunities and future directions in hydrogen sector development in Canada. *Int. J. Hydrogen Energy* **2022**, *47*, 9083–9102. [CrossRef]

polymers

MDPI

Article

Study of the Degradation of a TPS/PCL/Fique Biocomposite Material in Soil, Compost, and Water

Fabián Steven Mosquera Rodríguez [1], Alejandro Quintero Vélez [1], Estivinson Córdoba Urrutia [2], Howard Ramírez-Malule [1,*] and Jose Herminsul Mina Hernandez [3,*]

1 School of Chemical Engineering, Universidad del Valle, Calle 13 No. 100-00, Cali 760001, Colombia; mosquera.fabian@correounivalle.edu.co (F.S.M.R.); alejandro.quintero.velez@correounivalle.edu.co (A.Q.V.)
2 Group Investigación en Ciencia Animal y Recursos Agroforestales, Universidad Tecnológica del Chocó, Carrera 22 No. 18B-10, Quibdó 270001, Colombia; estivinson.cordoba@utch.edu.co
3 School of Materials Engineering, Group Materiales Compuestos, Universidad del Valle, Calle 13 No. 100-00, Cali 760001, Colombia
* Correspondence: howard.ramirez@correounivalle.edu.co (H.R.-M.); jose.mina@correounivalle.edu.co (J.H.M.H.)

Abstract: The degradability of the biocomposite produced from a binary mixture of thermoplastic banana starch (TPS) and polycaprolactone (PCL) reinforced with fique fibers (Fs) was evaluated in three different environments (soil, compost, water). An experimental design with two factors (soil and compost) and three levels (5, 10, and 20 cm) was used, with additional tests for a third aqueous environment (water from the lake of the Universidad del Valle) at a depth of 20 cm. The biocomposite was prepared from the implementation of a twin-screw extrusion process of the binary mixture TPS/PCL and fique fibers (54, 36, and 10% composition, respectively), followed by hot compression molding, and after that, generating ASTM D638 type V specimens using a stainless-steel die. The specimens were dried and buried according to the experimental design, for a total experimental time of 90 days, and removing samples every 30 days. After 90 days, all samples showed signs of degradation, where the best results were obtained in the compost at a depth of 20 cm (34 $\pm$ 4% mass loss and a decrease in tensile strength of 77.3%, which indicates that the material lost mechanical properties). TPS was the fastest disappearing component and promoted the degradation of the composite material as it disappeared. Finally, the aqueous media presented the lowest degradation results, losing only 20% of its initial mass after 90 days of the experiment, being the least effective environment in which the biocomposite can end up.

Keywords: degradation; thermoplastic starch; polycaprolactone; biocomposite; fique fibers; microorganisms

Citation: Mosquera Rodríguez, F.S.; Quintero Vélez, A.; Córdoba Urrutia, E.; Ramírez-Malule, H.; Mina Hernandez, J.H. Study of the Degradation of a TPS/PCL/Fique Biocomposite Material in Soil, Compost, and Water. *Polymers* **2023**, *15*, 3952. https://doi.org/10.3390/polym15193952

Academic Editors: Alexey L. Iordanskii and Alexandre Vetcher

Received: 30 August 2023
Revised: 20 September 2023
Accepted: 22 September 2023
Published: 30 September 2023

1. Introduction

The high accumulation of solid waste is one of the main problems for the environment today [1,2]. In Colombia, about 11.6 million tons of solid waste are generated annually, and a large part of this waste is single-use plastic, including bags, packaging, straws, and bottles made of fossil material. According to the Colombian Ministry of Environment and Sustainable Development, a large percentage of this plastic waste ends up in landfills and water sources [3], contributing to pollution due to the time it takes to degrade.

On a global scale, the situation is not improving; according to the Organization for Economic Co-operation and Development (OECD), global plastic production doubled from 2000 to 2019, adding 353 million tons (Mt) to the existing ones. Approximately two-thirds of this waste comes from plastic with a life cycle of less than five years, 40% of which comes from packaging, 12% from consumer goods, and 11% from clothing and textiles. At the end of their useful life, only 9% of plastics are recycled, 19% are incinerated, 50% are used in landfill, and 22% are not properly managed, being incinerated in open pits or ending up in land or water-based landfills. In 2019, 6.1 Mt of plastic waste leaked into aquatic media, and 1.7 Mt reached the oceans [4].

In accordance with the above, and to develop planet-friendly alternatives, numerous studies have been carried out on biocomposite and/or bioplastics, which seek to take advantage of biological material to replace conventional plastics [5–18], generating products with similar characteristics (high strength, durability, and robustness, among others), but with shorter degradation times and cleaner processes. However, the bioplastics industry faces challenges, such as high production costs, limited availability of raw materials, inferior mechanical properties compared to conventional plastics [19,20], the need for high-temperature composting facilities, and competition for land use. However, despite these challenges, the industry is growing and can contribute to global sustainability; as proof of the above, the production of bioplastics increased by 30% from 2015 to 2022 (being in 2015 1.7 Mt and reaching 2.217 Mt worldwide by 2022) [21,22].

Polycaprolactone (PCL) and starch blend fall within the group of biodegradable bioplastics; according to the latest data compiled by European Bioplastic in cooperation with the Nova-Institute, in 2020, over 20% of global production (443.1 thousand tons) corresponded to them, and these percentages are projected to grow by 8% by 2025, based on global production of 2.87 Mt of biopolymers. In addition, these biocomposites present a wide range of physical properties that make them attractive in the development and improvement of products, which, combined with other materials, can increase their physical characteristics and applications.

Due to the above, different works have been devoted to the study and development of substitutes for conventional plastics [23–27]. Mina et al. [28] have developed biocomposites based on a thermoplastic starch/polycaprolactone (TPS/PCL) matrix reinforced with fique fibers, focusing their work mainly on physicochemical, mechanical, and thermal characterizations, indicating the need to carry out degradability studies on these materials. In this sense, not necessarily all bio-based materials can be decomposed in microbiological processes; some may require specific conditions for their degradation, and, in case these conditions are not met, the permanence time of this biocomposite would increase, and it would not represent an efficient alternative to existing fossil materials.

Based on the above, the objective of this work is to evaluate the degradation of a biocomposite based on a thermoplastic banana starch/polycaprolactone (TPS/PCL) matrix reinforced with fique (F) fibers, in different media (soil, compost, and water), to obtain more information about the decomposition times under different environmental conditions. Thus, specific processes could eventually be developed to mitigate the environmental impact generated by the use and application of this material in consumer goods or others.

2. Materials

The native starch used to elaborate the TPS was obtained from the dried banana root of the Dominico Harton variety [29], from the Asociación de Productores de Finca Tradicional del Norte del Cauca (ASPROFINCA), located in the municipality of Villa Rica (Cauca, Colombia). The glycerol used as a plasticizer, from Químicos del Valle Uno A S.A.S. in Cali, Colombia, is of industrial grade, with a purity of 99.8%. The fique fiber was obtained from a market in the same city and was produced by the company Empaques del Cauca in the city of Popayan, Colombia; the material was used as reinforcement and was cut to an average length of 5 mm. The PCL used was acquired from Perstorp UK Limited of Warrington, UK, under reference CapaTM 6800, with a melting temperature (Tm) of 58–60 °C, moisture content < 1%, and an elongation at break of 800%. The commercial compost was purchased from Vivero Pasoancho in Cali, Colombia, and the soil and water used were extracted from an area near Villa Solar and the Lake Biology experimental station, respectively, located at the Universidad del Valle, Meléndez campus.

3. Experimental Procedure

3.1. Statistical Experimental Design

A 3^2 factorial design was used to evaluate the degradation of the biocomposite (TPS/PCL/F) in soil and compost during a period of 3 months; additionally, for comparison purposes, the tests were carried out in an aqueous media, for a single level. The response variable selected for the statistical analysis was the percent mass loss (%ML) of the biocomposite. The analysis of variance (ANOVA) with $\alpha = 0.05$ was performed using Minitab 18® software (Minitab Inc., State College, PA, USA), and to see the effects and interactions of the factors, the following were used: main effects diagram, contour plot, and surface plot. The factors and levels studied are shown in Table 1. This methodology was developed based on a bibliographic review of Rodríguez, et al. [30], Minchola [31], and Accinelli, C. et al. [32].

Table 1. Factors and levels in the experimental design.

Factor	Level
Degradation media	Soil Compost Water *
Depth	5 cm 10 cm 20 cm

Note: * Additional level; only 20 cm was studied in the experimental setup.

3.2. Selection, Sampling, and Characterization of Degradation Media (Soil, Compost, or Water)

To simulate the degradation of the biocomposite in the possible environments to which it would be exposed at the end of its useful life, soil and water samples were taken at the Meléndez campus of the Universidad del Valle, while the compost was purchased at the Pasoancho greenhouse (Figure 1). On one hand, each of the media samples was characterized by physicochemical analyses, such as moisture, soil texture, total organic matter, total nitrogen, pH, electrical conductivity, salt content, and microbial activity, in the case of compost and soil. On the other hand, the following were taken for the water sample: pH, electrical conductivity, microbial activity, chemical oxygen demand (COD), oxide reduction potential (ORP), and dissolved oxygen (DO). The analyses were carried out at the LASA agricultural water and soil laboratory of the School of Sanitary and Environmental Engineering, in accordance with the soil laboratory analytical methods proposed by the Agustin Codazzi Geographic Institute [33], and the methods of physical and chemical analysis of water quality [34].

(a) (b) (c)

Figure 1. Areas selected for sampling and collection of degradation media: (**a**) soil; (**b**) compost; and (**c**) water.

3.3. *Production of Thermoplastic Banana Starch*

The banana starch, previously dried at 60 °C for 24 h, was pre-mixed with glycerol at a ratio of 65 starch/35 glycerol (by mass), using a KitchenAid professional (KitchenAid®, Troy, OH, USA) mixer for 10 min, until the material was homogeneous and free of lumps. Then, the material was stored in polypropylene bags for 48 h. The resulting mixture was plasticized in a HAAKE twin-screw extruder machine (HAAKE Polylab OS RheoDrive 7, Thermo Scientific Inc., Waltham, MA, USA), using a temperature profile of 110, 110, 120, 120, 125, 125, 125, 135, 135, 140, 140, and 145 °C for the ten heating zones of the cylinder and of the head. Finally, the obtained TPS was pelletized using a blade pelletizer (Thermo Scientific Inc, Saint Louis, MO, USA).

3.4. *Preparation of the Binary Mixture TPS/PCL*

The pelletized TPS was manually mixed with the PCL at a ratio of 40 TPS/60 PCL (by mass) and subjected to an extrusion process, using the same equipment previously used to obtain the TPS. The temperature profile handled was 122, 124, 124, 126, 128, 130, 130, 130, 132, 132, 134, and 134 °C for the ten heating zones of the cylinder and of the head. Finally, the TPS/PCL binary mixture was pelletized.

3.5. *Preparation of the Biocomposite*

The TPS/PCL binary mixture was combined with the fique fibers at a ratio of 90 TPS-PCL/10 F (in mass), through an extrusion process following the same transformation conditions previously used to obtain the TPS/PCL binary mixture. Plates (composition: 54% PCL, 36% TPS, and 10% fibers) of 1 mm thickness were obtained with the biocomposite material by means of a hot-compression molding process, using 20 ± 0.05 g of material in a mold holder arranged in a semi-automatic hydraulic press Carver MH 4389-4021, with heating and cooling systems by water circulation (Carver Inc., Wabash, IN, USA): the pressure used was 3000 psi at a temperature of 170 ± 5 °C, and a heating-only time of 8 min was used, followed by another 8 min of heating under pressure. With the prepared plates and using a metal die, V-type pro-betas were obtained with the dimensions specified in the ASTM D-638-14 [35].

3.6. *Degradation–Assembly Configuration*

Wooden baskets of 53 × 48 × 35 cm in length, width, and height, respectively, were used in the set-up configuration. The degradation media, soil (preserving the depth profile), and compost were placed there. The samples necessary for a period of 90 days were placed equidistantly at the three depth levels (5, 10, and 20 cm), as shown in Figure 2a. In the case of water, the samples were placed inside a sample holder at a depth of 20 cm and protected by a wire-mesh cylinder, as shown in Figure 2b.

(**a**) (**b**)

Figure 2. Degradation setup: (**a**) soil and compost and (**b**) water.

3.7. *Analytical Characterization of Samples*

3.7.1. Fourier-Transform Infrared Spectroscopy (FT-IR)

For the determination of the main functional groups in the biocomposite material, PCL, and TPS, a Jasco spectrometer (FT/IR-4100) Type A (JASCO Manufacturing, Portland, OR, USA), operated at 100 scans and a resolution of 4 cm^{-1}, was used. Attenuated total reflectance (ATR) methodology was employed using an ATR PRO450-S accessory (JASCO Manufacturing, Portland, OR, USA).

3.7.2. Scanning Electron Microscopy (SEM)

To observe the surface morphology of the TPS/PCL/F biocomposite samples, a JEOL SEM scanning electron microscope model JSM-6490 (JEOL Ltd., Akishima, Tokyo, Japan) operated at 20 kV was used, where the samples were previously coated with a layer of gold using a Denton Vacuum Desk IV (Denton Vacuum, Moorestown, NJ, USA) cold spray coater model STANDAR. A PHENOM electron microscope model PROX (ThermoFisher Scientific, Waltham, MA, USA) operated at 15 kV, without any coating, was also used. Both devices were used to perform measurements at 500 magnifications.

3.7.3. Tensile Test

The tensile mechanical properties of the TPS/PCL/F biocomposite samples were determined for different sampling times (30, 60, and 90 days), at depth levels of 5, 10, and 20 cm. The tests were performed using a Tinius Olsen model H50KS (Tinius Olsen, Philadelphia, PA, USA) universal testing machine, with a 10 KN load cell and wedge-type grips. Type V specimens were used at a jaw displacement rate of 5 mm/min, in accordance with ASTM D-638 [35].

3.7.4. Mass Loss

Samples were previously dried at 40 °C for 24 h in a LabTech LDO-150F (Labtech S.R.L., Sorisole, Bergamo, Italy) forced convection oven, and the mass of the sample was determined before being subjected to the degradation assembly (M_I). Every 30 days, samples were removed from the degradation media, carefully cleaned with distilled water, and then dried at 40 °C for 24 h to record their mass (M_F). Finally, the model presented in Equation (1) was used to calculate the corresponding mass loss (ML).

$$\%ML = \left(\frac{M_I - M_F}{M_I}\right) \times 100 \tag{1}$$

where:

M_I = Initial mass of dry samples.
M_F = Final mass of dry samples subjected to the degradation environment.

4. Results and Discussions

4.1. *Characterization of Degradation Media (Soil, Compost)*

Table 2 shows the results obtained in the characterization of the soil and compost used as degradation environments.

Considering that soil is a system that integrally relates chemical, physical, and biological factors, the results were analyzed to determine the behavior of the media (soil and compost). Based on the pH values of the samples, they were classified according to the information reported by Jaramillo [36] in Table S1: with the above, the soil sample was classified as "strongly acidic", while the compost sample was classified as "very strongly acidic". Based on the above classification and according to the intervals established by the USDA in 1971 (Figure S1 [37]), the availability of nutrients for a strongly acid soil was high, compared to a very strongly acid soil, which can be confirmed by the values obtained for the percentage of nitrogen for the soil and compost (1.096 and 0.461%, respectively).

Table 2. Physicochemical characterization of soil and compost.

Parameter		Sample		Method
		Soil	**Compost**	**Gómez, I.D. [33]**
Humidity (%)		24.19	11.42	Gravimetric
pH		5.51 ± 0.07	4.75 ± 0.04	Potentiometric
Electric Conductivity (µS/cm)		95.1 ± 5.3	228.5 ± 3.5	Conductivity meter
Total Nitrogen (%)		1.096 ± 0.001	0.46 ± 0.07	Kjeldahl
Total Organic Matter (%)		5.1 ± 0.5	12.0 ± 0.1	Colorimetric
Salinity (dS/m)		0.095 ± 0.005	0.230 ± 0.004	Conductivity meter
Soil Texture (%)	Sand	38	66	Bouyoucos
	Clay	38.2	16.4	
	Silt	23.8	17.6	

With the results of electrical conductivity, the samples can be classified as "non-saline soil" due to the low content of salts present, as established by the Food and Agriculture Organization of the United Nations (FAO) (Table S2 [37]), so it has a negligible salinity effect.

Microbial activity in the compost is more than double that of the soil, according to Figure 3, a behavior that can be explained because of the higher amount of organic matter present in the compost (12%) compared to the soil (5.075%), which represents the consumable biomass for the microorganisms present, and although there is a lower amount of nitrogen in the compost, it is more available, according to Figure S1, where the solubility of different nutrients based on pH is related.

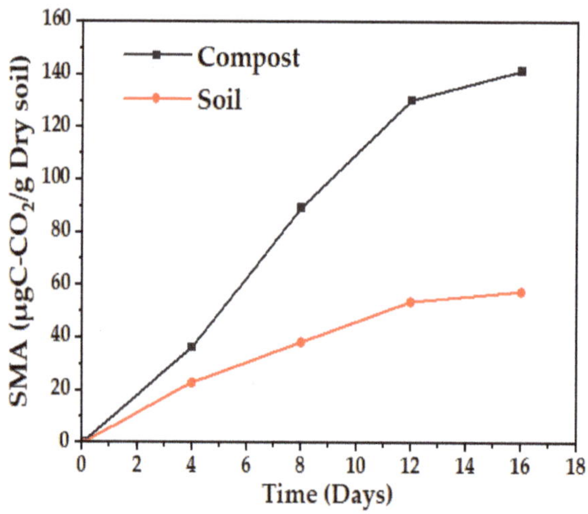

Figure 3. Soil microbial activity (SMA), soil sample, and compost.

Finally, in the case of texture analysis, this allows establishing the capacity of soil to maintain moisture, retain and release ions, and see the availability of nutrients and establishing its aeration capacity and permeability, characteristics that are due to its composition (clay, sand, and silt), considering the proportions receive a classification according to the triangle of textures presented in Figure S2. On one hand, the soil sample is classified as clay loam: its composition contains mostly clay and sand at 38.25 and 38%, respectively, which gives it qualities such as high moisture retention, as well as adhesion and stickiness. On the other hand, the compost sample is classified as sandy loam, with a majority composition of sand (66%): its particles are fine, which makes it saturated with little water, and it does not mold easily, dries quickly in the air, and is not sticky, which allows good aeration. In addition, loam soils are characterized by being fertile, with a high content of organic matter and minerals.

According to all of the above, it can be said that the selected media have good characteristics that would potentially favor the degradation of the biocomposite material.

In the characterization of the aqueous media (Table 3), several conditions were found that allow the proliferation of microorganisms that, together with the movement and friction processes generated by the agitation of the water, can help the decomposition of the biocomposite [38]. The optimum pH range for bacterial growth is between 6 and 8.5, and the sample presents a value of 6.62; therefore, it is an important factor that allows microbial growth, which can be observed in the presence of total coliforms and the number of colony-forming units of mesophilic organisms. Additionally, the values of dissolved oxygen, ORP, and COD are within life-supporting values for a large number of species [39,40].

Table 3. Physicochemical characterization of the lake water.

Parameter	Valor	Method
		Chacon, M.Y. [34]
pH	6.62 ± 0.05	Potentiometric
Electric Conductivity (μS/cm)	20.8	Conductivity meter
TDS (mg/L)	13.52	Potentiometric
COD (mg/L O_2)	0.55	Colorimetric
ORP (mV)	598.2	Potentiometric
DO (mg/L)	6.56	Potentiometric
Mesophiles (CFU/mL)	>300	Plate count
Total coliforms	Presence	Presence–Absence
E. coli	Presence	Presence–Absence

4.2. Analytical Characterization of Samples

4.2.1. FT-IR Spectroscopy

Figure 4 shows the IR spectra of the PCL, the TPS, and the biocomposite material (TPS/PCL/F). The last IR spectra are the reference for analysis of this study. In Table 4, the bands of higher intensity are related to the types of vibrations associated with the functional groups present [41,42]. Figures 5–7 show the FTIR spectra of the biocomposite samples in the three degradation media at different depths (5, 10, and 20 cm), in an evaluation period of 3 months.

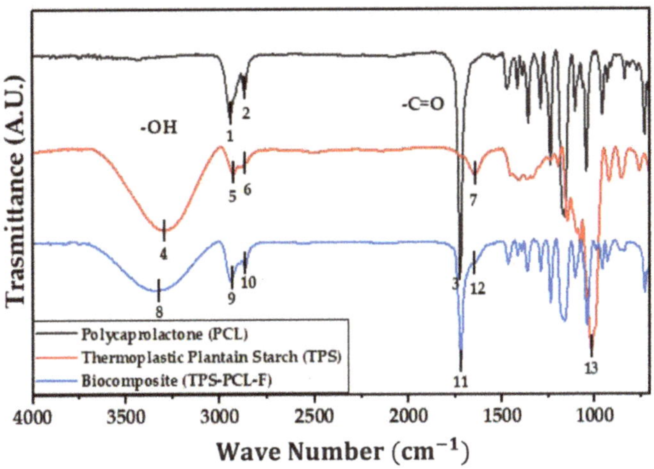

Figure 4. FTIR for PCL, TPS, and TPS/PCL/F biocomposite.

Table 4. Characteristic bands and binding type for PCL, TPS, and the initial sample of TPS/PCL/F biocomposite.

No.	Reference	Wave Number (cm^{-1})
1	Symmetric and asymmetric stretching of methylene groups (CH_2)	2942
2		2868
3	Carbonyl group stretching (C=O)	1721
4	Hydroxyl bond stretching vibration (-OH)	3291
5	Symmetric and asymmetric stretching of methylene groups (CH_2)	2930
6		2880
7	Hydroxyl bond bending vibration (-OH)	1645
8	Hydroxyl bond stretching vibration (-OH) (moved)	3335
9	Symmetric and asymmetric stretching of methylene groups (CH_2)	2941
10		2865
11	Carbonyl group stretching (C=O)	1721
12	Hydroxyl bond bending vibration (-OH)	1645
13	Stretching of starch glycosidic bonding (C-O-C)	1018

Signal 11 (1721 cm^{-1}) in the IR spectrum of Figure 4 corresponds to the stretching in tension of the carbonyl group of the polycaprolactone. This signal had no shift with respect to that found for PCL alone (signal 3), which suggests that hydrogen bonds were probably not significantly formed in the TPS/PCL mixture, thus indicating that it is predominantly immiscible, similar to what was previously reported by [41], who also worked with a binary mixture of TPS/PCL. However, a shift towards a higher wave number of signal 8 was evidenced, related to the stretching of the OH bond, so it can be inferred that intermolecular interactions were probably generated between the hydroxyl groups of TPS and cellulose from F fibers. This type of interactions between TPS and fibers from different sources (sisal, wood, straw, banana leaf, among others) has already been commented on in investigations carried out by Lubis M. et al., Jumaidin et al., and Wang et al. Additionally, Lubis et al., mention that the free hydroxyl groups of TPS that are mutually attracted to the hydroxyls of the macromolecular chain of the fiber are produced during the compression molding process, and, if the displacement is made at a lower wave number, the hydrogen bonds generated are more stable [43–45].

It can be observed in the signals belonging to the symmetric and asymmetric stretching of the (-CH_2) bond at 2941 and 2865 cm^{-1} (Figures 5–7) a greater decrease in the intensity of the bands at depths of 10 and 20 cm, compared with that of 5 cm. In addition, the band of the carbonyl group of the polycaprolactone at 1721 cm^{-1} does not undergo considerable modifications; it only presents a small decrease in the band intensity (it had an average decrease in its intensity of 6, 10, and 35% in soil, compost, and water, respectively, compared to the IR spectra of the day 0, black colored). However, there is no displacement in the wave number. This behavior is because PCL degrades slowly due to its high degree of crystallinity and low moisture absorption; therefore, it remains almost unchanged during the 3 months of the test. Nevertheless, the higher the humidity, the greater the decrease in the intensity of the carbonyl bond (CO), probably due to hydrolytic degradation [46].

Additionally, as shown by the infrared spectra in Figures 5–7, the representative signal of TPS (band from 3000 to 3600 cm^{-1} due to the stretching of the OH bond) decreased almost completely from the first 30 days of the test (at the end of the 90 days of testing, the intensities of these bands decreased on average by 83% for soil and compost and 66% in water). This is due to the susceptibility of TPS to degradative extraction and to the fact that it presents little stability when humidity conditions are high, added to the fact that the fibers of the biocomposite act as a support for the attack of microorganisms, which begins to generate channels in the matrix of the system that allow the entry of water and favor the disappearance of the plasticized starch more easily [47–49].

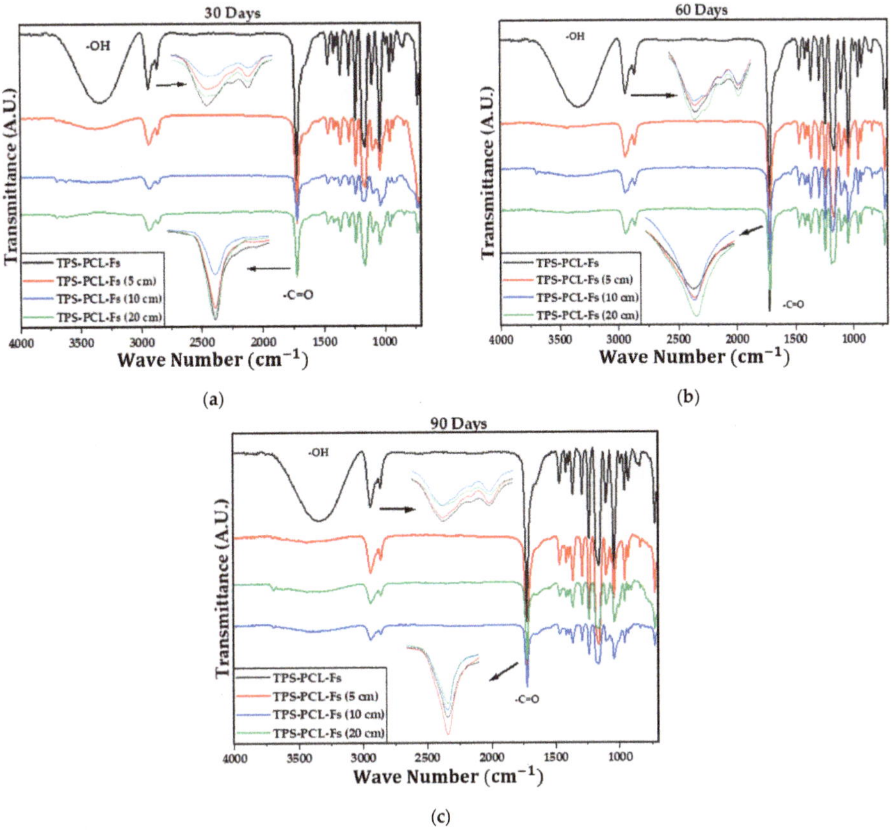

Figure 5. FTIR for TPS/PCL/F biocomposite in soil at (**a**) 30 days, (**b**) 60 days, and (**c**) 90 days.

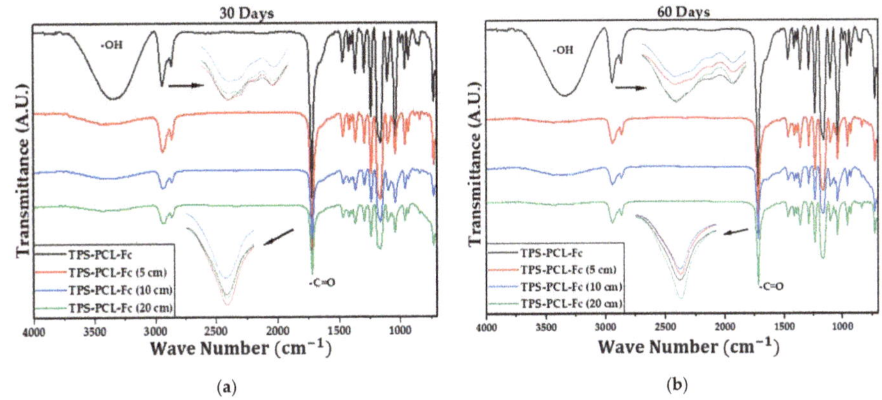

Figure 6. *Cont.*

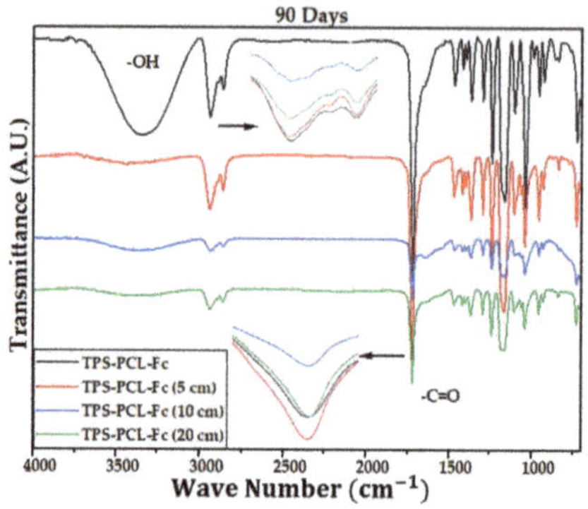

(c)

Figure 6. FTIR for the TPS/PCL/F biocomposite in compost at (**a**) 30 days, (**b**) 60 days, and (**c**) 90 days.

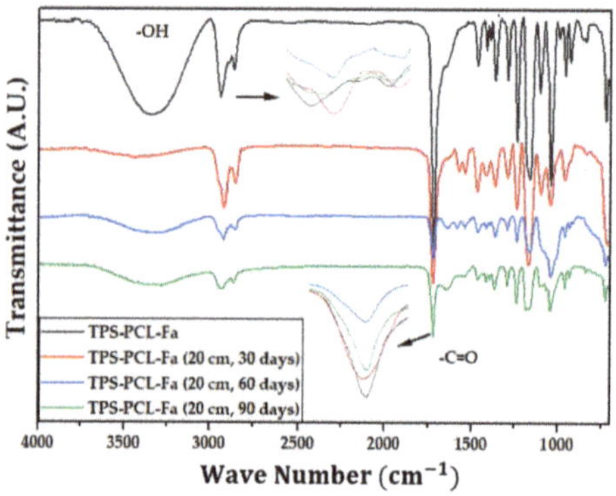

Figure 7. FTIR for TPS/PCL/F biocomposite in water at 20 cm at 30, 60, and 90 days.

4.2.2. Morphological Analysis

As the degradation process of the biocomposite advances, there is a change in the surface of the samples. In Figure 8, it can be observed that the initial samples had a mainly smooth surface, without perforations, being this the reference image to contrast with the results of the materials after conditioning at 30, 60, and 90 days in the different media (soil, compost, and water) and depth levels (Figures 9, 11, and 12).

Figure 9 clearly shows the deterioration of all the samples buried in the soil, compared to the reference image (Figure 8). For the first 30 days, at depths of 5 and 10 cm, there is evidence of loss of homogeneity in the surface and the presence of quite visible holes (red circles), in contrast to the depth of 20 cm, which can be attributed mainly to microbial action and fungal growth that were perceived when the samples were obtained (Figure 10). The characteristics of the worked clay loam soil allow the formation of conglomerates, and, as a consequence, its humidity conditions can vary at a few centimeters distance, since the soil dries more easily towards the surface and retains more water as the depth increases, generating a distinction in bacterial growth, which could explain the differences in the micrographs at 5, 10, and 20 cm depths, where the fibers of fique at the last level are more exposed (yellow circles, Figure 9). When the analysis is made for 60 and 90 days, it is observed that the external damage increases, with the presence of holes and craters, which means the surface was eroded again due to the microorganisms and the deterioration of the material, given the humidity conditions of the media.

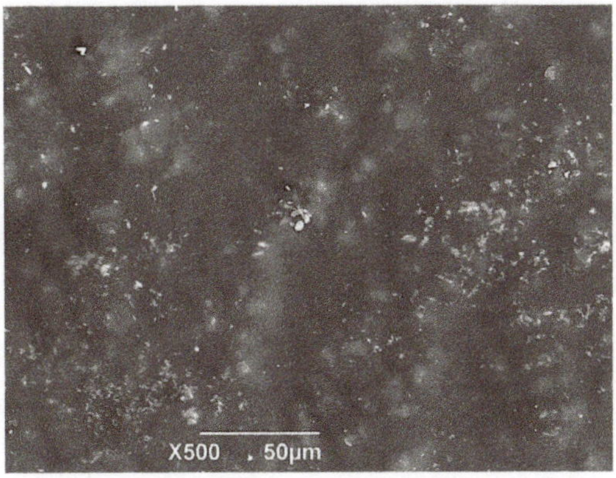

Figure 8. SEM micrograph of TPS/PCL/F Day 0 (standard sample).

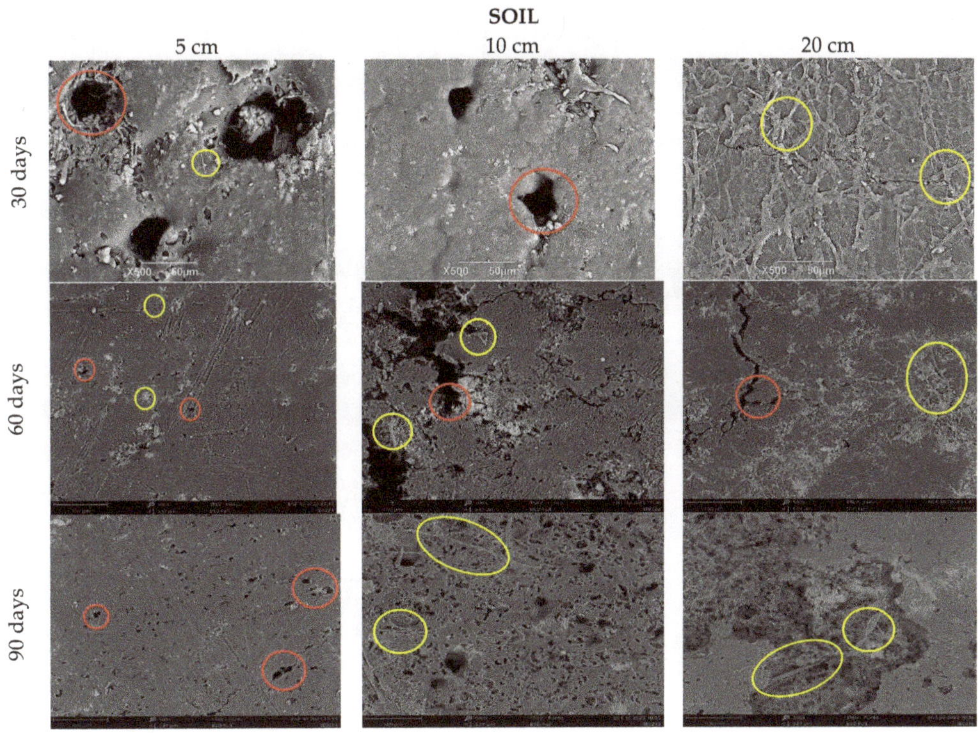

Figure 9. SEM micrograph of TPS/PCL/F in soil at three depth levels (5, 10, and 20 cm) at a time of (30, 60, and 90 days).

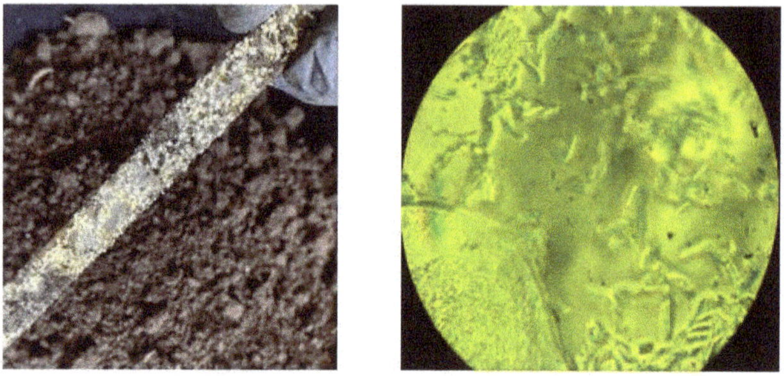

Figure 10. Photographic evidence of the presence of a fungus on the surface of the biocomposite; microscopic image of the fungus at 100× magnification.

Figure 11 shows the deterioration of the samples buried in the compost; as in the soil at the first 30 days, erosion is evident at 5 and 10 cm depths, there are holes (red circles), and the fiber begins to be visible (yellow circles), and at 20 cm, there seems to be less superficial damage. Campos A. et al. characterized a TPS/PCL matrix with 10% sisal fibers and studied its degradation. In their micrographs, they found the formation of agglomerates and the superficial appearance of fibers as a result of the poor interfacial adhesion of the matrix fiber, which contributed to the degradation of their mixture [50,51].

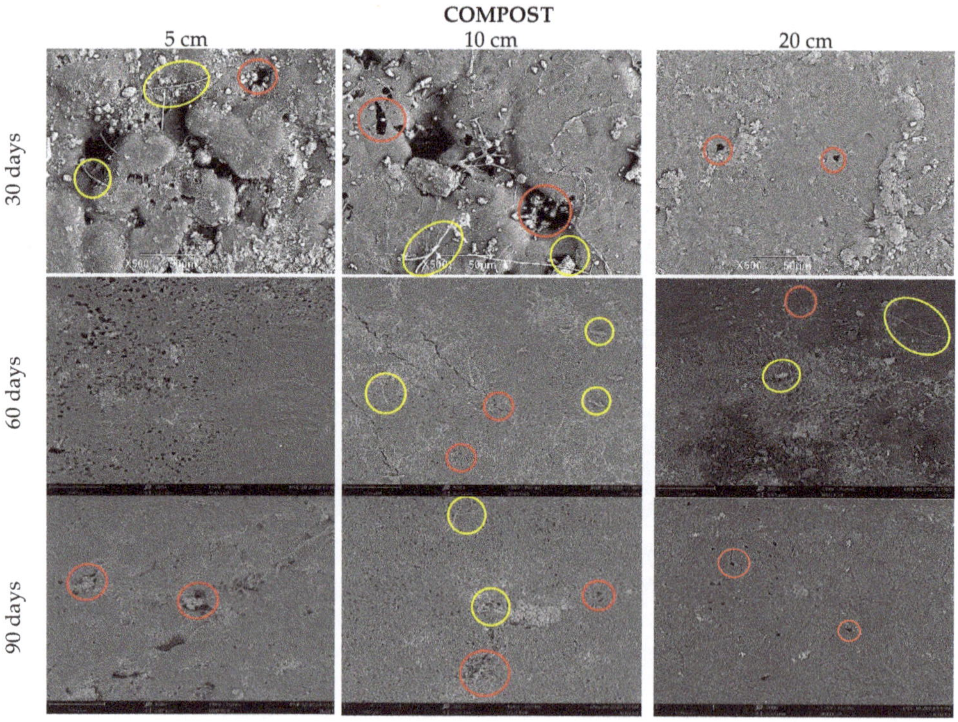

Figure 11. SEM micrograph of TPS/PCL/F in compost at three depth levels (5, 10, and 20 cm) at a time of (30, 60, and 90 days).

In the case of the aqueous media (Figure 12), the samples were kept at the same depth (20 cm), and thanks to the hydrolytic degradation, it is observed that the fibers are gradually exposed from 30 to 90 days of conditioning. Likewise, holes can be seen on the surface, probably due to the fact that the contact of the polymer with the aqueous media promotes water molecules penetrating into the matrix of the material, causing its swelling, rupture of the hydrogen bonds, and finally the hydrolysis of the unstable bonds, which ends up weakening the biocomposite material [52]. The images obtained with SEM equipment do not provide quantitative information about the degradation process.

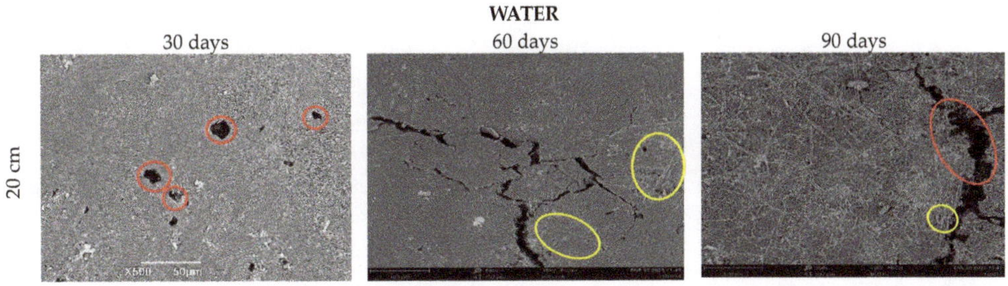

Figure 12. SEM micrograph of TPS/PCL/F in water, 20 cm depth at a time, of (30, 60, and 90 days).

4.2.3. Tensile Strength

The tensile test allows establishing the mechanical behavior of the biocomposite material and its change over time. Figures 13 and 14 show the variation in the maximum strength and the tensile modulus of elasticity of the TPS/PCL/F biocomposite, in the three degradation media, during the time of the experiment. The maximum strength presented, as expected, a progressive decrease over time, which was less marked in the case of conditioning in water in comparison with soil and compost at the three depth levels, showing the greatest changes for 10 and 20 cm. In the case of the modulus of elasticity, a decrease occurs, but with minimal changes, having more influence in the cases of soil and compost. When the TPS/PCL binary mixture is made, its processability and biodegradability improves, and the material has hydrophobic characteristics, decreasing plasticization by water absorption [53,54]; but, as described above, as degradation occurs, the structural damage in the surface morphology of the biocomposite increases (Figures 9, 11 and 12). This is possibly due to the loss of mass, allowing the absorption of water inside the TPS due to its hydrophilic nature, which can enable the action of the retrogradation process, allowing the rearrangement of the polymeric chains by the presence of OH groups, followed by recrystallization that increase the stiffness, altering the mechanical properties, decreasing the elongation, and affecting the quality of the biocomposite material, as described in the studies reported by Villada et al. [48], Mina [55], Cortes et al. [56], and López et al. [57].

4.2.4. Mass Loss (%ML)

The gravimetric analysis shows a gradual loss of mass, with an upward trend as time progresses, with compost being the media with the highest values obtained (34%) after comparing the respective curves at 5, 10, and 20 cm depths, while in the aqueous media, the loss of mass reached does not exceed 20% after 90 days of testing (Figure 15). This was to be expected, considering that PCL takes 2 to 4 years to achieve complete degradation in an aqueous media (hydrolytic degradation) [47] and that the biocomposite is composed of a majority proportion of this polymer (54%). The synergistic effect caused by the presence of bacteria and fungi in the biocomposite is what allows the degradation carried out on land to have higher kinetics.

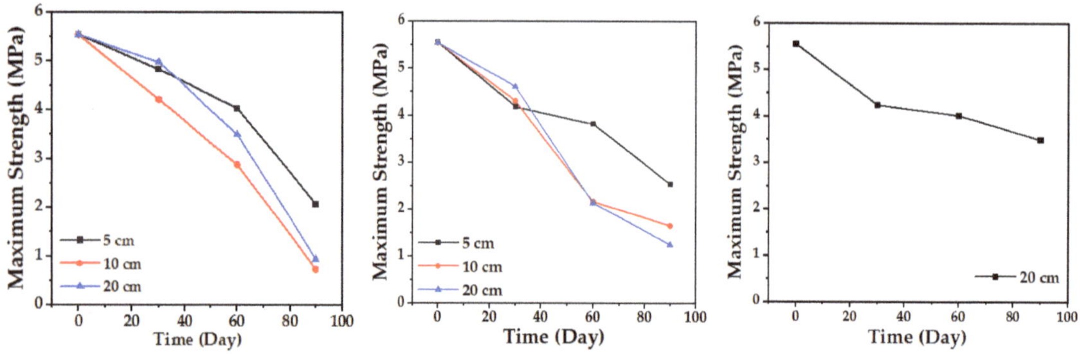

(a) (b) (c)

Figure 13. Maximum biocomposite tensile strength (TPS/PCL/F) over a period of 90 days in the degradation media: (**a**) Soil; (**b**) Compost; and (**c**) Water.

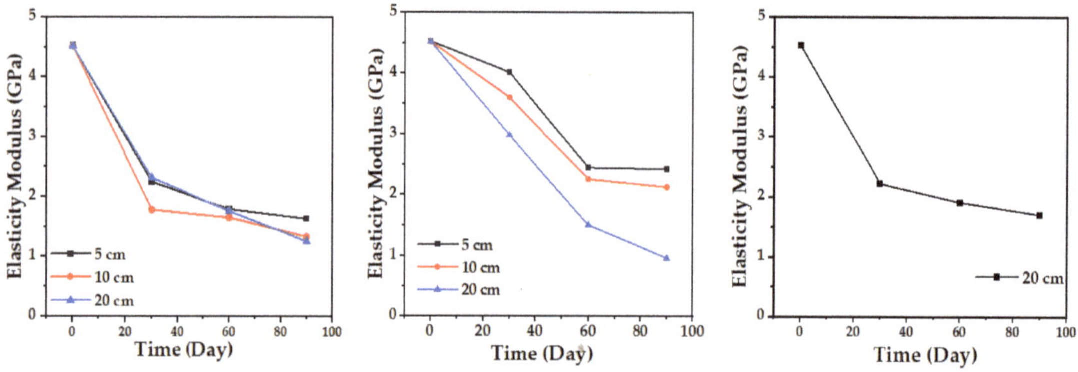

(a) (b) (c)

Figure 14. Tensile modulus of elasticity of the biocomposite (TPS/PCL/F) over a period of 90 days in the degradation media: (**a**) Soil; (**b**) Compost; and (**c**) Water.

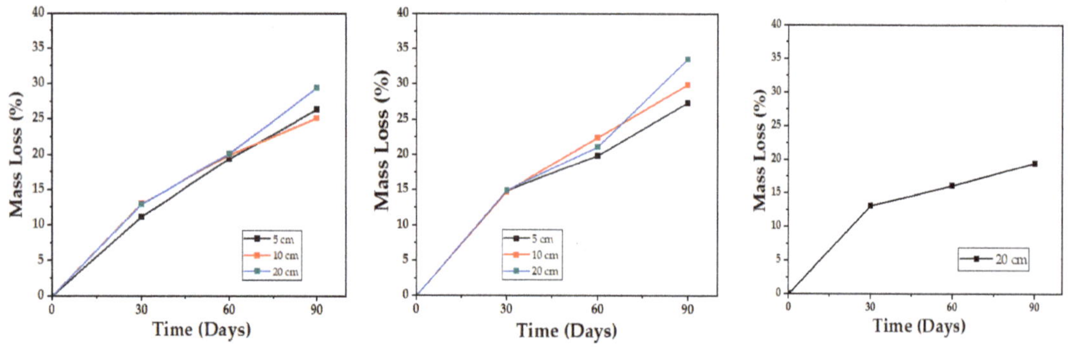

(a) (b) (c)

Figure 15. Mass loss profile for the biocomposite (TPS/PCL/F) over a period of 90 days in the degradation media: (**a**) Soil; (**b**) Compost; and (**c**) Water.

Mass losses can be attributed to a greater extent to the disappearance of the starch that is part of the biocomposite, since this is a more bio-susceptible material than PCL [48], and although its plasticization with glycerol reduces water absorption and gives it better mechanical properties, with the passage of time, the retrogradation and the decrease in free volume reduce its structural stability and make it fragile [58]. Consequently, two different effects are produced in the biocomposite: On the one hand, when the sample fractures, the surface area increases and more spaces are generated for fungi, bacteria, and even mites to degrade the organic matter, as was recorded in the study of aerobic biodegradation of starch plasticized with glycerol, carried out by Merchán J. et al. [59], where samples of equal mass, but with 4.65 times more surface area, achieved 3 times more degradation than those of smaller area in the same test time. On the other hand, the recrystallization of the material with time (mainly TPS) also makes it less prone to hydrolyze with moisture and hinders assimilation by microorganisms since the retrograded starch is less accessible to amylases, and digestion ceases or decreases [60]; according to [61], the crystallinity of starch samples increases with storage time, so that, as the test days pass, the degradation process slows down.

4.2.5. Statistical Analysis

With the mass loss data, an analysis of variance (ANOVA) was performed with a confidence level of 95%, comparing the results at 90 days for the three levels, soil, compost, and water (media 1, 2, and 3, respectively). Due to the mixture of effects contributing to degradation, there is a significant difference between the levels evaluated in soil (soil and compost) and those of the aqueous media, whose mass loss does not exceed the mean of 20%, this being the worst scenario for the final disposal of the biomaterial. In contrast, the best result was achieved with the compost at 20 cm depth, followed by the same media at 10 cm, as shown in the contour plots surface (Figure 16) and main effects (Figure 17) for the response variable. The above effect is probably a consequence of the higher microbial activity in media 2, as can be seen in Figure 3, together with its clay characteristics.

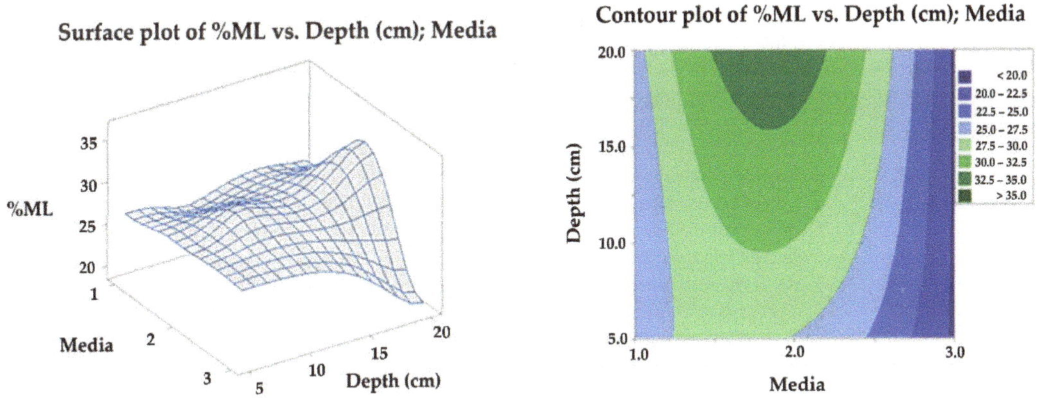

Figure 16. Surface and contour plot for TPS/PCL/F mass loss in degradation media at depth levels.

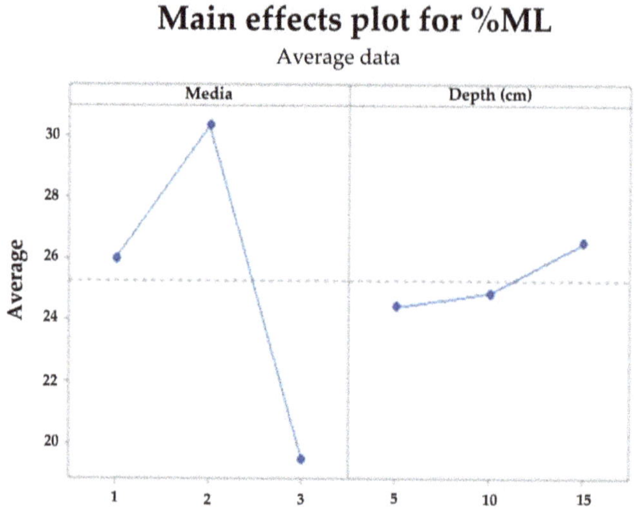

Figure 17. Main effects plot for TPS/PCL/F mass loss in degradation media at depth levels.

5. Conclusions

The degradation of the biocomposite based on TPS, PLC, and F of composition 36, 54, and 10%, respectively, was evaluated by subjecting it to different natural environments to simulate the conditions to which the material would be exposed once its useful life cycle is over. The samples evaluated showed signs of degradation from the first 30 days of testing, decreasing their initial mass and losing mechanical properties with respect to the reference values of day 0. At the end of the total time of the experiment, the biocomposite lost, on average, 26, 33, and 20% of its initial mass in the soil, compost, and water media, respectively. Additionally, mechanical properties, such as maximum strength and tensile modulus of elasticity, decreased considerably.

The SEM micrographs obtained on the biocomposite corroborated the surface deterioration of all the samples; the infrared spectra showed that the most bio-susceptible material of the biocomposite was TPS, since the signals corresponding to the main functional groups of starch decreased in intensity to a great extent for all depths. PCL showed greater permanence in the different environments.

The tensile test performed on the biocomposite, as well as the SEM micrographs, corroborated the deterioration of the biocomposite due to changes in its mechanical properties, decreasing the modulus of elasticity and maximum resistance, since the loss of mass and the deterioration on the surface of the biocomposite can make it susceptible to retrogradation, affecting its quality and behavior over time.

Finally, the best degradation conditions for the material were found in the compost media at 20 cm depth, followed by soil, and finally in the aqueous media, which is due to the synergistic effect produced by bacteria, mite fungi, and the humidity conditions to which they are subjected on land (soil and compost), while in water, the mechanisms for decomposing the biomaterial decrease and slow down the process. From the tests, it is also concluded that the degradation process is not linear, since in the mass loss, tensile test, and modulus of elasticity graphs, the slope varies as time goes by, being this variation smaller month by month. Nevertheless, it would be required to extend the time of the experiment to obtain the time required for the total degradation of the evaluated material. The results obtained in this study show a potential approach for the sustainable process to produce biodegradable biocomposite.

Supplementary Materials: The following supporting information can be downloaded at: https://www.mdpi.com/article/10.3390/polym15193952/s1.

Author Contributions: Conceptualization, F.S.M.R., A.Q.V., H.R.-M. and J.H.M.H.; Methodology, F.S.M.R., A.Q.V., H.R-M. and J.H.M.H.; Investigation, F.S.M.R. and A.Q.V.; Writing—original draft preparation, F.S.M.R., A.Q.V., H.R.-M., J.H.M.H. and E.C.U.; Writing—review and editing, F.S.M.R., A.Q.V., H.R.-M., J.H.M.H. and E.C.U. All authors have read and agreed to the published version of the manuscript.

Funding: This research was funded by Universidad Tecnológica del Chocó.

Institutional Review Board Statement: Not applicable.

Data Availability Statement: Not applicable.

Acknowledgments: The authors acknowledge the Laboratorio de Agua y Suelos Agricolas (LASA) of the Universidad del Valle, Catherine Gómez Paredes, and Diana Marcela Samboni for their contribution during the composition of this work.

Conflicts of Interest: The authors declare no conflict of interest.

References

1. Luyt, A.S.; Malik, S.S. Can Biodegradable plastics solve plastic solid waste accumulation? In *Plastics to Energy*; William Andrew Publishing: Norwich, NY, USA, 2019; pp. 403–423.
2. Momodu, N.S.; Dimuna, K.O.; Dimuna, J.E. Mitigating the impact of solid wastes in urban centres in Nigeria. *J. Hum. Ecol.* **2011**, *34*, 125–133. [CrossRef]
3. Ministerio de Ambiente y Desarrollo Sostenible. Plan Nacional para la Gestión Sostenible de los Plásticos de un Solo Uso. Colombia. 2021. Available online: https://www.minambiente.gov.co/wp-content/uploads/2022/02/plan-nacional-para-la-gestion-sostenible-de-plasticos-un-solo-uso-minambiente.pdf (accessed on 16 March 2023).
4. OCDE. La Contaminación por Plásticos Crece sin Cesar, en Tanto que la Gestión de Residuos y el Reciclaje se Quedan Cortos, dice la OCDE. 2023. Available online: https://www.oecd.org/espanol/noticias/perspectivas-globales-del-plastico.htm#:~:text=La%20producci%C3%B3n%20mundial%20de%20pl%C3%A1stico%20se%20duplic%C3%B3%20de%202000%20a,sum%C3%B3%20353%20millones%20de%20toneladas (accessed on 16 March 2023).
5. Rendón, R.R.; Lorenzo, M.A.; Olvera, R.; Trujillo, C.A. Bioplastic composed of starch and micro-cellulose from waste mango: Mechanical properties and biodegradation. *Polímeros Ciência Tecnol.* **2022**, *32*, e2022026. [CrossRef]
6. Krishnamurthy, A.; Amritkumar, P. Synthesis and characterization of eco-friendly bioplastic from low-cost plant resources. *SN Appl. Sci.* **2019**, *1*, 1432. [CrossRef]
7. Dzeikala, O.; Prochon, M.; Marzec, A.; Szczepanik, S. Preparation and characterization of Gelatin-Agose and Gelatin-Starch blends using alkaline solvent. *Int. J. Mol. Sci.* **2023**, *24*, 1473. [CrossRef]
8. Tarres, Q.; Aguado, R.; Domínguez, J.; Larraneta, E.; Delgado, M. Valorization of Kraft Lignin from black liquor in the production os composites materials with poly(caprolactane) and natural Stone groundwood fibers. *Polymers* **2022**, *14*, 5178. [CrossRef]
9. Scharnowski, J.B.; Rodriguez, A.; Pal, A.K.; Wang, T.; Snowdon, M.R.; Misra, M.; Mohanty, A.K. Biocomposites from thermoplastic postindustrial waste starches filled with mineral fillers for single-use flexible packaging. *Macromol. Mater. Eng.* **2022**, *307*, 2100960. [CrossRef]
10. Kim, G.M.; Chang, W.S.; Kim, Y.K. Biocomposites using whole or valuable component-extracted microalgae blended with polymers: A review. *Catalysts* **2022**, *12*, 25. [CrossRef]
11. Singh, S.S.; Thakur, A.; Sandilya, S.; Kumar, A. Recent advances in Bioplastics: Synthesis and emerging perspective. *Iran. J. Chem. Chem. Eng.* **2022**, *41*, 2704–2727.
12. Lee, T.H.; Yu, H.C.; Forrester, M.; Wang, T.P.; Shen, L.Y.; Liu, H.Z.; Li, J.Z.; Li, W.Z.; Kraus, G.; Cochran, E. Next-generation high-performance Bio-based naphthalate polymers drived from malic acid for sustainable food packaging. *ACS Sustain. Chem. Eng.* **2022**, *10*, 2624–2633. [CrossRef]
13. Chang, C.C.; Trinh, B.M.; Mekonnen, T.H. Robust multiphase and multilayer starch/polymer (TPS/PBAT) film with simultaneous oxygen/moisture barrier properties. *J. Colloid Interface Sci.* **2021**, *593*, 290–303. [CrossRef]
14. Ferrari, F.; Striani, R.; Fico, D.; Alam, M.M.; Greco, A.; Corcione, C.E. An overview on Wood waste valorization as biopolymers and biocomposites: Definition, classification, production, properties, and applications. *Polymers* **2023**, *14*, 5519. [CrossRef] [PubMed]
15. Meereboer, K.W.; Misra, M.; Mohanty, A.K. Review of recent advances in the biodegradability of polyhydroxyalkanoates (PHA) bioplastics and their composites. *Green Chem.* **2020**, *22*, 5519–5558. [CrossRef]
16. Endres, H.J. Bioplastics. *Andvances Biochem. Eng. Biotechnol.* **2017**, *166*, 427–468.
17. Kong, U.W.; Rawi, N.F.M.; Tay, G.S. The potential applications of reinforced bioplastics in various industries: A review. *Polymers* **2023**, *15*, 2399. [CrossRef] [PubMed]
18. Saharan, R.; Kharb, J. Exploration of bioplastics: A review. *Orient. J. Chem.* **2022**, *38*, 840–854. [CrossRef]

19. Barbato, A.G.; Pamplona, J.B. Os desafios para a difusão dos bioplásticos no Brasil. *Rev. Gestão Susten. Ambient.* **2022**, *11*, 365–390. [CrossRef]
20. Freire, E.; Novack, K.M.; Melo, T.M.S.; Finzi-Quintão, C.M.; Gualberto, M.S. Cenários Prospectivos para a Produção de derivados de Moringa oleifera: Uma Abordagem Sistêmica. *Rev. Virtual Química* **2022**, *14*, 529–536. [CrossRef]
21. European Bioplastics. What Are Bioplastics. 2023. Available online: https://www.european-bioplastics.org/bioplastics/ (accessed on 16 March 2023).
22. Springer Nature Group. *Greening the Plastics Industry: Advances and Challenges in Bioplastics Research*; Springer: London, UK, 2023; Available online: https://www.biomedcentral.com/collections/GPIACBR (accessed on 16 March 2023).
23. Tanasă, F.; Teacă, C.A.; Nechifor, M.; Zănoagă, M. Multicomponent Polymer Systems Based on Agro-Industrial Waste. In *Bioplastics for Sustainable Development*; Kuddus, M.R., Ed.; Springer: Singapore, 2021.
24. Dominici, F.; Luzi, F.; Benincasa, P.; Luigi, T.; Puglia, D. Biocomposites based on plasticized wheat flours: Effect on bran content on thermomechanical behavior. *Polymers* **2020**, *12*, 2248. [CrossRef]
25. Dang, K.M.; Yoksan, R.; Pollet, E.; Avérous, L. Morphology and properties of thermoplastic starch blended with biodegradable polyester and filled with halloysite nanoclay. *Carbohydr. Polym.* **2020**, *242*, 116392. [CrossRef]
26. Xiong, Z.; Zhang, Y.; Du, X.; Song, P.; Fang, Z. Green and scalable fabrication of core-shell biobased flame retardants for reducing flammability of polylactic acid. *ACS Sustain. Chem. Eng.* **2019**, *7*, 8954–8963. [CrossRef]
27. Chevali, V.S.; Srinivasamurthy, L.; Khan, B.; Wang, H. Experimental and molecular dynamics analysis of antimicrobial additive migration from biocomposite packaging. In *ICCM22 2019, Proceedings of the 22nd International Conference on Composite Materials, Melbourne, Australia, 11–16 August 2019*; Engineers Australia: Melbourne, Australia, 2019; pp. 803–808.
28. Mina, J.H.; González, A.V.; Muñoz, M.F. Micro- and macromechanical properties of a composite with a ternary PLA-PCL-TPS matrix reinforced with short fique fibers. *Polymers* **2020**, *12*, 58. [CrossRef] [PubMed]
29. Arcila, M.I.; Aranzazu, L.F.; Castrillon, C.; Valencia, J.A.; Bolaños, M.M.; Castellanos, P. *El Cultivo del Plátano*; Corporación colombiana de investigación Agropecuaria y Comité de Cafeteros del Quindío: Manizales, Colombia, 1999.
30. Rodríguez, P.; Medina, A.; Ariza, O. Evaluación de resultados de la degradación de un bioplástico sometido a procesos de compostaje según normas ASTM D-6400 y D-5988. *Inf. Técnico* **2018**, *82*, 112–115. Available online: http://revistas.sena.edu.co/index.php/inf_tec/issue/view/292/214 (accessed on 16 March 2023).
31. Minchola, G. Elaboración y Degradación de Bioplásticos de Residuos de *Solanum tuberosum y Tropaeolum tuberosum* en Lugares de Bajas Temperatura—Oyón. Bachelor's Thesis, Universidad César Vallejo, Lima, Perú, 2019.
32. Accinelli, C.; Saccà, M.; Mencarelli, M.; Vicari, A. Deterioration of bioplastic carrier bags in the environment and assessment of a new recycling alternative. *Chemosphere* **2012**, *89*, 136–143. [CrossRef] [PubMed]
33. Gómez, I.D. *Métodos Analíticos del Laboratorio de Suelos*, 6th ed.; Departamento Administrativo Nacional de Colombia, Instituto Geográfico Agustín Codazzi: Bogotá, Colombia, 2006.
34. Chacón, M.Y. *Análisis Físico y Químicos de la Calidad del Agua*, 1st ed.; Ospina, M.S., Andrade, A.F., Eds.; USTA: Bogotá, Colombia, 2017.
35. *ASTM D638-14*; Standard Test Method for Tensile Properties of Plastics. ASTM: West Conshohocken, PA, USA, 2014.
36. Jaramillo, D.F. *Introducción a la Ciencia del Suelo*; Universidad Nacional de Colombia: Medellín, Colombia, 2002.
37. Barbaro, L.A.; Karlanian, M.A.; Mata, D.A. *Importancia del pH y la Conductividad Eléctrica (CE) en los Sustratos para Plantas*; Instituto Nacional de Tecnología Agropecuaria (INTA): Buenos Aires, Argentina, 2002.
38. Moros, D.C. Presencia de Bacterias Mesófilas y Coliformes del agua de Riego en los Cultivos de Lechuga (*Lactuca sativa*) en la Finca El Rubí de la Vereda San José (Municipio Mosquera). Bachelor's Thesis, Pontificia Universidad Javeriana, Bogotá, Colombia, 2018.
39. Peña, E.; Chang, J. *Calidad de Agua Trabajo de Investigación Oxígeno Disuelto (OD)*; Escuela Superior Politécnica Del Litoral: Guayaquil, Ecuador, 2007.
40. Cervantes, J.; Orihuela, R.; Rutiaga, J.G. On the Development and Control of Microorganisms in the Paper Manufacture. *Concienc. Tecnol.* **2017**, *54*, 54–58.
41. Mina, J.H.; Valadez, A.; Toledano, T. Physico-Chemical studied of thermoplastic starch (TPS) y Polycaprolactone (PCL). *Biotecnol. Sect. Agropecu. Agroindustrial* **2013**, *11*, 31–40.
42. Cuevas, Z.B. Obtención y Caracterización de Almidones Termoplásticos Obtenidos a Partir de Almidones Injertados con Poliésteres Biodegradables. Ph.D. Thesis, Centro de Investigación Científica de Yucatán, Yucatán, México, 2017.
43. Lubis, M.; Harahap, M.B.; Ginting, M.H.S.; Sartika, M.; Azmi, H. Production of bioplastic from avocado seed starch reinforced with microcrystalline cellulose from sugar palm fibers. *J. Eng. Sci. Technol.* **2018**, *16*, 381–393.
44. Jumaidin, R.; Diah, N.A.; Ilyas, R.A.; Alamjuri, R.H.; Yusof, F.A.M. Processing and Characterisation of Banana Leaf Fibre Reinforced Thermoplactic Cassava Strach Composites. *Polymers* **2021**, *13*, 1420. [CrossRef]
45. Wang, C.; Li, F.; Wang, L.; Li, J.; Guo, A.; Zhang, C.; Liu, P. Research on Thermoplastic Starch and Different Fibers Reinforced Biomass Composites. *RSC Adv.* **2015**, *5*, 49824–49830. [CrossRef]
46. Puentes, M.C. Propiedades, Métodos de Sintesis y Aplicaciones de la Policaprolactona. Bachelor's Thesis, Universidad de los Andes, Bogotá, Colombia, 2020.
47. Cyras, V.P. *Relación Estructura, Propiedades y Procesamiento del Material Compuesto Biodegradable Obtenido a Partir de Policaprolactona/Almidón y Fibra Sisal*; Asociación Argentina de Materiales: Rosario, Argentina, 2001.

48. Villada, H.S.; Acosta, H.A.; Velasco, R.J. Research on thermoplastic starches, biodegradable products precursors. *Inf. Tecnol.* **2008**, *19*, 3–14. [CrossRef]
49. Lorenzo, A.T.; Sabino, M.A.; Müller, A.J. Estudio de la biodegradación de una polimezcla de Poli (E -Caprolactona) y almidón de maíz (Pcl/Alm) y su compatibilización con almidón plastificado. *Rev. Latinoam. Metal. Mater.* **2003**, *23*, 25–35.
50. De Campos, A.; Tonoli, G.H.D.; Marconcini, J.M.; Mattoso, L.H.C.; Klamczynski, A.; Greforski, K.S.; Wood, D.; Williams, T.; Chiou, B.; Imam, S.H. TPS/PCL Composite reinforced with treated Sisal Fibers: Property, biodegradation, and water-absorption. *J. Polym. Environ.* **2013**, *21*, 1–7. [CrossRef]
51. Campos, A.; Teodoro, K.B.R.; Teixeira, E.M.; Correa, A.C.; Marconcini, J.M.; Wood, D.F.; Williams, T.G.; Mattoso, L.H.C. Properties of thermoplastic starch and TPS/polycaprolactone blend reinforced with sisal whiskers using extrusion processing. *Polym. Eng. Sci.* **2012**, *53*, 800–808. [CrossRef]
52. Ramos, G. Desarrollo de Nuevos Poliésteres Alifáticos de Cristalinidad y Degradación Hidrolítica Controlada vía Polimerización por Apertura de Anillo (ROP) Utilizando Isopropóxido de Neodimio como Indicador. Master's Thesis, Centro de Investigación en Química Aplicada, Coahuila, México, 2021.
53. Valero, M.F.; Ortegón, Y.; Uscategui, Y. Biopolimeros: Avances y perspectivas. *Dyna* **2013**, *80*, 171–180.
54. Nara, S.; Komiya, T. Studies on the Relationship Between Water-satured State and Crystallinity by the Diffraction Method for Moistened Potato Starch. *Starch Stärke* **1983**, *35*, 407–410. [CrossRef]
55. Mina, J.H. Effect of the Incorporation of Polycaprolactone (PCL) on the Retrogradation of Binary Blends with Cassava Thermoplastic Starch (TPS). *Polymers* **2021**, *13*, 38. [CrossRef]
56. Cortés, J.F.; Fernández, A.L.; Mosquera, S.A.; Velasco, R. Evaluación de propiedades mecánicas, ópticas y de barrera en películas activas de almidón de yuca. *Biotecnol. Sect. Agropecu. Agroind.* **2014**, *12*, 88–97.
57. López, A.; Rivas, J.; Loaiza, M.; Sabino, M. Degradación de películas plastificadas de quitosano obtenidas a partir de conchas de camarón (L. *vannamei*). *Rev. Fac. Ing. UCV* **2010**, *25*, 133–143.
58. Iovino, R.; Zullo, R.; Rao, M.A.; Cassar, L.; Gianfreda, L.; Gianfreda, L. Biodegradation of poly(lactic acid)/starch/coir biocomposites under controlled composting conditions. *Polym. Degrad. Stab.* **2008**, *93*, 147–157. [CrossRef]
59. Merchán, J.P.; Ballesteros, D.; Jiménez, I.C.; Medina, J.A.; Álvarez, O. Estudio de la biodegradación aerobia del almidón termoplástico (TPS). *Lat. Am. J. Metall. Mater. S* **2009**, *1*, 39–44.
60. Zhang, P.; Hamaker, B.R. Banana starc structure and digestibility. *Carbohydr. Polym.* **2012**, *87*, 1552–1558. [CrossRef]
61. Bello, L.A.; Ottenhof, A.M.; Agama, E.; Farhat, I.A. Effect of storage time on the retrogradation of banana starch extrudate. *J. Agric. Food Chem.* **2005**, *53*, 1081–1086. [CrossRef] [PubMed]

polymers

MDPI

Article

Unfolding of Lignin Structure Using Size-Exclusion Fractionation

Audrey LaVallie [1,2], Anastasia A. Andrianova [1,3], Joshua Schumaker [1,4], Sarah Reagen [1,5], Shelly Lu [1], Irina P. Smoliakova [1], Evguenii I. Kozliak [1,*] and Alena Kubátová [1,*]

1 Department of Chemistry, University of North Dakota, 151 Cornell St., Mail Stop 9024, Grand Forks, ND 58202, USA; alavallie@nhsc.edu (A.L.); anastasia.andrianova@agilent.com (A.A.A.); josh.schumaker@hotmail.com (J.S.); sereagen@gmail.com (S.R.); shellylu01@gmail.com (S.L.); irina.smoliakova@und.edu (I.P.S.)
2 Nueta Hidatsa Sahnish College, 220 8th Ave. E, New Town, ND 58763, USA
3 Agilent Technologies, 2850 Centerville Rd., Wilmington, DE 19808, USA
4 SCIEX, 1201 Radio Rd., Redwood City, CA 94065, USA
5 North Dakota Office of the Attorney General, Crime Laboratory Division, 2641 E Main Ave., Bismarck, ND 58501, USA
* Correspondence: evguenii.kozliak@und.edu (E.I.K.); alena.kubatova@und.edu (A.K.)

Abstract: The heterogeneous and recalcitrant structure of lignin hinders its practical application. Here, we describe how new approaches to lignin characterization can reveal structural details that could ultimately lead to its more efficient utilization. A suite of methods, which enabled mass balance closure, the evaluation of structural features, and an accurate molecular weight (MW) determination, were employed and revealed unexpected structural features of the five alkali lignin fractions obtained with preparative size-exclusion chromatography (SEC). A thermal carbon analysis (TCA) provided quantitative temperature profiles based on sequential carbon evolution, including the final oxidation of char. The TCA results, supported with thermal desorption/pyrolysis gas chromatography–mass spectrometry (TD-Py-GC-MS) and ^{31}P NMR spectroscopy, revealed the unfolding of the lignin structure as a result of the SEC fractionation, due to the disruption of the interactions between the high- and low-MW components. The "unraveled" lignin revealed poorly accessible hydroxyl groups and showed an altered thermal behavior. The fractionated lignin produced significantly less char upon pyrolysis, 2 vs. 47%. It also featured a higher occurrence of low-MW thermal evolution products, particularly guaiacol carbonyls, and more than double the number of OH groups accessible for phosphitylation. These observations indicate pronounced alterations in the lignin intermolecular association following size-exclusion fractionation, which may be used for more efficient lignin processing in biorefineries.

Keywords: indulin AT/alkali/softwood kraft lignin characterization; lignin fractionation; preparative size-exclusion chromatography; lignin narrow molecular weight fractions

Citation: LaVallie, A.; Andrianova, A.A.; Schumaker, J.; Reagen, S.; Lu, S.; Smoliakova, I.P.; Kozliak, E.I.; Kubátová, A. Unfolding of Lignin Structure Using Size-Exclusion Fractionation. *Polymers* **2023**, *15*, 3956. https://doi.org/10.3390/polym15193956

Academic Editors: Alexandre Vetcher and Alexey L. Iordanskii

Received: 6 September 2023
Revised: 20 September 2023
Accepted: 25 September 2023
Published: 30 September 2023

1. Introduction

Lignin is one of the most abundant sources of renewable carbon: it builds up within the walls of plant cells and contributes up to 30% of biomass feedstocks [1,2]. As of 2019, the annual lignin market was estimated at USD 955 million with a projected 2% increase annually [3–5]. Due to its polyphenolic structure, a variety of value-added chemicals may be manufactured from lignin [6]; nonetheless, and despite significant efforts, the commercial use of lignin has not exceeded 2% of its annual production [3]. The disadvantage inherent to lignin is its irregular structure, which is composed of randomly cross-linked phenylpropanoid units [7]. An additional, and often underestimated, level of irregularity is caused by strong supramolecular aggregation, as lignin consists of multiple components having varied molecular weights (MW) [8–11]. The heterogeneous nature of lignin hinders a thorough characterization of its structure and, thus, limits its utilization [12]. The fractionation of lignin (desirably scalable) is considered to be a method whereby more homogenous subsections may be rendered: narrower molecular sizes and molecular weight (MW) distributions would greatly increase its homogeneity [5]. In addition, lignin fractionation is expected to break or reduce its aggregation, although this specific effect has not been documented.

Currently, three main methods have been used for lignin fractionation, either alone or in combination [13–16], i.e., selective solvent fractionation [5,12,17–26], differential precipitation [15,27–34], and membrane ultra- and nanofiltration [13–26,28–43]. A recent review also considered ionic liquid-assisted and enzyme-assisted fractionation; however, these new methods have not become mainstream protocols [44].

Selective solvent extraction is based on the differing partial solubility of lignin fractions in various solvents, which range greatly in their polarity and ability to solubilize lignin. Usually, lower-MW species are more soluble [5], so a sequential solvent application allows for obtaining fractions with increasing MW. Duval et al. provided a detailed analysis of the solvent parameters' influence on the MW of the dissolved fraction, leading to a suitable fractionation protocol [19]. A major advantage of this method is the application of relatively inexpensive instrumentation and a moderately easy scale-up process. On the other hand, solubility depends on both the structural features and the MW [40]. Similar issues are encountered when the selective precipitation of dissolved lignin is performed. Furthermore, a non-desired additional chemical alteration may take place [45].

Membrane filtration enables the direct fractionation of black liquor and has the advantage of a controlled separation determined by the MW through the variation in the membrane pore size [40–43,46–50]. A combination of solvent extraction and membrane-assisted ultrafiltration yielded consistent results [51–53]. However, the membranes tend to become foul and the filtration process is not readily scalable to satisfy industrial needs [22].

As early as 1969, Kirk et al. considered the preparative size-exclusion chromatography (SEC) fractionation of lignin using gel permeation chromatography (GPC) as an effective approach for collecting lignin fractions solely based on molecular size [45]. An apparent advantage of this method is that the molecular size cut-offs can readily be controlled by varying the retention-time windows for fraction collection post-SEC processing. Furthermore, SEC is known to be a scalable technique [54].

An SEC application for obtaining lignin fractions with a desired MW, i.e., the separation of nineteen fractions with the number-average MW ranging from 340 to 1250 Da, was reported by Botaro et al. [55] using Acetosolv sugarcane lignin, which had been precipitated with water addition. Fourier transform infrared spectroscopy (FTIR) was used for the analysis of the functional groups, while an analytical SEC demonstrated the differences in the MW distribution for each of the collected fractions [55]. The preparative SEC enabled fractionation, often producing fractions with low polydispersity indices; however, it yielded a fairly low mass range of only up to 1250 Da. While the obtained low MW could be due to the specific feedstock, the fractionation was performed with hydroxypropylated cross-linked dextran as the SEC stationary phase, which might have yielded a separation that was not based exclusively on the MW [56]. We and others have previously shown that the undesired non-SEC interactions may skew the MW-based separation, and that this problem is amplified when a hydroxylated stationary phase is used [56].

The non-SEC interactions between lignin and the stationary phase in SEC, purportedly arising from the heteropolymeric nature of lignin and the variety of its functional groups, have been considered as possible sources of error [57,58]. However, in our previous study, we demonstrated that a highly cross-linked porous polystyrene/divinylbenzene matrix-based (PSDVB) stationary phase enabled lignin separation based solely on its MW [56]. The accuracy of the MW determination was confirmed using independent methods, including matrix-assisted laser desorption ionization with time-of-flight high-resolution mass spectrometry (MALDI-TOF HR MS) and direct infusion electrospray ionization (ESI) with TOF HR MS, which we also developed for lignin MW analysis [59,60].

In this study, capitalizing on the development of a suite of methods for lignin characterization, we set out to determine whether lignin's chemical structure correlates with its MW.

Following lignin fractionation using a preparative SEC, with a scaling-up of the previously validated conditions [56], a detailed structural characterization of the obtained fractions was performed. The hypothesis tested herein was that the fractions of different MWs may also be structurally different. We also postulated that the suite of analytical protocols used in this study, which included both thermal and spectroscopic methods, could provide insights into the lignin structure and reveal significant supramolecular interactions, which may contribute to lignin recalcitrance.

Of those, the TCA application was the significant innovation enabling accurate carbon quantification and providing specific thermal evolution profiles. We have successfully applied this method, in combination with TD-Py-GC-MS, analytical SEC, and ESI-TOF HR MS, to lignin analysis, observing the changes in its chemical structure while accurately closing the mass balance of carbon [61,62]. However, this study, for the first time, uses TCA to explain the changes in the polymer thermal properties observed due to the lignin components' interactions, an outcome that was not anticipated when developing this method.

2. Materials and Methods

2.1. Materials and Reagents

Alkali lignin was purchased from Sigma-Aldrich (St. Louis, MO, USA). It was determined to have an elemental composition of C (64.14%), H (5.79%), S (1.39%), and N (0.46%) by Atlantic Microlab, Inc. (Norcross, GA, USA). HPLC grade unstabilized tetrahydrofuran (THF), containing no preservatives, was obtained from Fisher Scientific (Fair Lawn, NJ, USA). Deionized water was obtained using a Direct-Q® 3 system from Millipore (Billerica, MA, USA).

For the preparative SEC, the alkali lignin was dissolved entirely in a 1:1 (v/v) THF/water mixture at a concentration of 50 mg/mL and further diluted with THF to form a lignin solution with a final concentration of 10 mg/mL, containing 10% of water. When the water content was decreased, no precipitation was observed.

For column calibration, two sets of narrow-range polymeric standards were used, i.e., a polystyrene (PS) standard set with a MW peak maxima (M_p) of 580–19,760 Da, and polymethyl methacrylate (PMMA) standards (M_p 550–56,600 Da) from Agilent Technologies (Santa Clara, CA, USA). Pinoresinol (≥95% purity, Sigma Aldrich, Burlington, MA, USA) was used as the lignin structure model compound to verify the calibration, as suggested in the previous SEC validation study [56].

For the ^{31}P NMR analysis, a set of lignin model compounds, including phenol, guaiacol, methyl guaiacol, ethyl guaiacol, propyl guaiacol, vanillin, acetovanillin, syringaldehyde, vanillic acid, homovanillic acid, and bicreosol (all ≥95% purity, Sigma Aldrich) were analyzed with respect to the chemical shift, for identification and quantitation. Pyridine (≥99.8%), cyclohexanol (≥99%), 2-chloro-4,4,5,5-tetramethyldioxaphospholane (TMDP) (≥95%), deuterochloroform (≥99.8%), and chromium acetylacetonate (≥97%) (all from Sigma Aldrich) were used for sample preparation.

2.2. Lignin Fractionation via Preparative SEC

To confirm that the SEC separation was primarily controlled by size exclusion, SEC calibration was performed with two sets of standards differing in their functional groups (PS, PMMA), as was demonstrated to be effective in our previous work (Figure S1 and Table S1) [56]. Additional details can be found elsewhere [60].

The preparative SEC fractionation was performed on an Agilent 1100 Series HPLC system utilizing a preparative PLgel column (300 × 25 mm, with 10 μm particle size and a 1000 Å pore size) (Agilent Technologies, St. Cloud, MN USA). The system was equipped with a diode array detector (DAD). For this work, the analytical flow cell was replaced by a preparative flow cell (Agilent Technologies). An unstabilized THF was used as the mobile phase at a flow rate of 5.0 mL/min; it was essential to use unstabilized THF to obtain pure lignin fractions without butylated hydroxytoluene or other additives used for

THF stabilization. An extended loop capillary was installed in the injection loop to perform a 500 µL injection of a 10 mg/mL lignin solution.

Several fractionation trials (with the SEC chromatograms shown in Figure S2), which obtained 4–6 fractions, were performed by slightly varying the collection-time windows, yet provided comparable results (the SEC data from the preliminary fractionations are shown in the Supplementary Information). The final protocol targeted the collection of representative fractions with differing MW. For example, the pre-eluate (i.e., the fraction collected before fraction 1, which is henceforth labeled as that with the highest MW) was not further considered because it contained mere traces of organic carbon.

In the final fractionation protocol, the pre-eluate was collected first, during the retention times when no increase in the DAD signal was observed. Then, MW fractions 1–5 were obtained during the following elution-time windows: 14–16, 16–18, 18–20, 20–22, and 22–24 min. The fraction collection was performed manually. The procedure was repeated ten times, resulting in a final volume of 100 mL for each of the six collected fractions. Each fraction was concentrated using evaporation under a stream of nitrogen to a final volume of 2 mL. As the control, an aliquot of THF (100 mL) was also concentrated to a final volume of 2 mL.

2.3. Analytical SEC of Lignin Fractions

The obtained preparative SEC fractions, the control sample (concentrated THF), an aliquot of pure THF, and an untreated lignin solution (50 mg/mL) were analyzed using an analytical-scale SEC on an Agilent 1100 Series HPLC system equipped with a DAD with an analytical high-pressure flow cell, utilizing a PLgel analytical column (300 × 7.5 mm, with a 5 µm particle size and a 1000 Å pore size, 500–60,000 Da separation range) equipped with a PLgel guard column (50 × 7.5 mm) (Agilent Technologies). The SEC column with a PSDVB stationary phase was calibrated with PS standards (Figure S1). Unstabilized THF was used as the mobile phase at a flow rate of 1.0 mL/min. The injection volume for all samples was set to 20 µL.

The SEC analysis DAD abundance profile was based on the absorbance in a UV-Vis range of 220–700 nm. The SEC determination of the MW as M_n (number-average MW), M_w (mass-average MW), and dispersity index (DI) values were based on standard SEC equations [56].

Standard deviations are not provided in pertinent Tables and Figures because the results are reported for a single preparative SEC separation. The results were confirmed by four similar prior experiments (shown in the Supplementary Materials), with some of the fractions being collected on a slightly modified time scale, which prevented an exact statistical evaluation. Nevertheless, when the outcomes were compared, the variance did not exceed 10% for similar fractions.

2.4. Thermal Carbon Analysis

A thermal optical analyzer from Sunset Laboratory Inc. (Portland, OR, USA) was employed to obtain the quantitative thermal carbon evolution profiles (TCA), which enabled a comprehensive carbon fractionation and characterization [62,63]. TCA is a relatively novel method of accounting for all the carbon mass in an organic sample (converting all carbon to CO_2 and then methane, with the uniform quantification of various carbonaceous species), with the added benefit that the temperature programs provide the fractionation by volatility, i.e., higher molecular weight compounds evolve at higher temperatures. Further, the evolved fractions are separated into two kinds: (1) those due to thermal physical desorption, from an ambient temperature up to 300 °C, and (2) pyrolytic products which evolve at higher temperatures, reflecting those lignin components that cannot volatilize without chemical decomposition—i.e., higher-MW compounds. The remaining material, recalcitrant char, is then evolved and recovered via combustion, thus closing the mass balance on carbon [62].

A sample (20 μL) was introduced into a Pall Flex 2500QAT-UP tissue quartz filter (Pall Corp, East Hills, NY, USA), dried on a hot plate at 40 °C for 4 min, and placed into the oven. The sample was desorbed/pyrolyzed at selected temperature steps for specific time durations. A detailed description of the applied TCA protocol can be found elsewhere [62,63]. Briefly, the thermal desorption temperatures were 30, 200, and 300 °C, while pyrolysis took place at 400, 500, and 890 °C in a helium atmosphere. This sequence was followed by oven-cooling to 550 °C, the introduction of an oxidizing carrier gas mixture of He with 10% O_2, and heating to 890 °C in order to evolve the coked carbon fraction. All the evolved species were converted to CO_2 and then to methane, thus allowing for quantification using a single standard with a flame ionization detector [62].

2.5. ESI-TOF HR MS Analysis

The HR MS mass distribution of the different lignin fractions was obtained using an Agilent 6210 LC/TOF with a mass resolution of >13,000 (at *m*/*z* 2722) and mass accuracy <2 ppm (*m*/*z* 609.2807) with ESI [59]. The samples were introduced via direct infusion with a syringe pump at a flow rate of 5.0 μL/min. The analysis was performed in the positive ion mode with electrospray ionization (i.e., the capillary potential) and collision-induced dissociation (the fragmentor potential) set at 3500 and 150 V, respectively. Nitrogen, at a flow rate of 4 L/min, was used as the nebulizing gas. The nebulization temperature and pressure were set to 250 °C and 20 psi, respectively. The TOF MS system was calibrated with $[(CsI)_n + Cs]^+$ clusters formed by an introduction of cesium iodide (30 mmol/L solution in ACN/water 1:1 (*v*/*v*)) via direct infusion at a flow rate of 5 μL/min.

The Mass Hunter software package B.07.00 (Agilent Technologies, St. Cloud, MN, USA was used for data processing. The mass spectra of lignin were deconvoluted using a built-in tool utilizing an unbiased isotope model with a peak spacing tolerance of 0.0025 *m*/*z*. The maximal assigned charge state was not limited. Hydrogen (proton) was considered as the charge carrier. The peaks selected for deconvolution were filtered based on their absolute height (≥100 counts) and the relative height of the largest peak, which was set to ≥0.1% of the largest peak unless otherwise stated. The maximum number of peaks was not specified.

2.6. TD-Py-GC-MS Analysis

TD-Py-GC-MS was performed on a CDS Analytical Inc. 5200 pyroprobe (Oxford, PA, USA), connected to an Agilent GC 7890 with 5975C MS. The quartz tube with quartz wool was cleaned outside of the probe at 1200 °C for 5 s. The sample was introduced at a 5.0–10.0 μL volume onto the quartz wool filter before the probe was inserted and, once inserted, the probe was heated sequentially through 200, 300, 400, 500, and 890 °C steps, just as in our original publication [62]. The probe was held at each temperature for 30 s, except at the 890 °C step, which was held for 10 s. The transfer line and valve oven were kept at 300 and 320 °C, respectively, and the pyroprobe assembly was held at 300 °C. The temperature steps of 200, 300, 400, 500, and 890 °C were repeated twice during the runs to ensure that all potential polymers had evaporated for the online GC-MS analysis. Additional details of this experimental setup are provided elsewhere [62,63].

The GC-MS was equipped with a 51 m HP 5MS column (0.25 μm film thickness and 0.25 mm inner diameter). The GC inlet was kept at 300 °C, with a 10:1 split ratio. The GC oven temperature program started at a 50 °C hold for 1 min, followed by a 40 °C/min gradient up to 80 °C, and then used the second gradient of 25 °C/min, up to 320 °C, with a final hold of 7 min. The MS analysis was performed with no solvent delay in the 10–550 *m*/*z* mass range. The resulting GC-MS data showed that the second run for each pyroprobe temperature yielded no residual species before the next increased temperature step, hence, indicating no carryover. The total ion current (TIC) chromatograms of the fractions and blank sample were analyzed for lignin-derived compounds; the peaks were labeled when the compounds were identified with ≥80% accuracy for the NIST library search results or based on confirmation with the standards [62].

2.7. ^{31}P NMR Analysis

A Bruker AVANCE 500 NMR spectrometer was used to record the ^{31}P spectra. Samples for the proton-decoupled ^{31}P NMR (to be called just ^{31}P NMR henceforth), were obtained using a mixture of pyridine and $CDCl_3$ (ratio of 1.6:1). For the quantitative ^{31}P NMR studies, the pulse width was optimized to give the 90° flip angle at approximately 10 μs. The optimized pulse delay was 20 s. The ^{31}P NMR spectra of the TMDP and its hydrolysis product were obtained at 256 scans, while the spectra of the phosphitylated lignin, lignin degradation products, and other analytes were obtained using 1024 scans.

The general procedure for the phosphitylation reaction was as follows, using a combination of the published protocols [64,65]: A mixture of pyridine and $CDCl_3$ (400 μL of 1.6:1 (v/v)) was added to a 4.0 mL vial with a magnetic stir bar. Then, a sample to be phosphitylated was introduced into the vial. During the quantification studies, chromium acetylacetonate (1.0 mg) and cyclohexanol (10 μL, the internal standard for integration) were added to the vial before introducing the phosphitylation reagent. Two molar equivalents of TMDP were added dropwise to the solution. After stirring at room temperature for 5 min, the phosphitylated sample was transferred to an NMR tube. The ^{31}P NMR spectra were recorded within 1 h after preparation of the sample. Prior to phosphitylation, the samples that were dissolved in the water/DCM were dried using a rotary evaporator at 20 torr for 60 s to remove the solvents from the system, since the presence of hydroxyl groups in the solvents was not conducive to the phosphitylation reaction.

The ^{31}P NMR signal of the phosphitylated cyclohexanol was observed at δ 145.2 ppm. A sample of hydroxylated TMDP, 2-hydroxy-4,4,5,5-tetramethyl-1,3,2-dioxaphospholane, was produced by adding two droplets of water to a solution of 250 μL pyridine, 150 μL $CDCl_3$, and 15 μL TMDP, which was stirred for 5 min. The ^{31}P NMR signal of the hydroxylated TMDP was observed at δ 132.2 ppm.

The number of hydroxyl groups present in each sample was calculated through integration relative to the cyclohexanol peak found at 145.2 ppm with a known concentration. The following regions were used for the classification of the hydroxyl groups detected in the weight-fractionated samples: 143–150 ppm for alcohols, 138–143 ppm for phenols, and 135–138 ppm for carboxylic acids.

Prior to conducting the NMR studies of the fractions, we verified the method used by matching its results to the amounts of the hydroxyl groups in indulin reported by Meadwest Vaco in the manufacturer's certificate of analysis (Table 1). It is of note that the phosphitylated benzylic and aliphatic hydroxyl groups are indistinguishable using ^{31}P NMR spectroscopy, and the number of aliphatic moieties given in Table 1 includes the hydroxyl groups at the benzylic position. While the number of aliphatic and, especially, phenolic hydroxyl groups determined using ^{31}P NMR spectroscopy is somewhat lower, the difference between our data and those of Meadwest Vaco for the total amount of OH-containing groups was negligible when the carboxylic acids (not reported by Meadwest Vaco) were added.

Table 1. Comparison of the number of mmoles of hydroxyl groups present per g of alkali lignin reported by Meadwest Vaco (for indulin) vs. the results obtained in the present study using ^{31}P NMR spectroscopy for alkali lignin and indulin.

		^{31}P NMR Data	
	Meadwest Vaco Data	**Alkali Lignin**	**Indulin**
Phenolic OH	3.6	2.85 ± 0.04	2.4 ± 0.3
Benzylic OH	0.06	2.7 ± 0.2	2.4 ± 0.3
Aliphatic OH	2.9		
Carboxylic Acid OH	Not Reported	0.78 ± 0.01	0.7 ± 0.1
Total	6.5	6.3 ± 0.2	5.5 ± 0.6

3. Results

As demonstrated in our previous work, the application of a highly cross-linked porous PSDVB stationary phase enabled lignin separation, which was truly based on MW, according to the calibration using both polar and non-polar standards [56]. So, in this study, lignin was effectively separated into five main fractions via preparative SEC (Figure 1). As a result, a narrower MW distribution of the species within the fractions, as compared to the original, intact lignin, was achieved.

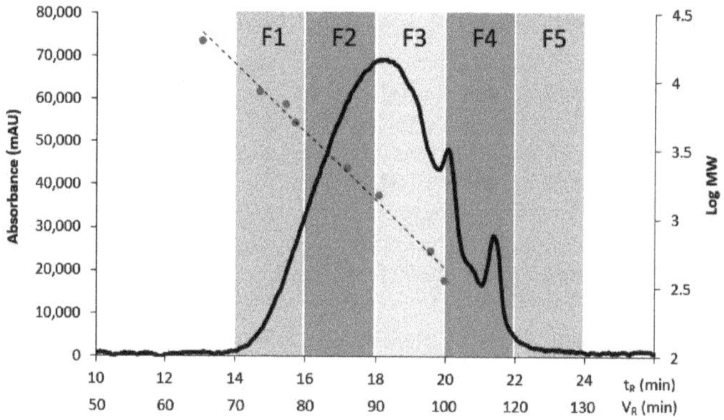

(a)

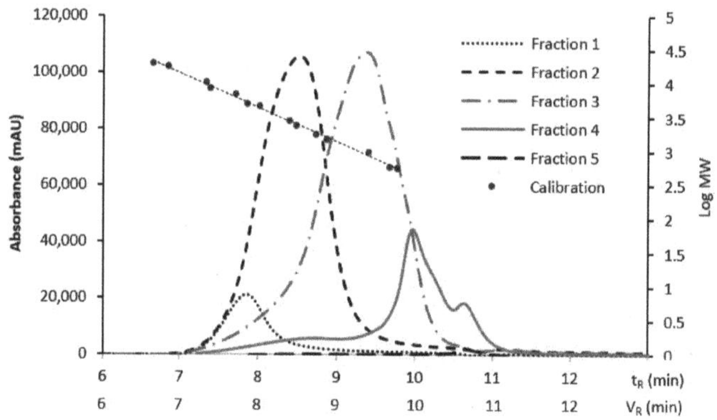

(b)

Figure 1. SEC analysis of (**a**) chromatogram denoting alkali lignin fractionation using preparative SEC showing the collection of five fractions (see experimental section for detail) and (**b**) chromatograms of fractions analyzed with analytical SEC. The dotted straight line with data points, as well as the right vertical axes, refer to the SEC calibration. (**a**) Preparative SEC and (**b**) analytical SEC.

3.1. Mass and MW Distribution in SEC Fractions

The distribution of lignin among the fractions obtained using the preparative SEC (Figure 1a) was assessed using two methods based on (i) the TCA results enabling the accurate quantification of carbon and (ii) an analytical SEC with UV/Vis detection (Figure 1). Similar distribution profiles were obtained using both methods (Table 2), suggesting that the UV/Vis response was mostly based on the contribution of lignin aromatic moieties, which made up the bulk of the lignin mass. It is of note that the TCA provided the inherently non-biased carbon abundance, as all the carbon was evolved, and then all of the evolved carbon was quantified as a single analyte, methane. Thus, the fraction abundance obtained using TCA presents the true values, which are absolute rather than relative. Given this feature, the obtained TCA data are re-visited in Section 3.4, where they are matched to the TD-Py-GC-MS profiles.

Table 2. Distribution of lignin sample across fractions as measured using TCA (the data are proportional to mass) and DAD (based on UV-Vis absorption) following the preparative SEC.

Fraction #	1	2	3	4	5
TCA % distribution of lignin across fractions	3.9	32	43	20	0.47
DAD % distribution of lignin across fractions	5.6	37	43	14	0.30

The general trends of the analytical SEC based on the average MW values, M_p, M_n, and M_w listed in Table 3, were as expected, showing sequentially decreasing values. Large-MW species eluted first, and the following fractions featured successively lower-MW species, presumably phenolic oligomers, then dimers with monomers.

Table 3. MW characterization using the analytical SEC of both the unfractionated alkali lignin and the five fractions obtained using the preparative SEC.

			Analytical SEC			
Fractions	**1**	**2**	**3**	**4**	**5**	**Alkali Lignin**
M_n	4698	2824	1342	802	248	1631
M_w	5862	3823	2626	2595	313	2740
M_z	6647	4943	4979	4932	396	3723
DI	1.2	1.4	2.0	3.2	1.3	1.7
		Estimates based on preparative SEC				
Expected MW range (g/mol)	4300–13,300	1400–4300	440–1400	140–440	<140	
% species: in the expected range	58	69	61	50	5	
of MW > than expected	0	15	29	44	95	
of MW < than expected	41	15	10	5	0	

However, a closer look at the analytical SEC profiles acquired for each fraction revealed rather broad peaks of a non-Gaussian shape (Figure 1b) for each fraction, indicating a possible carryover of high-MW lignin components into the late fractions, as well as the occurrence of low-MW components in the early fractions.

This observation indicated strong interactions between the unfractionated lignin components of varied MWs within the original, intact lignin structure. We hypothesized that lignin fractionation broke some of these interactions, leading to a partial "unraveling" of the aggregated lignin structure. Evidence of the supramolecular structure of lignin was reported earlier [10,11]. As an extreme case, Crestini et al. showed that milled wood lignin components are associated so strongly that the resulting polymer was actually an aggregate of linear oligomers [9]. The rest of this study reports and discusses the evidence confirming the correctness of this assumption obtained using various analytical methods.

The comparison of the expected MW (obtained using the preparative SEC) and the actual M_n, M_w, and PI (based on the analytical SEC) presented in Table 3 supports the occurrence of intermolecular interactions between the unfractionated lignin components of varied MWs and their disruption as a result of their run through the preparative and then the analytical SEC columns. Presumably, these aggregates partially disassembled and became separated from each other. The contributions of the species with higher-than-expected MWs were rather sizable, particularly for fractions 4 and 5 (Table 3). Note that while fraction 5 is small in abundance, the obvious contribution of the higher-MW species cannot be ignored.

Interactions between lignin components have been observed earlier; the interest in studying them stems from the fact that they are responsible for the formation and high stability of lignin nanoparticles [10,11]. The tendency toward aggregation and particle size have been shown to depend on the extracting solvent [12,66], suggesting some dependence on the MW. However, to the best of our knowledge, no direct evidence of the particular interactions between low- and high-MW lignin components with defined MWs has been reported until this study.

The observed peak broadening (in Figure 1) is unlikely to have been caused by common chromatographic artifacts, such as peak fronting, caused by the overloading of separation sites on the preparative SEC. This phenomenon was not observed in our earlier study [56], nor did it become apparent in the first, preparatory SEC run in this study, even though the column loading was significantly higher. Similarly, column interactions, besides size exclusion, which could potentially cause peak tailing were not observed during the first SEC run.

However, verification using independent methods was still warranted. The average MW values obtained with SEC calibration were verified using a complementary method, ESI-TOF HR MS (developed earlier) [59], as shown in the next section. Then, corroboration was sought for the occurrence of interactions between the high- and low-MW lignin components by comparing the thermal behavior of the SEC fractions and native lignin using two independent methods, TCA and TD-Py-GC-MS.

3.2. ESI-TOF HR MS Analysis of SEC Fractions

The ESI-TOF HR MS, just as the SEC did, showed sequentially smaller number-average and weight-average molecular weights (M_n and M_w) values as the fractions proceeded from 1 to 5, while the unfractionated lignin showed M_n and M_w values in the mid-range of the fractions' values (Figure 2, Table S2). The actual ESI-TOF HR MS profiles of fractions 1–5, shown in Figure S3, also reflect this expected trend through the sequence of the mass spectra: fraction 1 clearly shows an abundance of mid- and high-MW values for the compounds, while fractions 2–4 feature steadily fewer high-MW peaks and then fewer mid-MW species and, finally, by fraction 5, only low-MW compounds were present in the amounts detectable using MS.

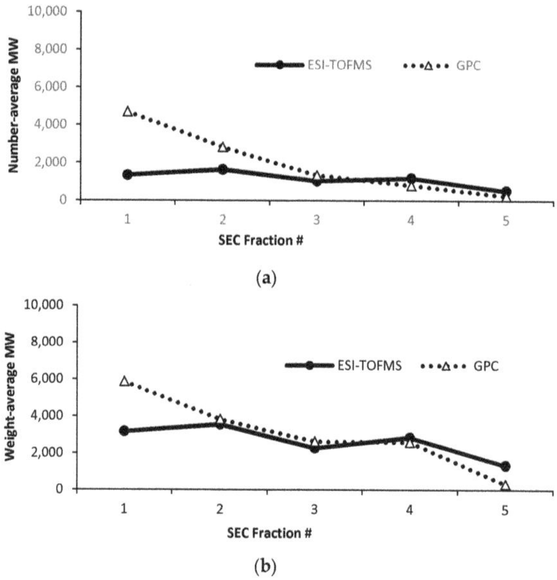

Figure 2. Comparison of SEC vs. ESI-TOF HR MS number-average and weight-average molecular weights for all lignin fractions. (**a**) M_n and (**b**) M_w.

The analytical SEC and ESI-TOF HR MS yielded similar, average-MW values for fractions 2–4, which comprise the bulk of the fractionated lignin. The difference between the values obtained using MS and SEC was pronounced only for fraction 1, which is consistent with the limitation of MW determined using MS, as large-MW species may not volatilize and may excessively fragment in comparison to lower-MW species. It is of note that the low abundance or even apparent absence of high-MW peaks in fractions 3–5 does not necessarily mean that they are absent: unlike the UV detection used in the SEC, the HR MS signals of those species present below their detection limits do not accumulate.

Thus, the carryover of the high- and low-MW species observed in the SEC could not be confirmed by the ESI-TOF HR MS due to this method's limitations. Confirmation of this was obtained by using thermal methods, as shown in the next section.

3.3. *Quantitative Profiles of SEC Fractions*

To explain the observed broad-peak profiles suggesting interactions between the low- and high-MW species (Figure 1b and Table 3), fractions 1–5 were subjected to a further detailed quantitative and qualitative chemical characterization. A TCA was used for quantification, while a TD-Py-GC-MS and ^{31}P NMR provided the structural characterization.

The TCA, in addition to determining the total carbon amount in each fraction (which was used for the verification of fraction abundance, shown in Table 2), provided the carbon distribution among the portions which evolved at different temperatures. As mentioned in Section 2.4, those were: (1) the TD directly evolved via evaporation, below 300 °C, (2) the products of the subsequent higher-temperature pyrolysis and, finally, (3) the char or coke, i.e., the material that was not broken down pyrolytically, even at 890 °C, which subsequently combusted in the presence of oxygen. Accounting for all the types of carbon enabled a comprehensive sample quantification.

Fractions 1–3 showed pronounced pyrolysis at 890 °C, which was expected due to their relatively high MW (Figure 3a). Less expected was the occurrence of sizable TD portions in their TCA profiles. The TCA thus confirmed that these SEC fractions contained some low-MW compounds which volatilized at TD temperatures, which was the observation made when using the analytical SEC. As mentioned earlier, the occurrence of species with unexpected MWs is possibly due to the strong adsorption of high- and low-MW compounds on each other, i.e., an aggregation specific to the mismatch in size of the interacting species.

The TCA profiles of fractions 4 and 5 corroborate this conclusion. It is of note that while fraction 4 is abundant, fraction 5 is not, so its magnified profile is shown in Figure 3a as an insert. Contrary to the expectation of their increasing volatilization at TD temperatures, their TCA profiles actually showed rather large portions which evolved at pyrolytic temperatures. This observation is consistent with the high-MW lignin components' carryover into these fractions, as shown in Figure 1b.

Some high-MW chemicals may be formed as a result of heating, as the presence of adsorbed high-MW compounds may enhance polymerization. This phenomenon of the re-polymerization of low-MW lignin components during thermal treatment is well known [67–70]. Thus far, various methods, including TCA, have been used merely to confirm the MW profiles obtained using SEC. However, TCA, besides closing the mass balance on carbon, provides additional opportunities for revealing the structural changes occurring in lignin as a result of its fractionation by MW, particularly in combination with TD-Py-GC-MS.

Namely, the modes of thermal evolution were rather different for the fractionated lignin and its unfractionated precursor. The latter, unlike the former, appeared to have stronger interactions, most likely via hydrogen bonding, creating a network similar to cross-linking. This intermolecular cross-linking may hinder the pyrolytic evolution of small-MW products, i.e., causing polymerization/charring. This is apparent when comparing the TCA profile of intact lignin to that of the sum of all the fractions in Figure 3b, where the unfractionated lignin yielded about 47% char (coke), while 24% of the carbon was volatilized at 890 °C. In contrast, the amount of carbon which evolved at 890 °C without oxygen increased in the fractionated lignin, at the expense of the final carbon portion, which evolved only with oxygen.

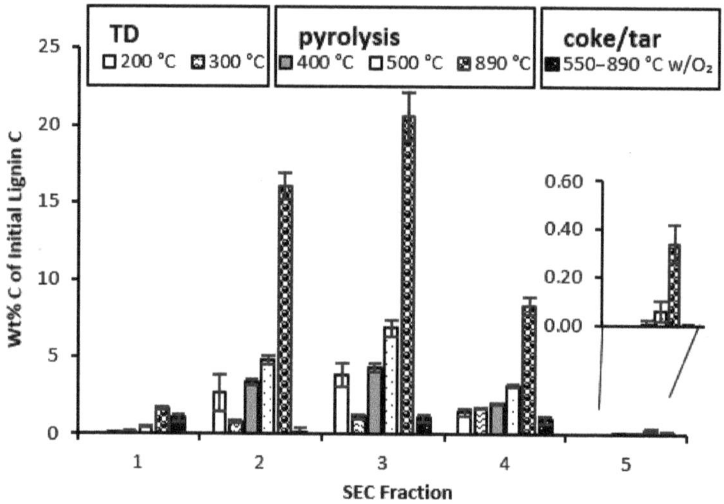

(**a**)

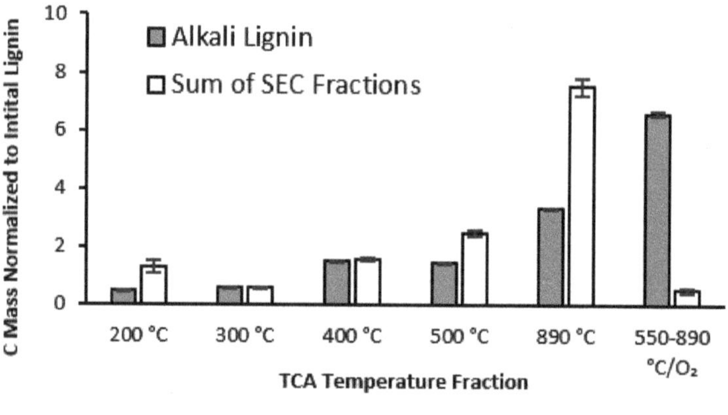

(**b**)

Figure 3. TCA analysis: (**a**) Carbon % distribution among the TCA temperature portions for each lignin fraction; the insert in panel (**a**) shows magnified fraction 5. (**b**) Carbon mass sums for all SEC fractions at each temperature step of TCA compared to the corresponding alkali lignin carbon masses.

This striking difference indicates that unfractionated lignin has a much greater propensity to charring, compared to any of its MW fractions, including even that with the highest MW. Apparently, charring is enhanced when polymers interact with abundant and accessible small-MW fragments, as is the case in the unfractionated lignin—this is a novel conclusion resulting from the current work. Presumably, the lignin molecules fold with the hydrogen bonds serving as a "glue", with the smaller-MW components "cementing" the fold. Even though such interactions still skew the expected TCA pattern in fractions 4 and 5, the extent of their charring is nowhere near that of the original lignin, presumably because such interactions are much stronger in the latter.

These results are in agreement with the report of Chua et al., who made a crude separation by lignin dissolution in THF. They observed that both the THF-soluble and -insoluble fractions, presumably of low and high MWs, respectively, produced significantly less char upon pyrolysis than the original lignin, i.e., similar to what we observed for all MW fractions [71].

These results are also consistent with the study of Li et al., who reported that magnesium and calcium hydroxides tend to reduce lignin agglomeration, i.e., the interactions between the lignin components. As a result, more organic phenolic compounds and CO_2 were formed upon the pyrolysis of this material with reduced interactions, at the expense of the hard-shelled char characteristic of the untreated lignin [72].

The next question was what type and size of compounds volatized at each temperature step, and whether those profiles were different for the fractionated and native lignin. This relationship was investigated using the TD-Py-GC-MS speciation data.

3.4. TD-Py-GC-MS: Speciation of Fractionated Lignin

The five SEC fractions were analyzed via TD-Py-GC-MS to identify the various compounds which evolved sequentially at increasing temperatures. The corresponding profiles, including the non-fractionated lignin, are shown in Figure 4 (supported by Figures S4 and S5).

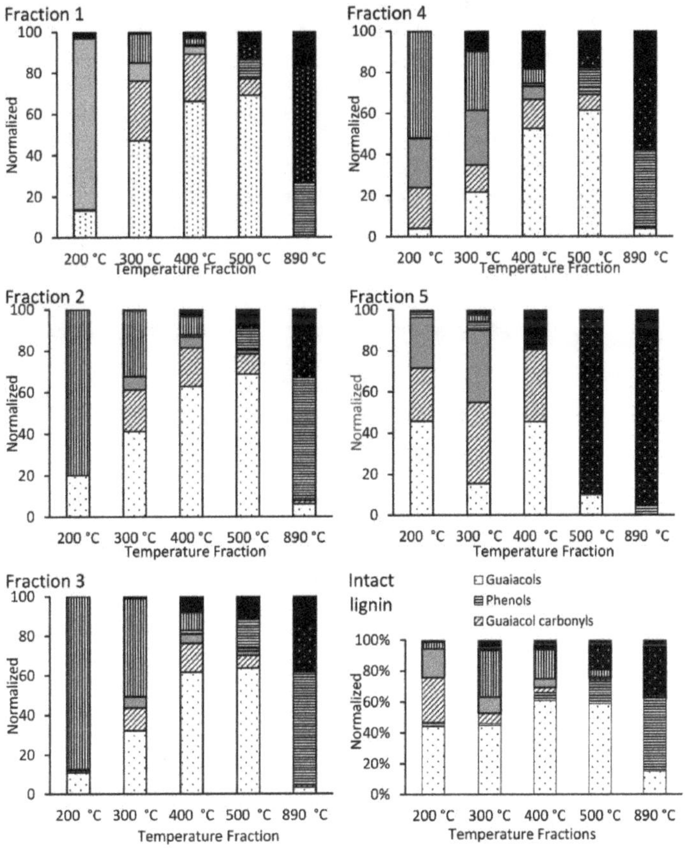

Figure 4. Normalized distribution of different classes of organic compounds for each TD-Py-GC-MS temperature for SEC fractions 1–5 and unfractionated lignin.

The semi-quantitative TD-Py-GC-MS of fractions 1–5 corroborated the TCA results, showing a similar total abundance as well as the premature elution of some low-MW compounds in the early fractions (1–3), and high-MW lignin components in the later fractions, 4 and 5 (Figure S4).

The identified volatile compounds (i.e., those thermally desorbed up to 300 °C) in the fractionated lignin consisted mainly of guaiacyl dimers, guaiacol carbonyls and acids, and guaiacols. In contrast, guaiacols as the homology series were the most abundant class of compounds in the early pyrolytic fractions (400 and 500 °C), followed by guaiacyl carbonyls and dimers. These observations were consistent with those observed earlier [61,62]. The general trends and mechanistic considerations can be found elsewhere [61]. Here, the focus will be on the differences between the SEC fractions.

With a further increase in temperature, other pyrolytic products—phenols, aromatic hydrocarbons and, ultimately, polycyclic aromatics (mostly retene and phenanthrenes)—appeared as immediate structural char precursors that became most abundant in the final, 890 °C temperature step. Approaching this temperature, the aromatic and polycyclic hydrocarbons became dominant in fraction 5, although they were abundant in all the fractions. These results are in agreement with those of Shao et al., who observed a more abundant formation of phenols and aromatic hydrocarbons upon the high-temperature (800 °C) pyrolysis of high-MW lignin components [73]. On the other hand, they are also consistent with the observation of Li et al., that polycyclic hydrocarbons tend to form as a result of the pyrolysis of low-MW lignin components [74].

The observation of abundant polycyclic aromatics in fraction 5 is not a common feature, though, because low-MW chemicals characteristic of this fraction are expected to volatilize rather than undergo pyrolysis. Thus, the observed "hard" pyrolysis is evidence of the presence of high-MW components. The fact that PAH formation is actually the most pronounced in fraction 5 (even more than in the unfractionated lignin) indicates that the formation of char precursors is most likely when a mixture of high- and low-MW components is pyrolyzed, suggesting their interaction.

The TD-Py profiles in the SEC fractions appear to be additive and, thus, qualitatively similar when compared to the native lignin (Figure 4). Thus, the SEC fractions are similar in their basic chemical structure, including the original lignin, corroborating a similar conclusion made in the literature for a larger-scale fast pyrolysis of lignin [73]. Nonetheless, some sizable differences, both quantitative and even qualitative, were observed, particularly in the TD fractions. The acetic and propionic acids which evolve at 200 °C were surprisingly abundant in fraction 1 and nowhere else, suggesting their preferential adsorption in high-MW species, presumably via hydrogen bonding. The other qualitative difference was the abundance of guaiacyl acids (predominantly homovanillic) and dimers in fraction 4 mentioned earlier.

In comparison to the original lignin, the volatile guaiacol carbonyls which evolve at TD temperatures were in a much higher abundance in the SEC fractions, the effect of which was also most pronounced in fraction 4. This difference may be explained by the stronger adsorption of these compounds (perhaps, with multiple hydrogen bonding) within the non-fractionated lignin, hindering their evolution, with a partial cleavage of these intermolecular bonds as a result of fractionation by MW. This explanation is consistent with the hypothesis of strong interactions between the lignin components of different MWs. For further confirmation of this hypothesis, and to reveal the chemical mechanism of such interactions, another independent method was used, ^{31}P NMR, for the detection of hydroxyl groups potentially involved in hydrogen bonds formation.

3.5. Hydroxyl Group Quantitation using ^{31}P NMR Spectroscopy

The ^{31}P NMR spectra of the SEC fractionated lignin samples enabled the determination of the main functional groups (Figures S6 and S7). The number of mmoles found in each sample are summarized in Table 4. Then, the number of moles of the post-SEC hydroxyl groups was determined for all five fractions. Surprisingly, their sum turned out to be significantly larger (>2 times) than the number determined for the alkali lignin before fractionation.

Table 4. Number of mmoles determined in NMR samples in the SEC weight fractions per g of alkali lignin.

	Alcohols	Phenols	Acids	Unidentified	Total
Fraction 1	0.62	0.35	0.00	0.02	0.99
Fraction 2	2.11	1.73	0.12	0.24	4.19
Fraction 3	2.81	2.03	0.20	0.09	5.13
Fraction 4	1.50	1.77	0.33	0.30	3.90
Fraction 5	0.31	0.08	0.04	0.00	0.43
Total of fractions	7.35	5.96	0.69	0.65	14.65
Intact lignin	2.67	2.85	0.78	ND	6.29

It is noteworthy that the spectra of all the fractions also contained a cluster of unidentified signals of low intensity centered around 150.6 ppm. These signals (labeled as "unidentified" henceforth) have not been reported in other studies of phosphitylated lignin-based samples, nor were they observed in the spectra of the other lignin and lignin-derived samples investigated by our group. It is possible that the TMDP reacted with some other highly electronegative species such as sulfur, which is present in alkali lignin.

The most plausible explanation is that some hydroxyl groups in the unfractionated lignin before SEC were inaccessible for phosphitylation, due to steric hindrance. This observation corroborates the TCA results, which showed that the unfractionated lignin is much more recalcitrant toward pyrolysis, presumably due to the lignin fragments' folding. The fractionation process possibly caused modifications of the complex three-dimensional structure of lignin in such a way as to allow more hydroxyl groups to undergo phosphitylation. This explanation is consistent with the observed, strong interactions between the low- and high-MW lignin fractions that may be responsible for lignin association. Li et al. observed that lignin fractionated by partial dissolution in acetone significantly increased the adsorption capacity toward a cationic dye, methylene blue [75], thus corroborating our findings.

It is of note that all the fractions made a sizable contribution to the observed increase in the number of phosphitylation-accessible OH groups, as compared to the original lignin. The contribution of the middle fractions was the greatest in terms of the absolute numbers. For example, the single fraction 3 showed an amount similar to that of the whole unfractionated lignin, with fractions 2 and 4 also making an impact.

Nonetheless, fractions 1 and 5 stood out in terms of their relative contribution to the OH group pool. Based on Table 4 (the rightmost column), fractions 1 and 5 contain sizable percentages of the hydroxyl groups, 11.2% and 8.8%, respectively. However, according to Table 2, these two fractions account for only 3.9 and 0.7% of the total carbon (determined using TCA), respectively. Thus, they, and particularly fraction 5, with the lowest MW, contain disproportionally large hydroxyl group contents (mostly at the expense of the most abundant fraction 3). Furthermore, fractions 1 and, in particular, 5, showed an enrichment in aliphatic hydroxyl groups at the expense of phenolic ones (Table 4). For the reader's convenience, the relative distributions of the different types of hydroxyl groups are shown explicitly in Figure S6. The enrichment of fraction 5 with aliphatic OH groups contradicts the earlier observed enrichment of lower-MW fractions with phenolic OH groups [39,53]. However, these earlier data were obtained when using a different MW fractionation method, ultrafiltration. We also observed the enrichment with phenolic OH groups up to fraction 4—but not in fraction 5 (Table 4, Figure S6). This exception can be readily explained by the carryover of some fraction 1 material into it. Note that these two fractions, though of low abundance, contain the species with the greatest mismatch in size, i.e., those expected to form the strongest interactions prior to SEC fractionation.

These findings may explain the observed strong interactions between the low- and high-MW lignin components, resulting in the significant broadening of the SEC peaks (Figure 1b). They may also explain the difference in the composition of the TD fractions discussed in the previous section. The aforementioned interactions may be enabled by strong hydrogen bonding, which would reduce the effective charges. It is well known that lignin aggregation is more pronounced in non-hydrogen-bonding polar solvents, e.g., THF, where the solvent does not interfere with such interactions [9,10]. In turn, unfractionated lignin dissolution in a polar solvent may cause the transfer of such interacting hydroxyl groups into the internal domains of lignin-agglomerated particles, making them less accessible and less amenable to analysis.

Alternative explanations were also considered for the observed increased amount of OH groups in the fractionated lignin. Theoretically, the THF used in the SEC method could be a source of some of the additional OH groups identified in our experiments. For example, THF could have reacted with lignin under the conditions used during the SEC analysis, resulting in the cleavage of the THF ring which allowed for the formation of a new hydroxyl group. THF peroxide formation was also a possibility, as the product of this reaction had a new OH group and, therefore, would readily react with the TMDP. Finally, the ether bond cleavage in the THF could occur either before or during the SEC application. However, these qualitative considerations did not pass the quantitative test. Namely, the ^{31}P NMR spectrum of the phosphitylated-concentrated THF contained two phospholane signals. The first was a signal in the aliphatic hydroxyl group region at 147.4 ppm, whereas the other was located in the phenolic region at 142.0 ppm. However, both signals had relatively small integrations and an insignificant impact on the final total. Furthermore, to eliminate any potential artifacts, the pre-eluate was used as a blank—and even after this treatment the sum of the MW fractions still yielded a greater amount of OH groups than the non-fractionated lignin (Table 4).

Thus, the unraveling of the lignin supramolecular structure appears to be the most plausible explanation of the observed mismatch in the OH group quantitation between the original lignin and the sum of its MW fractions, and is supported by their altered thermal behavior.

4. Conclusions

Lignin fractionation by MW using SEC showed strong interactions between its low- and high-MW components. The occurrence of these interactions was observed after the re-run of the separated MW fractions through the same SEC column, resulting in pronounced peak broadening. Such interactions were confirmed using several methods, including ESI-TOF HR MS, TCA, and TD-Py-GC-MS. The application of these methods, particularly TCA, for lignin MW fraction analysis was fruitful, having shown that fractionation by MW caused several significant and unexpected changes in lignin properties. In particular, the fractionated lignin showed a smaller propensity to charring as compared to the original lignin, as shown by the combination of TCA and TD-Py-GC-MS. Furthermore, the component structure was shown to be different among the obtained MW fractions, with an uneven distribution of phenolic and aliphatic hydroxyl groups. Namely, the number of accessible hydroxyl groups, e.g., those amenable to ^{31}P NMR analysis, increased more than two-fold. As an apparent result, several volatile compounds, in particularl guaiacol carbonyls, were much more abundant in the TD fractions of the TD-Py-GC-MS of the fractionated lignin, at the expense of charring. These phenomena can be explained by the association between high- and low-MW lignin components in unfractionated lignin, presumably via hydrogen bonding, thus making a stable and recalcitrant particle that folds upon itself, forming supramolecular particles. After the SEC separation, lignin appears to unravel/unfold due to a partial loss of such an association, resulting in a reduction in its structural integrity. The observed reduction in lignin recalcitrance by fractionation may be used for its more efficient processing in biorefineries.

Supplementary Materials: The following supporting information can be downloaded at: https://www.mdpi.com/article/10.3390/polym15193956/s1, Table S1—Calibration data for PS and PMMA standards obtained for analytical SEC. Table S2—MW characterization of the five fractions obtained using preparative SEC and unfractionated alkali lignin conducted with ESI-TOF HR MS. Figure S1—Calibration curve for PS and PMMA standards obtained for analytical SEC. Figure S2—SEC preparative and analytical chromatograms for preliminary fractionation experiment. Figure S3—ESI-TOF HR MS spectra for SEC lignin fractions 1–5. Figure S4—Aggregated TD-Py-GC-MS peak areas based on the GC-MS TIC response. Figure S5—TD-Py-GC-MS profiles for each SEC fraction based on the TIC peak-areas response. Figure S6—^{31}P NMR spectra of SEC weight-fractionated lignin samples. Figure S7—Relative abundance of characteristic functional groups determined using ^{31}P NMR.

Author Contributions: Conceptualization, A.K. and E.I.K.; methodology, A.L., A.A.A., J.S., I.P.S. and A.K.; formal analysis, A.L., A.A.A. and J.S.; investigation, A.L., A.A.A., J.S., S.R. and S.L.; writing—original draft preparation, A.L.; writing—review and editing, A.L., I.P.S., A.K. and E.I.K.; visualization, A.L. and A.K.; supervision, I.P.S. and A.K.; project administration, A.K.; funding acquisition, A.K. All authors have read and agreed to the published version of the manuscript.

Funding: The authors would like to acknowledge the funding from the National Science Foundation (NSF) (ND EPSCoR CSMS) and A.A.A.'s doctoral dissertation award (IIA-1355466). S.L. was supported through the NSF Research Experience for Undergraduates (Grant no. CHE 1460825). This work was completed under the USDA-NIFA-NCRSGC-Bioeconomy—Advancing the Bioeconomy through Regional Sun Grant Centers: Production of 3D Graphene from Renewable Lignin Through Flash Catalytic Thermochemical Processes (Grant. No. 202-35502-32916/3TJ372). Any opinions, findings, and conclusions or recommendations expressed in this material are those of the authors and do not necessarily reflect the views of the NSF or USDA.

Institutional Review Board Statement: Not applicable.

Data Availability Statement: Additional data supporting the reported results can be found in the Supplementary Materials.

Conflicts of Interest: Author Anastasia A. Andrianova is currently employed by Agilent Technologies. Joshua Schumaker is currently employed by SCIEX. The remaining authors declare that the research was conducted in the absence of any commercial or financial relationships that could be construed as a potential conflict of interest.

References

1. Kang, S.; Li, X.; Fan, J.; Chang, J. Hydrothermal conversion of lignin: A review. *Renew. Sustain. Energy Rev.* **2013**, *27*, 546–558. [CrossRef]
2. Pu, Y.; Hu, F.; Huang, F.; Davison, B.H.; Ragauskas, A.J. Assessing the molecular structure basis for biomass recalcitrance during dilute acid and hydrothermal pretreatments. *Biotechnol. Biofuels* **2013**, *6*, 1–13. [CrossRef]
3. Zhang, X.; Tu, M.; Paice, M. Routes to potential bioproducts from lignocellulosic biomass lignin and hemicelluloses. *Bioenergy Res.* **2011**, *4*, 246–257. [CrossRef]
4. Gouveia, S.; Fernandez-Costas, C.; Sanroman, M.A.; Moldes, D. Enzymatic polymerisation and effect of fractionation of dissolved lignin from *Eucalyptus globulus* Kraft liquor. *Bioresour. Technol.* **2012**, *121*, 131–138. [CrossRef]
5. Jääskeläinen, A.; Liitiä, T.; Mikkelson, A.; Tamminen, T. Aqueous organic solvent fractionation as means to improve lignin homogeneity and purity. *Ind. Crop. Prod.* **2017**, *103*, 51–58. [CrossRef]
6. Ragauskas, A.; Williams, C.; Davison, B.; Britovsek, G.; Cairney, J.; Eckert, C.; Frederick, W., Jr.; Hallett, J.; Leak, D.; Liotta, C.; et al. The path forward for biofuels and biomaterials. *Science* **2006**, *311*, 484–789. [CrossRef]
7. Brebu, M.; Cazacu, G.; Chirila, O. Pyrolysis of lignin—A potential method for obtaining chemicals and/or fuels. *Cell. Chem. Technol.* **2011**, *45*, 43–50.
8. Gidh, A.V.; Decker, S.R.; See, C.H.; Himmel, M.E.; Williford, C.W. Characterization of lignin using multi-angle laser light scattering and atomic force microscopy. *Anal. Chim. Acta* **2006**, *555*, 250–258. [CrossRef]
9. Crestini, C.; Melone, F.; Sette, M.; Saladino, R. Milled wood lignin: A linear oligomer. *Biomacromolecules* **2011**, *12*, 3928–3935. [CrossRef]
10. Zwilling, J.D.; Jiang, X.; Zambrano, F.; Venditti, R.A.; Jameel, H.; Velev, O.D.; Rojas, O.J.; Gonzalez, R. Understanding lignin micro- and nanoparticle nucleation and growth in aqueous suspensions by solvent fractionation. *Green Chem.* **2021**, *23*, 1001–1012. [CrossRef]
11. Chen, L.; Luo, S.-M.; Huo, C.-M.; Shi, Y.-F.; Feng, J.; Zhu, J.-Y.; Xue, W.; Qiu, X. New insight into lignin aggregation guiding efficient synthesis and functionalization of a lignin nanosphere with excellent performance. *Green Chem.* **2022**, *24*, 285–294. [CrossRef]

12. Li, M.-F.; Sun, S.-N.; Xu, F.; Sun, R.-C. Sequential solvent fractionation of heterogeneous bamboo organosolv lignin for value-added application. *Sep. Purif. Technol.* **2012**, *101*, 18–25. [CrossRef]
13. Helander, M.; Theliander, H.; Lawoko, M.; Henriksson, G.; Zhang, L.; Lindström, M.E. Fractionation of technical lignin: Molecular mass and pH effects. *Bioresources* **2013**, *8*, 2270–2282. [CrossRef]
14. Niemi, H.; Lahti, J.; Hatakka, H.; Kärki, S.; Rovio, S.; Kallioinen, M.; Mänttäri, M.; Louhi-Kultanen, M. Fractionation of organic and inorganic compounds from black liquor by combining membrane separation and crystallization. *Chem. Eng. Technol.* **2011**, *34*, 593–598. [CrossRef]
15. Wang, G.; Chen, H. Fractionation of alkali-extracted lignin from steam-exploded stalk by gradient acid precipitation. *Sep. Purif. Technol.* **2013**, *105*, 98–105. [CrossRef]
16. García, A.; Spigno, G.; Labidi, J. Antioxidant and biocide behaviour of lignin fractions from apple tree pruning residues. *Ind. Crop. Prod.* **2017**, *104*, 242–252. [CrossRef]
17. Boeriu, C.; Fiţigău, F.; Gosselink, R.; Frissen, A.; Stoutjesdijk, J.; Peter, F. Fractionation of five technical lignins by selective extraction in green solvents and characterisation of isolated fractions. *Ind. Crop. Prod.* **2014**, *62*, 481–490. [CrossRef]
18. Cui, C.; Sun, R.; Argyropoulos, D.S. Fractional precipitation of softwood kraft lignin: Isolation of narrow fractions common to a variety of lignins. *ACS Sustain. Chem. Eng.* **2014**, *2*, 959–968. [CrossRef]
19. Duval, A.; Vilaplana, F.; Crestini, C.; Lawoko, M. Solvent screening for the fractionation of industrial kraft lignin. *Holzforschung* **2016**, *70*, 11–20. [CrossRef]
20. Li, H.; McDonald, A.G. Fractionation and characterization of industrial lignins. *Ind. Crop. Prod.* **2014**, *62*, 67–76. [CrossRef]
21. Sadeghifar, H.; Wells, T.; Le, R.; Sadeghifar, F.; Yuan, J.; Ragauskas, A. Fractionation of organosolv lignin using acetone:water and properties of the obtained fractions. *ACS Sustain. Chem. Eng.* **2017**, *5*, 580–587. [CrossRef]
22. Guo, G.; Li, S.; Wang, L.; Ren, S.; Fang, G. Separation and characterization of lignin from bio-ethanol production residue. *Bioresour. Technol.* **2013**, *135*, 738–741. [CrossRef] [PubMed]
23. Xiao, L.; Xu, F.; Sun, R.-C. Chemical and structural characterization of lignins isolated from *Caragana sinica*. *Fibers Polym.* **2011**, *12*, 316–323. [CrossRef]
24. Dodd, A.; Kadla, J.; Straus, S. Characterization of fractions obtained from two industrial softwood kraft lignins. *ACS Sustain. Chem. Eng.* **2015**, *3*, 103–110. [CrossRef]
25. Wang, S.; Wang, Y.; Cai, Q.; Wang, X.; Jin, H.; Luo, Z. Multi-step separation of monophenols and pyrolytic lignins from the water-insoluble phase of bio-oil. *Sep. Purif. Technol.* **2014**, *122*, 248–255. [CrossRef]
26. Wang, G.; Chen, H. Fractionation and characterization of lignin from steam-exploded corn stalk by sequential dissolution in ethanol–water solvent. *Sep. Purif. Technol.* **2013**, *120*, 402–409. [CrossRef]
27. Mussatto, S.; Fernandes, M.; Roberto, I. Lignin recovery from brewer's spent grain black liquor. *Carbohydr. Polym.* **2007**, *70*, 218–223. [CrossRef]
28. Ház, A.J.M.; Dubinyová, L.; Sládková, A.; Šurina, I. Thermal properties and size distribution of lignins precipitated with sulphuric acid. *Wood Res.* **2015**, *60*, 375–384.
29. Šurina, I.; Jablonský, M.; Ház, A.; Sladková, A.; Briškárová, A.; Kačík, F.; Šima, J. Characterization of non-wood lignin precipitated with sulphuric acid of various concentrations. *Bioresources* **2015**, *10*, 1408–1423. [CrossRef]
30. Li, X.-H.; Wu, S.-B. Chemical structure and pyrolysis characteristics of the soda-alkali lignin fractions. *Bioresources* **2014**, *9*, 6277–6289. [CrossRef]
31. Lourençon, T.; Hansel, F.; da Silva, T.; Ramos, L.; de Muniz, G.; Magalhães, W. Hardwood and softwood kraft lignins fractionation by simple sequential acid precipitation. *Sep. Purif. Technol.* **2015**, *154*, 82–88. [CrossRef]
32. Liang, X.; Liu, J.; Fu, Y.; Chang, J. Influence of anti-solvents on lignin fractionation of eucalyptus globulus via green solvent system pretreatment. *Sep. Purif. Technol.* **2016**, *163*, 258–266. [CrossRef]
33. Jiang, X.; Savithri, D.; Du, X.; Pawar, S.; Jameel, H.; Chang, H.-M.; Zhou, X. Fractionation and characterization of kraft lignin by sequential precipitation with various organic solvents. *ACS Sustain. Chem. Eng.* **2017**, *5*, 835–842. [CrossRef]
34. Lange, H.; Schiffels, P.; Sette, M.; Sevastyanova, O.; Crestini, C. Fractional precipitation of wheat straw organosolv lignin: Macroscopic properties and structural insights. *ACS Sustain. Chem. Eng.* **2016**, *4*, 5136–5151. [CrossRef]
35. Mörck, R.; Reimann, A.; Kringstad, K. Fractionation of Kraft lignin by successive extraction with organic solvents. III. Fractionation of Kraft lignin from birch. *Holzforschung* **1988**, *42*, 111. [CrossRef]
36. Ropponen, J.; Räsänen, L.; Rovio, S.; Ohra-aho, T.; Liitiä, T.; Mikkonen, H.; van de Pas, D.; Tamminen, T. Solvent extraction as a means of preparing homogeneous lignin fractions. *Holzforschung* **2011**, *65*, 543–549. [CrossRef]
37. Yuan, T.-Q.; He, J.; Xu, F.; Sun, R.-C. Fractionation and physico-chemical analysis of degraded lignins from the black liquor of Eucalyptus pellita KP-AQ pulping. *Polym. Degrad. Stab.* **2009**, *94*, 1142–1150. [CrossRef]
38. Toledano, A.; Garcia, A.; Mondragon, I.; Labidi, J. Lignin separation and fractionation by ultrafiltration. *Sep. Purif. Technol.* **2010**, *71*, 38–43. [CrossRef]
39. Sevastyanova, O.; Helander, M.; Chowdhury, S.; Lange, H.; Wedin, H.; Zhang, L.; Ek, M.; Kadla, J.F.; Crestini, C.; Lindström, M.E. Tailoring the molecular and thermo-mechanical properties of kraft lignin by ultrafiltration. *Sep. Purif. Technol.* **2014**, *71*, 38–43. [CrossRef]
40. Toledano, A.; Serrano, L.; Garcia, A.; Mondragon, I.; Labidi, J. Comparative study of lignin fractionation by ultrafiltration and selective precipitation. *Chem. Eng. J.* **2010**, *157*, 93–99. [CrossRef]

41. Werhan, H.; Farshori, A.; von Rohr, P. Separation of lignin oxidation products by organic solvent nanofiltration. *J. Memb. Sci.* **2012**, *423–424*, 404–412. [CrossRef]
42. Busse, N.; Fuchs, F.; Kraume, M.; Czermak, P. Treatment of enzyme-initiated delignification reaction mixtures with ceramic ultrafiltration membranes: Experimental investigations and modeling approach. *Sep. Sci. Technol.* **2016**, *51*, 1546–1565. [CrossRef]
43. Zinovyev, G.; Sumerskii, I.; Korntner, P.; Sulaeva, I.; Rosenau, T.; Potthast, A. Molar mass-dependent profiles of functional groups and carbohydrates in kraft lignin. *J. Wood Chem. Technol.* **2017**, *37*, 171–183. [CrossRef]
44. Xu, J.; Li, C.; Dai, L.; Xu, C.; Zhong, Y.; Yu, F.; Si, C. Biomass fractionation and lignin fractionation towards lignin valorization. *ChemSusChem* **2020**, *13*, 4284–4295. [CrossRef]
45. Kirk, T.; Brown, W.; Cowling, E. Preparative fractionation of lignin by gel-permeation chromatography. *Biopolymers* **1969**, *7*, 135–153. [CrossRef]
46. Jönsson, A.-S.; Nordin, A.-K.; Wallberg, O. Concentration and purification of lignin in hardwood kraft pulping liquor by ultrafiltration and nanofiltration. *Chem. Eng. Res. Des.* **2008**, *86*, 1271–1280. [CrossRef]
47. Brodin, I.; Sjöholm, E.; Gellerstedt, G. Kraft lignin as feedstock for chemical products: The effects of membrane filtration. *Holzforschung* **2009**, *63*, 290–297. [CrossRef]
48. Norgren, M.; Lindström, B. Physico-chemical characterization of a fractionated kraft lignin. *Holzforschung* **2000**, *54*, 528–534. [CrossRef]
49. Wallberg, O.; Jönsson, A.-S.; Wimmerstedt, R. Ultrafiltration of kraft black liquor with a ceramic membrane. *Desalination* **2003**, *156*, 145–153. [CrossRef]
50. Aminzadeh, S.; Lauberts, M.; Dobele, G.; Ponomarenko, J.; Mattsson, T.; Lindström, M.E.; Sevastyanova, O. Membrane filtration of kraft lignin: Structural characteristics and antioxidant activity of the low-molecular-weight fraction. *Ind. Crops. Prod.* **2018**, *112*, 200–209. [CrossRef]
51. Zheng, L.; Lu, G.; Pei, W.; Yan, W.; Li, Y.; Zhang, L.; Huang, C.; Jiang, Q. Understanding the relationship between the structural properties of lignin and their biological activities. *Int. J. Biol. Macromol.* **2021**, *190*, 291–300. [CrossRef] [PubMed]
52. Costa, C.A.E.; Pinto, P.C.R.; Rodrigues, A.E. Lignin fractionation from E. Globulus kraft liquor by ultrafiltration in a three stage membrane sequence. *Sep. Purif. Technol.* **2018**, *192*, 140–151. [CrossRef]
53. Allegretti, C.; Boumezgane, O.; Rossato, L.; Strini, A.; Troquet, J.; Turri, S.; Griffini, G.; D'Arrigo, P. Tuning lignin characteristics by fractionation: A versatile approach based on solvent extraction and membrane-assisted ultrafiltration. *Molecules* **2020**, *25*, 2893. [CrossRef] [PubMed]
54. Janson, J.-C. *Process Scale Liquid Chromatography*; Wiley: Hoboken, NJ, USA, 2007.
55. Botaro, V.; Curvelo, A. Monodisperse lignin fractions as standards in size-exclusion analysis: Comparison with polystyrene standards. *J. Chromatogr. A* **2009**, *1216*, 3802–3806. [CrossRef]
56. Andrianova, A.; Yeudakimenka, N.; Lilak, S.; Kozliak, E.; Ugrinov, A.; Sibi, M.; Kubátová, A. Size exclusion chromatography of lignin: The mechanistic aspects and elimination of undesired secondary interactions. *J. Chromatogr. A* **2017**, *1534*, 101–110. [CrossRef]
57. Banoub, J.; Delmas, G.H.J.; Joly, N.; Mackenzie, G.; Cachet, N.; Benjelloun-Mlayah, B.; Delmas, M.A. Critique on the structural analysis of lignins and application of novel tandem mass spectrometric strategies to determine lignin sequencing. *J. Mass Spectrom.* **2015**, *50*, 5–48. [CrossRef]
58. Uliyanchenko, E.; van der Wal, S.; Schoenmakers, P. Challenges in polymer analysis by liquid chromatography. *Polym. Chem.* **2012**, *3*, 2313–2335. [CrossRef]
59. Andrianova, A.; DiProspero, T.; Geib, C.; Smoliakova, I.; Kozliak, E.; Kubátová, A. Electrospray ionization with high-resolution mass spectrometry as a tool for lignomics: Lignin mass spectrum deconvolution. *J. Am. Soc. Mass Spectrom.* **2018**, *29*, 1044–1059. [CrossRef]
60. Andrianova, A.A. Novel Approaches in Lignomics Employing Liquid Chromatography and Mass Spectr aphy And Mass Spectrometry. Ph.D. Thesis, University of North Dakota, Grand Forks, ND, USA, 2017.
61. LaVallie, A.L.; Bilek, H.; Andrianova, A.; Furey, K.; Voeller, K.; Yao, B.; Kozliak, E.; Kubátová, A. Quantitative insights on de/repolymerization and deoxygenation of lignin in subcritical water. *Bioresour. Technol.* **2021**, *342*, 125974. [CrossRef]
62. Voeller, K.; Bilek, H.; Kreft, J.; Dostalkova, A.; Kozliak, E.; Kubátová, A. Thermal carbon analysis enabling comprehensive characterization of lignin and its degradation products. *ACS Sustain. Chem. Eng.* **2017**, *5*, 10334–10341. [CrossRef]
63. Voeller, K.M. Characterization of Kraft Alkali Lignin and Products of Its Thermal Degradation by Fractional Pyrolysis Method. Master Thesis, University of North Dakota, Grand Forks, ND, USA, 2016.
64. Joffres, B.; Lorentz, C.; Vidalie, M.; Laurenti, D.; Quoineaud, A.A.; Charon, N.; Daudin, A.; Quignard, A.; Geantet, C. Catalytic hydroconversion of a wheat straw soda lignin: Characterization of the products and the lignin residue. *Appl. Car. Environ.* **2014**, *145*, 167–176. [CrossRef]
65. Pu, Y.; Cao, S.; Ragauskas, A.J. Application of quantitative 31P NMR in biomass lignin and biofuel precursors characterization. *Energy Environ. Sci.* **2011**, *4*, 3154–3166. [CrossRef]
66. Hamzah, M.H.; Bowra, S.; Cox, P. Organosolv lignin aggregation behaviour of soluble lignin extract from Miscanthus x giganteus at different ethanol concentrations and its influence on the lignin esterification. *Chem. Biol. Technol. Agricult.* **2021**, *8*, 65. [CrossRef]
67. Hu, J.; Shen, D.; Wu, S.; Zhang, H.; Xiao, R. Effect of temperature on structure evolution in char from hydrothermal degradation of lignin. *J. Anal. Appl. Pyrol.* **2014**, *106*, 118–124. [CrossRef]

68. Yang, S.; Yuan, T.; Li, M.; Sun, R. Hydrothermal degradation of lignin: Products analysis for phenol formaldehyde adhesive synthesis. *Int. J. Biol. Macromol.* **2015**, *72*, 54–62. [CrossRef]
69. Yong, T.; Matsumura, Y. Kinetic analysis of lignin hydrothermal conversion in sub- and supercritical water. *Ind. Eng. Chem. Res.* **2013**, *52*, 5626–5639. [CrossRef]
70. Zhao, Y.; Li, X.; Wu, S.; Li, Y. Temperature impact on the hydrothermal depolymerization of *Cunninghamia lanceolata* enzymatic/mild acidolysis lignin in subcritical water. *Bioresources* **2016**, *11*, 21–32. [CrossRef]
71. Chua, Y.W.; Wu, H.; Yu, Y. Interactions between low- and high-molecular-weight portions of lignin during fast pyrolysis at low temperatures. *Energy Fuels* **2019**, *33*, 11173–11180. [CrossRef]
72. Li, J.; Bai, X.; Dong, Z.; Chen, Y.; Yang, H.; Wang, X.; Chen, H. Influence of additives on lignin agglomeration and pyrolysis behavior. *Fuel* **2020**, *263*, 116629. [CrossRef]
73. Shao, L.; Zhang, X.; Chen, F.; Xu, F. Fast pyrolysis of Kraft lignins fractionated by ultrafiltration. *J. Anal. Appl. Pyr.* **2017**, *128*, 27–34. [CrossRef]
74. Li, S.; Luo, Z.; Wang, W.; Lu, K.; Yang, Y.; Liang, X. Characterization of pyrolytic lignin and insight into its formation mechanisms using novel techniques and DFT method. *Fuel* **2020**, *262*, 116516. [CrossRef]
75. Li, H.; Yuan, Z.; Xing, Y.; Li, J.; Fang, J.; Chang, L.; Li, C. Acetone fractionation: A simple and efficient method to improve the performance of lignin for dye pollutant removal. *RSC Adv.* **2019**, *9*, 35895–35903. [CrossRef] [PubMed]

polymers

MDPI

Article

Biodegradation Potential of Polyethylene Terephthalate by the Two Insect Gut Symbionts *Xanthomonas* sp. HY-74 and *Bacillus* sp. HY-75

Jong-Hoon Kim †, So-Hye Lee †, Byeong-Min Lee, Kwang-Hee Son and Ho-Yong Park *

Microbiome Convergence Research Center, Korea Research Institute of Bioscience and Biotechnology, Daejeon 34141, Republic of Korea; kjh1018@kribb.re.kr (J.-H.K.); sohye2202@kribb.re.kr (S.-H.L.); ghsaor@kribb.re.kr (B.-M.L.); sonkh@kribb.re.kr (K.-H.S.)

* Correspondence: hypark@kribb.re.kr

† These authors contributed equally to this work.

Abstract: Polyethylene terephthalate (PET) is a plastic material that is widely used in beverage bottles, food packaging, and other consumer products, which is highly resistant to biodegradation. In this study, we investigated the effects of two insect gut symbionts, *Xanthomonas* sp. HY-74 and *Bacillus* sp. HY-75, during PET biodegradation. Both strains degraded PET-containing agar plates, and the sole nutrition source assay showed that HY-74 had different degradation rates depending on the presence of specific carbon and nitrogen sources, whereas HY-75 exhibited comparable degradation across all tested conditions. The two strains biodegraded the PET film with 1.57 ± 0.21% and 1.42 ± 0.46% weight loss after 6 weeks, respectively. Changes in the morphology and structure of the PET films, such as erosion, scratching, and surface roughening, were determined using scanning electron microscopy (SEM). Further, the two strains biodegraded PET powder, broke it into its degradation products, and changed the surface functional groups. This is the first study to investigate the biodegradation of PET by Hymenoptera gut-derived microbes and offers promising insights into the potential applications of insect gut symbionts in PET waste management.

Keywords: polyethylene terephthalate; biodegradation; insect gut symbiont; plastic waste management

Citation: Kim, J.-H.; Lee, S.-H.; Lee, B.-M.; Son, K.-H.; Park, H.-Y. Biodegradation Potential of Polyethylene Terephthalate by the Two Insect Gut Symbionts *Xanthomonas* sp. HY-74 and *Bacillus* sp. HY-75. *Polymers* **2023**, *15*, 3546. https://doi.org/10.3390/polym15173546

Academic Editors: Alexandre Vetcher and Alexey L. Iordanskii

Received: 24 July 2023
Revised: 17 August 2023
Accepted: 23 August 2023
Published: 25 August 2023

1. Introduction

Plastics are widely used in human life because of their utility and convenience; however, they cause serious environmental problems. Plastic waste has recently attracted worldwide attention, and many countries have enacted laws and countermeasures. The recycling rate is approximately only 9% despite the fact that global plastic consumption has reached 460 million tons, quadrupling over 30 years [1]. Only 19% of plastics are incinerated, generating toxic products, whereas the remaining 72% are discarded in the environment [2]. Therefore, biodegradation strategies are becoming more crucial than incineration and chemical strategies.

Polyethylene terephthalate (PET) is a widely used plastic material that is commonly used in beverage bottles, food packaging, and other consumer products. It is a thermoplastic polymer that is composed of repeating units of terephthalic acid (TPA) and ethylene glycol (EG), which are linked together through ester bonds. It is lightweight and convenient, making it a popular choice for sustainable packaging. However, PET's physical resistance to degradation contributes to it being a major source of plastic pollution as it can take hundreds of years to decompose in the environment [3]. Therefore, researchers are exploring ways to biodegrade PET using microorganisms, enzymes, and other methods to reduce its environmental impact.

Scientists have discovered several bacteria that are effective plastic-degrading agents. For example, in the case of PET degradation, *Ideonella sakaiensis* was discovered in 2016 and is capable of breaking down PET into its monomers [4]. Other bacteria, such as *Bacillus* sp. [5–8], *Pseudomonas* sp. [9,10], *Thermobifida fusca* [11], and *Rhodococcus ruber* [12], have been found to degrade PET through the production of various enzymes. Plastic-degrading microorganisms are found in diverse environments, including soil, seawater, and plastic waste. In the case of insects, their potential role in breaking down plastic was first discovered when they consumed packaging materials [13]. Since then, many studies have been conducted to identify insect gut-related microbes with rare plastic-degradation properties. Yang et al. [14] demonstrated that gut-associated bacteria in mealworms (*Tenebrio molitor*) could effectively degrade various types of plastics. Similarly, other studies have shown that the gut microbes of waxworms (*Galleria mellonella*) can degrade various types of plastics, including low-density polyethylene and polypropylene (PP) [15,16]. However, most studies have been limited to specific insects and plastic materials, and PET biodegradation by insect-derived microbes has not yet been reported. Therefore, there is an urgent need to secure microbial resources based on biodiversity and promote scientific advancement in PET biodegradation.

In this study, two PET-degrading bacterial strains, *Xanthomonas* sp. HY-74 and *Bacillus* sp. HY-75, were isolated from the intestines of two Hymenoptera species: *Xylocopa appendiculata* and *Eumenes decoratus*, respectively. The PET degradation abilities of HY-74 and HY-75 were evaluated using various physicochemical techniques, including weight loss measurements, scanning electron microscopy (SEM), Fourier transform infrared spectroscopy (FTIR), and high-performance liquid chromatography (HPLC).

2. Materials and Methods

2.1. Plastic Materials

The amorphous PET film was purchased from Goodfellow (Huntingdon, UK), cut into pieces (20 mm × 10 mm × 0.25 mm), and weighed. After soaking the film in 70% (*v*/*v*) ethanol overnight, the PET film was air-dried on a clean bench. Commercial PET powder (particle size <300 μm) was purchased from Goodfellow (Huntingdon, UK), soaked in 70% (*v*/*v*) ethanol, and air-dried on a clean bench.

2.2. Screening, Isolation, and Identification of PET Degrading Bacteria

The adult bees, *Xylocopa appendiculata* and *Eumenes decorates*, collected from the mountain Myongdo-bong (Jinan-gun, Jeollabuk-do, Republic of Korea), were first washed with 70% (*v*/*v*) ethanol to remove contaminants on the body surface before being washed twice with sterilized water. The digestive tract was dissected, and the intestinal contents were carefully recovered. The intestinal contents were diluted in phosphate-buffered saline and spread onto a double-layer agar plate containing 0.1% (*w*/*v*) PET, as described by Charnock et al. (2021) with minor modifications. The top layers were 1.5% agar and 0.1% PET powder dissolved in 40 mL dimethyl sulfoxide (Duchefa, Haarlem, Netherlands), and Reasoner's 2A agar (R2A; MBcell, Seoul, Republic of Korea) was used as the bottom layer in the PET agar plate. After incubation at 30 °C for 14 days, the bacteria that formed a translucent halo around the areas of bacterial growth were selected for further study.

The two strains were assessed for their extracellular enzyme activity as described by Bhagobaty and Joshi [17] with minor modifications. The activated bacterial cultures were streaked onto each R2A agar plate containing a specific substrate and incubated at 30 °C for 3 days. The ability to produce protease and polycaprolactone (PCL) degrading enzyme was qualitatively confirmed by observing transparent halo zones surrounding the colonies on the R2A agar containing 2% (*w*/*v*) skim milk and 0.1% (*w*/*v*) PCL, respectively. The extracellular lipase activity was qualitatively determined using R2A agar with 1% (*w*/*v*) Tween 80 and 0.01% (*w*/*v*) $CaCl_2$ and confirmed by the formation of crimson dots around the colonies, which were further measured.

Strain identification was performed using poly chain reaction amplification and nucleotide sequence analysis of the 16S rRNA gene. Genomic DNA was extracted using a DNA extraction kit (Bioneer, Daejeon, Republic of Korea) according to the manufacturer's guidelines. A partial 16S rRNA sequence from the isolated strain was amplified for phylogenetic identification using the universal primers 24F (5′-AGAGTTTGATCCTGGCTCAG-3′) and 1492R (5′-AAGTCGTAACAAGGTAACC-3′). The 16S rRNA sequences of closely related strains were retrieved from the National Center for Biotechnology Information (http://www.ncbi.nlm.nih.gov/GenBank/index.html; access on 11 September 2022) and aligned using CLUSTAL X [18]. The molecular phylogeny of 16S rRNA was inferred using the neighbor-joining method in MEGA software version X.

2.3. Sole Nutrition Source Assay

The ability of HY-74 and HY-75 to utilize PET as their sole nutrient source was assessed. These strains' growth and PET-degradation capabilities were investigated using a minimal salt medium (MSM; 0.1% NH_4NO_3, 0.07% K_2HPO_4, 0.02% KH_2PO_4, 0.02% $CaCl_2{\cdot}2H_2O$, 0.005% KCl, 0.001% $FeSO_4{\cdot}7H_2O$, 0.001% $ZnSO_4{\cdot}7H_2O$, 0.001% $MnSO_4{\cdot}7H_2O$, 15 g agar, and 1 L water) with PET as the sole nutrient source. The PET plates containing double-layered agar were prepared as described above. Four variations of MSM were prepared as the bottom layers: MSM, MSMG (0.5% glucose supplementation), MSMN (0.5% ammonium sulfate supplementation), and MSMGN (0.5% glucose and 0.5% ammonium sulfate supplementation). A total of 10 μL of activated cultures of HY-74 and HY-75 was added to each prepared plate and incubated at 30 °C for 4 weeks. After 4 weeks, translucent halos around the areas of bacterial growth in each medium were monitored and calculated using image J version 1.53t (US National Institute of Health, MD, USA). HY-74 and HY-75 were collected from each agar plate by scraping the surface with a cell scraper (SPL Life Science Inc., Pochen, Republic of Korea), and the total cell dry weight was measured to evaluate growth in each medium.

2.4. Degradation of PET Film

A total of 100 μL of HY-74 and HY-75 bacterial suspensions (the mid-log phase) was inoculated into each 10 mL of MSM, R2A, and LB (BD Difco, Franklin Lakes, NJ, USA) media in 50 mL Erlenmeyer flasks containing pre-weighed pieces of PET film (as described in Section 2.1), and the flasks were incubated on a rotary shaker (180 rpm) at 30 °C for 6 weeks, 8 mL of incubated media was replaced with 8 mL of fresh media every week. A negative control without bacterial inoculation was maintained under the same conditions. After incubation, each film was rinsed with 1% SDS and then thoroughly washed with distilled water five times to remove residual cells. The completely air-dried film was weighed and collected to observe the changes in the surface morphology and polymer bond formation. The experiments were performed in triplicate under the same conditions. For the SEM observations, the washed PET films were coated with gold using a sputter coater (Q15ORS, Quorum, East Sussex, UK) and analyzed using the FEI Quanta 250 FEG (FEI, Hillsboro, OR, USA). A PET film without inoculation was used as a negative control. The FTIR spectra of the PET films treated with or without bacteria were recorded at wavenumbers ranging from 400 to 4000 cm^{-1} (4 cm^{-1} resolution) using a Nicolet iS50 infrared spectrometer (Thermo Fisher Scientific Instrument, Waltham, MA, USA).

2.5. PET Powder Degradation Assay

HY-74 and HY-75 were cultured on a rotary shaker at 30 °C for 4 weeks in 250 mL Erlenmeyer flasks containing 50 mL of R2A with 1% (*w*/*v*) PET powder (<300 μm). After incubation, the supernatants were centrifuged at 13,000× *g* for 15 min at 4 °C and subsequently filtered using a 0.22 μm filter (Millipore, Burlington, MA, USA). The supernatant was freeze-dried using a freeze dryer (Alpha 1–4 LD plus, Martin Christ, Osterode, Germany), extracted with methanol, and filtered using a 0.22 μm pore filter.

An Agilent 1260 Infinity II Quaternary LC (Agilent, Santa Clara, CA, USA) with a Brownlee SPP C18 column (4.6 mm × 50 mm × 2.7 μm, PerkinElmer, Waltham, MA, USA) was used for HPLC analysis. Methanol and 50 mM phosphoric acid were used as the mobile phases with a gradient of 20% methanol for 0–2 min and 40% methanol for 12 min. The column was maintained at 40 °C, and the flow rate was 1 mL/min. The wavelength of the UV detector was set at 254 nm. Bis(2-hydroxyethyl) terephthalic acid (BHET) (Sigma Aldrich, St. Louis, MO, USA), mono(2-hydroxyethyl) terephthalic acid (MHET) (Advanced ChemBlocks Inc., Hayward, CA, USA), and TPA (Sigma Aldrich) were dissolved in methanol.

2.6. Statistical Analysis

One-way analysis of variance was performed using SPSS software (version 24; SPSS, Inc., Chicago, IL, USA). To compare the mean values, Scheffé's method was employed, and p-values less than or equal to 0.05 were deemed statistically significant.

3. Results

3.1. Isolation and Identification of PET Degrading Strain

Two bacterial strains isolated from the intestines of two Hymenoptera, *Xylocopa appendiculata* and *Eumenes decorates*, had a high degree of transparent halo zones on the test plate containing 0.1% (*w*/*v*) PET (Figure S1, see in Supplementary Materials). According to phylogenetic profiling based on the partial 16S rRNA region, the HY-74 strain was most closely related to the *Xanthomonas sontii* strain PPL1 (NQYO01000058) with 99.92% 16S rRNA nucleotide sequence similarity (Figure 1A). The HY-75 strain was most closely related to the *Bacillus siamensis* strain KCTC13613 (AJVF01000043) with 99.57% 16S rRNA nucleotide sequence similarity (Figure 2A). The 16S rRNA nucleotide sequences of the two strains were deposited in GenBank under accession numbers OQ921840 and OQ921871. The two strains were selected as potential candidates for PET degradation and were stored for further studies.

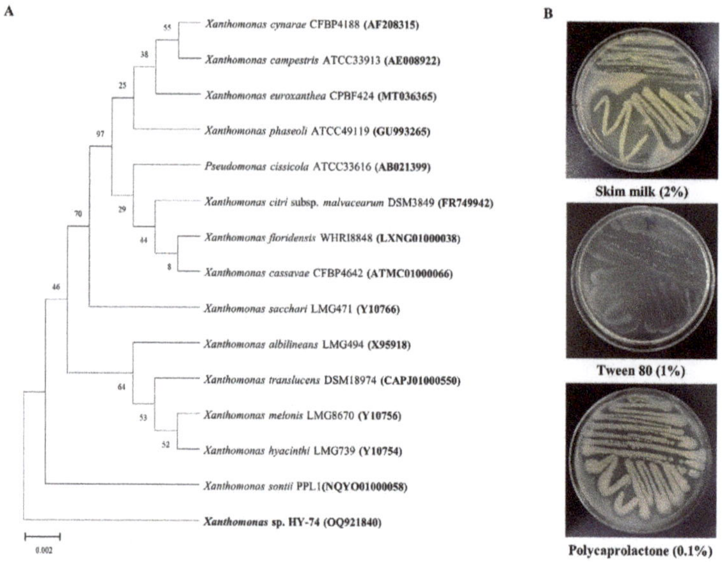

Figure 1. Phylogenetic relationship of the HY-74 strain from the intestine of *Xylocopa appendiculata* based on partial 16S rRNA gene sequencing (**A**). Neighbor-joining phylogenetic tree based on 16S rRNA gene sequences and closely related species constructed using MEGA X software. Numbers at branch nodes indicate the bootstrap percentages of 1000 replications. Extracellular enzymatic activities of the HY-74 strain (**B**).

Two bacterial strains were evaluated for the production of extracellular enzymes (Figures 1B and 2B). The results indicated that both strains have the ability to produce extracellular protease, PCL-degrading enzyme (the formation of a transparent halo zone), and lipase (the formation of a calcium complex).

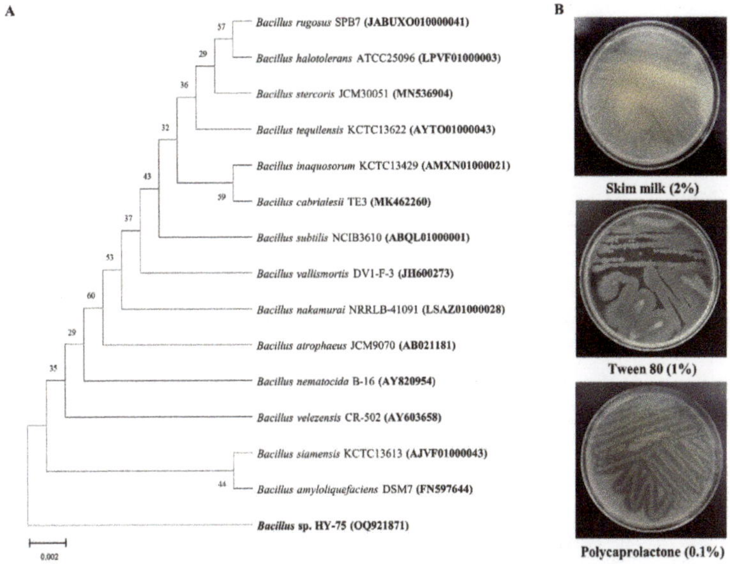

Figure 2. Phylogenetic relationship of the HY-75 strain from the intestine of *Eumenes decorates* based on partial 16S rRNA gene sequencing (**A**). Neighbor-joining phylogenetic tree based on 16S rRNA gene sequences and closely related species constructed using MEGA X software. Numbers at branch nodes indicate the bootstrap percentages of 1000 replications. Extracellular enzymatic activities of the HY-75 strain (**B**).

3.2. Sole Nutrition Source Assay

The ability of HY-74 and HY-75 to use PET as the sole nutritional source was assessed under four different carbon and nitrogen conditions in MSM using a double-layer PET agar plate (Figure 3). Over the course of 28 days, the HY-74 strain exhibited the following dry cell weight at the four different conditions: (i) 0.18 ± 0.05 g/L with MSM, (ii) 0.62 ± 0.10 g/L with MSMG, (iii) 0.60 ± 0.15 g/L with MSMN, and (iv) 1.36 ± 0.10 g/L with MSMGN. The HY-75 strain displayed the following dry cell weight at the four different conditions: (i) 0.13 ± 0.05 g/L with MSM, (ii) 0.57 ± 0.13 g/L with MSMG, (iii) 0.35 ± 0.08 g/L with MSMN, and (iv) 0.98 ± 0.14 g/L with MSMGN. The growth of the two strains on PET as the sole carbon and nitrogen source was obvious, but the growth was lower than that on minimal medium with an additional carbon and nitrogen source. The two strains exhibited the formation of degradation halos, as monitored by visual inspection. Under the same conditions, the PET degradation ability of the HY-74 strain was significantly higher than that of the HY-75 strain. The halo zones of the HY-74 strain were 319 ± 50 mm^2 with MSM, 357 ± 63 mm^2 with MSMG, 356 ± 30 mm^2 with MSMN, and 287 ± 51 mm^2 with MSMGN. HY-75 exhibited transparent halo zones of (i) 238 ± 32 mm^2 with MSM, (ii) 169 ± 27 mm^2 with MSMG, (iii) 473 ± 66 mm^2 with MSMN, and (iv) 201 ± 63 mm^2 with MSMGN.

3.3. PET Film Degradation

The ability of the HY-74 and HY-75 strains to degrade the PET film was determined using weight loss measurements, SEM observations, and FTIR analysis (Figure 4). The dry weight of the PET film after six weeks of incubation was significantly lower in all the tested media. In the HY-74 strain, the weight losses of the treated films were the following: MSM (0.087 ± 0.017 mg), R2A (0.197 ± 0.063 mg), and LB (0.196 ± 0.018) (Figure 4E). The HY-75 strain-degraded film contained MSM (0.097 ± 0.018 mg), R2A (0.210 ± 0.029 mg), and LB (0.163 ± 0.041 mg) (Figure 4D). With a weight loss of 1.57 ± 0.21% (for HY-74) and

1.42 ± 0.46% (for HY-75), R2A media showed the highest PET film degradation. The dry weight of the PET film of the untreated sample did not change after six weeks of incubation.

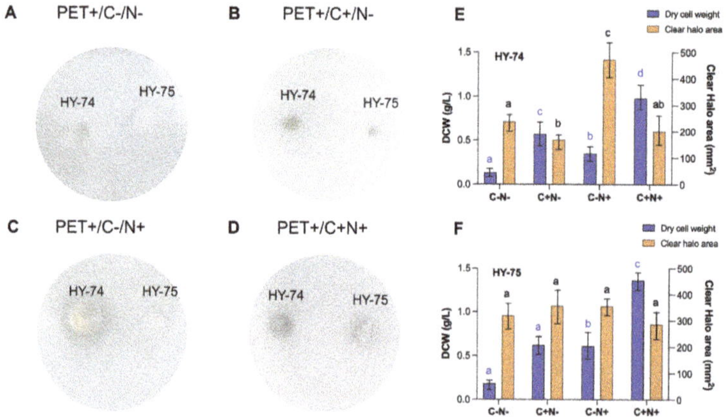

Figure 3. Growth and PET degradation of the HY-74 and HY-75 strains on a PET-containing agar plate as a sole nutrition source. A 0.1% PET-containing minimal salt agar medium was used to grow the strains for 4 weeks. (**A**) Minimal salt medium (MSM); (**B**) MSM supplemented with 0.5% glucose (MSMG); (**C**) MSM supplemented with 0.5% ammonium sulfate (MSMN); (**D**) MSM supplemented with 0.5% glucose and 0.5% ammonium sulfate (MSMGN). Dry cell weight was measured to assess bacterial growth, and PET biodegradation activity was demonstrated by visible clear halo zones surrounding the colonies for HY-74 (**E**) and HY-75 (**F**). Data are presented as means ± SD (n = 3). Different letters above the error bars indicate a significant difference by Scheffé's test ($p < 0.05$).

The surface morphological changes of the PET film were analyzed using SEM at 20,000× and 40,000× magnifications after six weeks of incubation (Figure 4A–C). The PET films incubated with both strains showed clear morphological changes due to microbial degradation. Scanning electron micrographs of the PET films treated with both strains showed significant damage to the surface morphology, such as erosion, scratches, and roughness, compared to the control.

The FTIR spectra of the PET film degradation by strains HY-74 and HY-75 showed changes in functional groups accompanied by more subtle changes at other wave numbers (Figure 4F,G). The peaks associated with aromatic C-H stretching (723 cm^{-1}), C-O stretching (1097 cm^{-1}), C-O-C stretching (1240 cm^{-1}), and C=O stretching (1714 cm^{-1}) decreased compared to the control. For HY-74, the O-H stretching (3200–3600 cm^{-1}) increased, whereas no significant difference was observed for HY-75. Specifically, the C-H stretching (723 cm^{-1}) corresponds to the out-of-plane bending vibrations in the aromatic and aliphatic C-H bonds, while the C-O stretching (1097 cm^{-1}) indicates the ester linkages between the terephthalate and EG units. Furthermore, the C-O-C stretching (1240 cm^{-1}) represents the vibrations in the ester group, and the C=O stretching (1714 cm^{-1}) signifies the cleavage of the ester bonds in PET. The increased hydroxyl group (OH) peak (3200–3600 cm^{-1}) suggested the formation of OH in the degradation products. These changes indicated that the PET film underwent chemical alterations, resulting in the formation of new functional groups and a modified composition.

3.4. Degradation of PET Powder

After 4 weeks of incubation with HY-74 and HY-75, the supernatant was analyzed using HPLC to confirm degradation in the presence of degradation products. In the standard HPLC spectra, the PET degradation products, including TPA, MHET, and BHET, showed clear peaks at 3.08, 5.08, and 6.75 min, respectively. Consistently, the supernatants treated with HY-74 and HY-75 showed TPA peaks at a retention time of 3.08 min after

4 weeks as shown in Figure 5, while MHET and BHET were not detected. The absence of these peaks indicates that they were subsequently hydrolyzed to TPA. No degradation products were detected in the control group.

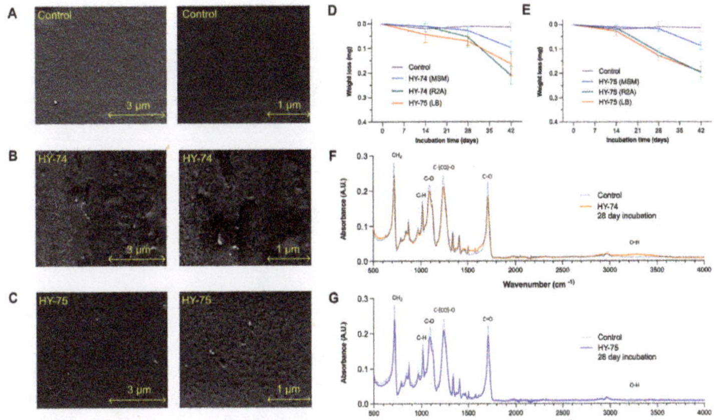

Figure 4. Degradation of PET film by HY-74 and HY-75 strains after 6 weeks of incubation. Scanning electron micrographs of non-treated control films (**A**), films treated with the HY-74 strain (**B**), and films treated with the HY-75 strain (**C**). Time course of degradation of PET film by the HY-74 strain (**D**) and HY-75 strain (**E**) in MSM, R2A, and LB media. Comparison of FTIR spectra of PET film degradation of the HY-74 strain (**F**), HY-75 strain (**G**), and non-treated control. Data are presented as means ± SD (n = 3).

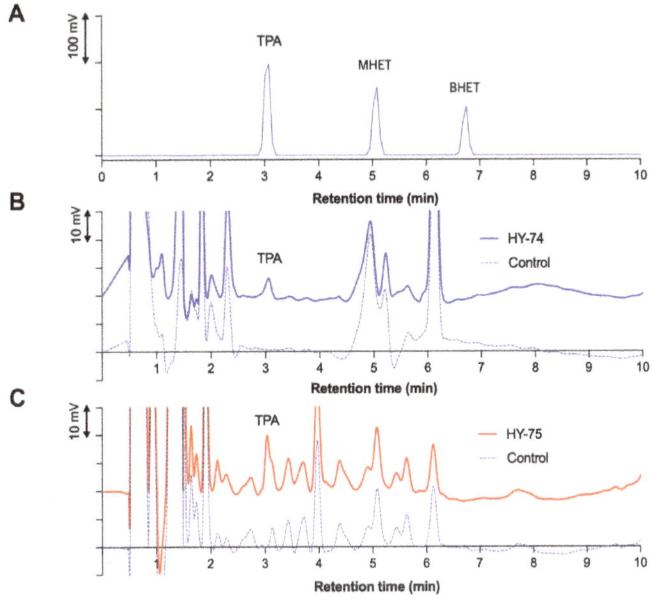

Figure 5. Determination of PET powder degradation by the HY-74 and HY-75 strain using high-performance liquid chromatography (HPLC) analysis after 4 weeks of incubation. (**A**) HPLC spectra of a standard sample displaying peaks for degradation products MHET, BHET, and TPA. HPLC spectra showing PET biodegradation by HY-74 strain (**B**) and HY-75 strain (**C**) after 4 weeks of incubation.

4. Discussion

Over the last decade, there has been growing interest in the plastic degradation capabilities of insect larvae for PP, polystyrene (PS), and polyvinyl chloride (PVC) because herbivorous insects may be a valuable resource for microorganisms that can break down synthetic plastics. Insects that can digest plastics using their gut microbiota are gaining interest for use in bioremediation, although their environmental benefits remain unknown [19]. These microbes exploit the enzymatic pathways of decomposing plant materials to degrade synthetic plastic polymers because of their similar chemical structures [20]. However, most plastics biodegraded by insect gut microbes are polyethylene [21–23], PS [24], PVC [25,26], and polyurethane [27,28] with little knowledge available on the gut microbiome of insects capable of degrading PET. In this study, the PET degradation abilities of two insect gut-derived bacterial strains, HY-74 and HY-75, were confirmed using PET-containing agar plates, PET powder, and a PET film. To the best of our knowledge, this is the first report on the isolation and characterization of PET-degrading bacteria of insect intestinal origin.

HY-75 demonstrated differential degradation activities under minimal and nutrient-rich conditions. The nitrogen source appeared to be a crucial factor in promoting PET degradation in HY-74, whereas the presence of glucose as a carbon source seemed to have a suppressive effect on degradation. Differences in the degrading abilities of the different media suggest that HY-74 produces enzymes in response to the limited availability of nutrients. In minimal media, microorganisms often need to synthesize specific enzymes to break down and utilize the limited available resources, whereas in nutrient-rich media, they may not need to produce these enzymes because of the abundance of readily available nutrients [29,30]. Enzyme production is usually subjected to catabolite repression in which the presence of a preferred carbon or nitrogen source in rich media inhibits the expression of certain enzymes [31]. These preferred sources may not be present in MSM without glucose, leading to the induction of enzyme production that utilizes alternative substrates. The degradation activity of HY-74 was similar or comparable for all the different nutritional sources. With controlled carbon and nitrogen sources, the degradation activity was not significantly affected by the specific carbon or nitrogen sources present. The degradation-related enzymes of the HY-74 strain may be constitutively expressed and continuously produced by the organism regardless of the presence of a specific substrate. This finding indicates that HY-74 can efficiently degrade PET regardless of the nutritional source.

Quantitative and visual analyses of the PET film degraded by strains HY-74 and HY-75 indicated that they could also degrade PET film types. After 6 weeks of incubation, the highest degradation rates of PET film for HY-74 and HY-75 were $1.57 \pm 0.21\%$ and $1.42 \pm 0.46\%$, respectively. In the scanning electron images, the PET film exhibited disruption and changes in the surface morphology. Although similar observations have been previously reported, HY-74 and HY-75 showed similar or higher weight losses than other bacteria on PET films in a shorter period [8,32,33], while their degradation efficiency was not as high as the whole-cell degradation exhibited by *Clostridium thermocellum* [34]. Further investigation of the factors that influence bacterial growth and PET degradation, including the physical properties of PET, temperature, pH, and nutrient availability, is required to develop effective bioremediation strategies.

Many hydrolases from microorganisms, such as cutinase and esterase, have been discovered and modified to primarily target and break down PET ester bonds. Hydrolase can degrade PET into MHET and BHET or ultimately into TPA and EG. As a result, some microorganisms can use the TPA and EG monomers generated in this degradation process, either by incorporating them into the tricarboxylic acid cycle or transforming them into valuable chemicals that enable the biodegradation and bioconversion of PET. In this study, the bacterial strains HY-74 and HY-75 demonstrated the ability to break down PET into TPA, as confirmed with HPLC. We hypothesized that these microorganisms produce a specific type of hydrolase that degrades PET. However, the comprehensive understanding of the specific enzymes involved and the intricate process of TPA metabolism is limited in this study. Further research, such as determining the specific enzymes involved and elucidating the degradation process of TPA metabolism, is required to understand the subsequent PET biodegradation by HY-74 and HY-75 and their bioconversion steps. The findings of this study provide promising prospects for the use of insect gut bacteria for PET biodegradation. Further research should focus on optimizing the conditions for the growth and activity of the identified PET-degrading microorganisms as well as exploring genetic engineering methods to enhance the efficiency of PET biodegradation. For instance, recent studies have focused on engineering and optimizing PET-degrading enzymes to enhance their performance [35], investigating microbial communities for potential PET degradation [36], and developing genetically engineered whole-cell catalysts, such as *E. coli*, expressing PET-degrading enzymes to enhance PET-degradation efficiency [37].

5. Conclusions

The PET biodegradable bacteria *Xanthomonas* sp. HY-74 and *Bacillus* sp. HY-75 were isolated from the intestines of Hymenoptera, *Xylocopa appendiculata* and *Eumenes decorates*, respectively. Both strains demonstrated the ability to degrade PET as evidenced by the observed changes in chemical functional groups and a distinctive TPA peak in the HPLC results. Additionally, morphological alterations of the PET film, surface modifications, and PET film weight reduction (1.57 ± 0.21% for HY-74, 1.42 ± 0.46% % for HY-75) were observed after a 6-week period. Further research employing multi-omics approaches and synthetic microbial communities is needed to elucidate the specific enzymes involved and their metabolic pathways, and to optimize the degradation conditions for practical applications in PET waste management. This study highlights the potential of insect-derived bacterial strains for PET biodegradation and highlights the importance of further exploration of their practical applications in mitigating the global plastic pollution crisis.

Supplementary Materials: The following supporting information can be downloaded at: https://www.mdpi.com/article/10.3390/polym15173546/s1, Figure S1. Screening of PET degrading bacteria, *Xanthomonas* sp. HY-74 strain from *Xylocopa appendiculata* (A) and *Bacillus* sp. HY-75 from *Eumenes decorates* (B).

Author Contributions: Conceptualization, J.-H.K.; methodology, K.-H.S.; validation, J.-H.K.; formal analysis, S.-H.L.; investigation, S.-H.L. and B.-M.L.; resources, J.-H.K.; data curation, J.-H.K. and K.-H.S.; writing—original draft preparation, J.-H.K.; writing—review and editing, J.-H.K.; visualization, S.-H.L.; supervision, H.-Y.P.; funding acquisition, H.-Y.P. All authors have read and agreed to the published version of the manuscript.

Funding: This research was supported by the Korea Research Institute of Bioscience and Biotechnology (KRIBB) Research Initiative Program (KGM5492322) of the Ministry of Science and ICT of the Republic of Korea.

Institutional Review Board Statement: Not applicable.

Data Availability Statement: The data presented in this study are available on request from the corresponding author.

Conflicts of Interest: The authors declare no conflict of interest.

References

1. OECD. Plastic Pollution Is Growing Relentlessly as Waste Management and Recycling Fall Short, Says OECD. 2022. Available online: https://www.oecd.org/environment/plastic-pollution-is-growing-relentlessly-as-waste-management-and-recycling-fall-short.htm (accessed on 22 December 2022).
2. Geyer, R.; Jambeck, J.R.; Law, K.L. Production, use, and fate of all plastics ever made. *Sci. Adv.* **2017**, *3*, e1700782. [CrossRef]
3. Andrady, A.L.; Neal, M.A. Applications and societal benefits of plastics. *Philos. Trans. R. Soc. B Biol. Sci.* **2009**, *364*, 1977–1984. [CrossRef]
4. Yoshida, S.; Hiraga, K.; Takehana, T.; Taniguchi, I.; Yamaji, H.; Maeda, Y.; Toyohara, K.; Miyamoto, K.; Kimura, Y.; Oda, K. A bacterium that degrades and assimilates poly (ethylene terephthalate). *Science* **2016**, *351*, 1196–1199. [CrossRef]
5. Maheswaran, B.; Al-Ansari, M.; Al-Humaid, L.; Raj, J.S.; Kim, W.; Karmegam, N.; Rafi, K.M. In vivo degradation of polyethylene terephthalate using microbial isolates from plastic polluted environment. *Chemosphere* **2023**, *310*, 136757. [CrossRef]

6. Ribitsch, D.; Heumann, S.; Trotscha, E.; Herrero Acero, E.; Greimel, K.; Leber, R.; Birner-Gruenberger, R.; Deller, S.; Eiteljoerg, I.; Remler, P.; et al. Hydrolysis of polyethyleneterephthalate by *p*-nitrobenzylesterase from *Bacillus subtilis*. *Biotechnol. Prog.* **2011**, *27*, 951–960. [CrossRef]
7. Jadaun, J.S.; Bansal, S.; Sonthalia, A.; Rai, A.K.; Singh, S.P. Biodegradation of plastics for sustainable environment. *Bioresour. Technol.* **2022**, *347*, 126697. [CrossRef]
8. Roberts, C.; Edwards, S.; Vague, M.; León-Zayas, R.; Scheffer, H.; Chan, G.; Swartz, N.A.; Mellies, J.L. Environmental consortium containing *Pseudomonas* and *Bacillus* species synergistically degrade polyethylene terephthalate plastic. *mSphere* **2020**, *5*, e01151-20. [CrossRef]
9. Bollinger, A.; Thies, S.; Knieps-Grünhagen, E.; Gertzen, C.; Kobus, S.; Höppner, A.; Ferrer, M.; Gohlke, H.; Smits, S.H.; Jaeger, K.E. A novel polyester hydrolase from the marine bacterium *Pseudomonas aestusnigri*–structural and functional insights. *Front. Microbiol.* **2020**, *11*, 114. [CrossRef]
10. Vague, M.; Chan, G.; Roberts, C.; Swartz, N.A.; Mellies, J.L. *Pseudomonas* isolates degrade and form biofilms on polyethylene terephthalate (PET) plastic. *bioRxiv* **2019**, 647321. [CrossRef]
11. Kawai, F.; Kawabata, T.; Oda, M.; Murase, Y. Discovery of a new bacterial PET hydrolase and synthetic PET degradation by combined culture of *Ideonella sakaiensis* and a hydrolytic bacterium. *Sci. Rep.* **2019**, *9*, 13845.
12. Li, J.; Gu, J.D.; Pan, L. Transformation of dimethyl phthalate, dimethyl isophthalate and dimethyl terephthalate by *Rhodococcus rubber* Sa and modeling the processes using the modified Gompertz model. *Int. Biodeterior. Biodegrad.* **2005**, *55*, 223–232. [CrossRef]
13. Bowditch, T.G. Penetration of polyvinyl chloride and polypropylene packaging films by *Ephestia cautella* (lepidoptera: Pyralidae) and *Plodia interpunctella* (Lepidoptera: Pyralidae) Larvae, and *Tribolium confusum* (Coleoptera: Tenebrionidae) Adults. *J. Econ. Entomol.* **1997**, *90*, 1028–1031. [CrossRef]
14. Yang, Y.; Yang, J.; Wu, W.M.; Zhao, J.; Song, Y.; Gao, L.; Yang, R.; Jiang, L. Biodegradation and mineralization of polystyrene by plastic-eating mealworms: Part 1. Chemical and physical characterization and isotopic tests. *Environ. Sci. Technol.* **2015**, *49*, 12080–12086. [CrossRef] [PubMed]
15. Yang, Y.; Chen, J.; Wu, W.M.; Zhao, J.; Yang, J. Complete genome sequence of *Bacillus* sp. YP1, a polyethylene-degrading bacterium from waxworm's gut. *J. Biotechnol.* **2015**, *200*, 77–78. [CrossRef]
16. Nyamjav, I.; Jang, Y.; Park, N.; Lee, Y.E.; Lee, S. Physicochemical and structural evidence that *Bacillus cereus* isolated from the gut of waxworms (*Galleria mellonella* larvae) biodegrades polypropylene efficiently in vitro. *J. Polym. Environ.* **2023**, 1–14. [CrossRef]
17. Bhagobaty, R.K.; Joshi, S.R. Enzymatic activity of fungi endophytic on five medicinal plant species of the pristine sacred forests of Meghalaya, India. *Biotechnol. Bioprocess.* **2012**, *17*, 33–40. [CrossRef]
18. Thompson, J.D.; Gibson, T.J.; Plewniak, F.; Jeanmougin, F.; Higgins, D.G. The CLUSTAL_X windows interface: Flexible strategies for multiple sequence alignment aided by quality analysis tools. *Nucl. Acids Res.* **1997**, *25*, 4876–4882. [CrossRef]
19. Jang, S.; Kikuchi, Y. Impact of the insect gut microbiota on ecology, evolution, and industry. *Curr. Opin. Insect Sci.* **2020**, *41*, 33–39. [CrossRef]
20. Sanchez-Hernandez, J.C. A toxicological perspective of plastic biodegradation by insect larvae. *Comp. Biochem. Physiol. C Toxicol. Pharmacol.* **2021**, *248*, 109117. [CrossRef]
21. Lou, Y.; Li, Y.; Lu, B.; Liu, Q.; Yang, S.S.; Liu, B.; Ren, N.; Wu, W.M.; Xing, D. Response of the yellow mealworm (*Tenebrio molitor*) gut microbiome to diet shifts during polystyrene and polyethylene biodegradation. *J. Hazard. Mater.* **2021**, *416*, 126222. [CrossRef]
22. Peng, B.Y.; Sun, Y.; Wu, Z.; Chen, J.; Shen, Z.; Zhou, X.; Wu, W.M.; Zhang, Y. Biodegradation of polystyrene and low-density polyethylene by *Zophobas atratus* larvae: Fragmentation into microplastics, gut microbiota shift, and microbial functional enzymes. *J. Clean. Prod.* **2022**, *367*, 132987. [CrossRef]
23. Latour, S.; Noël, G.; Serteyn, L.; Sare, A.R.; Massart, S.; Delvigne, F.; Francis, F. Multi-omics approach reveals new insights into the gut microbiome of *Galleria mellonella* (Lepidoptera: Pyralidae) exposed to polyethylene diet. *bioRxiv* **2021**, 6. [CrossRef]
24. Bulak, P.; Proc, K.; Pytlak, A.; Puszka, A.; Gawdzik, B.; Bieganowski, A. Biodegradation of different types of plastics by *Tenebrio molitor* insect. *Polymers* **2021**, *13*, 3508. [CrossRef] [PubMed]
25. Peng, B.Y.; Chen, Z.; Chen, J.; Yu, H.; Zhou, X.; Criddle, C.S.; Wu, W.M.; Zhang, Y. Biodegradation of Polyvinyl Chloride (PVC) in *Tenebrio molitor* (Coleoptera: Tenebrionidae) larvae. *Environ. Int.* **2020**, *145*, 106106. [CrossRef] [PubMed]
26. Zhang, Z.; Peng, H.; Yang, D.; Zhang, G.; Zhang, J.; Ju, F. Polyvinyl chloride degradation by a bacterium isolated from the gut of insect larvae. *Nat. Commun.* **2022**, *13*, 5360. [CrossRef]
27. Kim, J.H.; Choi, S.H.; Park, M.G.; Park, D.H.; Son, K.H.; Park, H.Y. Biodegradation of polyurethane by Japanese carpenter bee gut-associated symbionts *Xanthomonas* sp. HY-71, and its potential application on bioconversion. *Environ. Technol. Innov.* **2022**, *28*, 102822. [CrossRef]
28. Kim, J.H.; Choi, S.H.; Park, M.G.; Park, D.H.; Son, K.H.; Park, H.Y. Polyurethane biodegradation by *Serratia* sp. HY-72 isolated from the intestine of the Asian mantis *Hierodula patellifera*. *Front. Microbiol.* **2022**, *13*, 1005415. [CrossRef]
29. Harder, W.; Dijkhuizen, L. Physiological responses to nutrient limitation. *Annu. Rev. Microbiol.* **1983**, *37*, 1–23. [CrossRef]
30. Allison, S.D.; Vitousek, P.M. Responses of extracellular enzymes to simple and complex nutrient inputs. *Soil Biol. Biochem.* **2005**, *37*, 937–944. [CrossRef]
31. Görke, B.; Stülke, J. Carbon catabolite repression in bacteria: Many ways to make the most out of nutrients. *Nat. Rev. Microbiol.* **2008**, *6*, 613–624. [CrossRef]
32. Denaro, R.; Aulenta, F.; Crisafi, F.; Di Pippo, F.; Viggi, C.C.; Matturro, B.; Tomei, P.; Smedile, F.; Martinelli, A.; Di Lisio, V.; et al. Marine hydrocarbon-degrading bacteria breakdown poly (ethylene terephthalate) (PET). *Sci. Total Environ.* **2020**, *749*, 141608. [CrossRef]

33. Kumari, A.; Bano, N.; Bag, S.K.; Chaudhary, D.R.; Jha, B. Transcriptome-guided insights into plastic degradation by the marine bacterium. *Front. Microbiol.* **2021**, *12*, 751571. [CrossRef]
34. Yan, F.; Wei, R.; Cui, Q.; Bornscheuer, U.T.; Liu, Y.J. Thermophilic whole-cell degradation of polyethylene terephthalate using engineered *Clostridium thermocellum*. *Microb. Biotechnol.* **2021**, *14*, 374–385. [CrossRef]
35. Joo, S.; Cho, I.J.; Seo, H.; Son, H.F.; Sagong, H.Y.; Shin, T.J.; Choi, S.Y.; Lee, S.Y.; Kim, K.J. Structural insight into molecular mechanism of poly (ethylene terephthalate) degradation. *Nat. Commun.* **2018**, *9*, 382. [CrossRef]
36. Danso, D.; Chow, J.; Streit, W.R. Plastics: Environmental and biotechnological perspectives on microbial degradation. *Appl. Environ. Microbiol.* **2019**, *85*, e01095-19. [CrossRef]
37. Gercke, D.; Furtmann, C.; Tozakidis, I.E.; Jose, J. Highly crystalline post-consumer PET waste hydrolysis by surface displayed PETase using a bacterial whole-cell biocatalyst. *ChemCatChem* **2021**, *13*, 3479–3489. [CrossRef]

polymers

Article

Degradation of Polylactic Acid Polymer and Biocomposites Exposed to Controlled Climatic Ageing: Mechanical and Thermal Properties and Structure

Adam Vašíček *, Petr Lenfeld and Luboš Běhálek

Faculty of Mechanical Engineering, Technical University of Liberec, Studentska 1402/2, 46117 Liberec, Czech Republic
* Correspondence: adam.vasicek@tul.cz; Tel.: +420-604-134-654

Abstract: This paper deals with the study of the degradation of polylactic acid (PLA) material structures and biocomposite systems with a PLA matrix containing ground natural particulate waste fillers, buckwheat husks and egg shells. Waste fillers were used without difficult cleaning operations to describe the effect of the raw waste material on PLA. Biocomposites with raw waste materials are increasingly coming to the forefront in car interiors and packaging products. The prepared material systems were exposed to controlled climatic ageing simulating long-term solar radiation and cyclic outdoor conditions. The degradation of the biocomposite systems was evaluated via thermal (differential scanning calorimetry) and mechanical properties (tensile and flexural tests, Charpy impact toughness). In addition to evaluating the degradation of the material structures using standardized tests, the influence and effect of controlled climatic ageing was visually assessed using SEM images (electron microscopy) of the surfaces and fracture surfaces of the test specimens.

Keywords: polylactic acid; biocomposite; buckwheat husks; egg shells; degradation; controlled climate ageing

Citation: Vašíček, A.; Lenfeld, P.; Běhálek, L. Degradation of Polylactic Acid Polymer and Biocomposites Exposed to Controlled Climatic Ageing: Mechanical and Thermal Properties and Structure. *Polymers* **2023**, *15*, 2977. https://doi.org/10.3390/polym15142977

Academic Editors: Alexey Iordanskii and Alexandre Vetcher

Received: 9 June 2023
Revised: 30 June 2023
Accepted: 5 July 2023
Published: 8 July 2023

1. Introduction

There is now a growing interest in environmental sustainability, and thus, in biopolymers to replace synthetic polymers that are not biodegradable and have a negative environmental impact [1,2]. Biopolymers are degraded into low molecular weight natural substances when exposed to environmental influences, microorganisms, elevated temperatures or ambient climates [3]. In 2022, 2.17 million tonnes of bioplastics were produced worldwide, with polylactic acid accounting for the largest production (20%) [4]. Polylactic acid (PLA) is considered a promising successor to commonly used synthetic polymers with similar properties to, for example, polyethylene terephthalate (PET) [5]. Among the primary advantages of PLA is precisely its biodegradability and the possible formation of both amorphous and semicrystalline phases, as the ratio of L,D-isomers affects the crystallinity or biodegradation rate of the polymer [5–8]. PLA has been applied in packaging materials such as fruit boxes, pouches or beverage cups [9]. PLA in copolymerization, e.g., with methacrylic acid as a cross-linked polymer, has been applied as a thickener for printed fabrics [10]. Studies have also looked at the branched PLA structure, which offers excellent processing and injection moulding properties [11]. The primary source for PLA production is starch-rich agricultural products such as corn or potatoes [12]. PLA production from corn stover or seaweed has also been investigated [13].

PLA changes its mechanical and optical properties due to thermal, oxidation or photodegradation processes during the product's lifetime [14]. The marine ecosystem can lead to biopolymer degradation, manifested by swelling or hydrolysis [15]. Polymers undergo hydrolytic degradation due to moisture, as water molecules diffuse into the polymer and cause ester and primary chain cleavage. The degradation occurs mainly in the amorphous section; the crystalline section is more stable [16]. Exposure to solar UV radiation leads to bond breaking, i.e., ageing and possible degradation of PLA [17,18]. Under the influence of incident photons of UV radiation, polymer chains undergo cleavage or disintegration of the crystal structure of the polymer [16]. The effect of UV radiation can be simulated by outdoor exposure, where samples are placed in exposure racks and exposed to sunlight for 12–18 months, according to ASTM D5272-08 [19]. Currently, xenon test chambers are often used, where photodegradation is simulated with accelerations of 2:1, 5:1 and 63:1 [18].

In terms of PLA performance, fillers are added to the matrix. This creates a composite structure that can lead to better mechanical properties and lower product costs. In order to maintain the biodegradability of PLA, natural waste fillers are increasingly being used, which can be degraded by environmental conditions [20]. Natural fillers are either fibrous or particulate [21,22]. Among the fibrous ones, sisal, banana, bamboo, hemp and pineapple fibres are used [21]. Suitable particulate fillers are, e.g., buckwheat husks and onion skins. Buckwheat husks are a waste product of buckwheat production that cannot be used as livestock feed. They are used, for example, as fuel for boilers. Buckwheat husks have both antioxidant and bacteriostatic properties and are composed of organic and inorganic substances [22,23]. Egg shells have also been studied as particulate filler for PLA [24]. Egg shells are a widespread waste product of food, bakery and poultry processing plants [24,25]. Surveys in 2002 reported that egg shells can be used as fertilizers or as an ingredient in animal feed, and about one-third of egg shells end up in landfills [26]. They can also be applied as adsorbents for various unwanted organic and inorganic compounds from the waste segment [25].

2. Materials and Methods

The PLA matrix used was commercially available PLA Ingeo 3001D [27], suitable for injection moulding, which was purchased (NatureWorks, Ingeo 3001D, Nakhon Sawan, Thailand). The PLA polymer contains more than 99% L-lactic acid and up to 1% D-lactic acid. The biocomposite materials were prepared from PLA biopolymer matrix and natural waste fillers, buckwheat husks and egg shells. Buckwheat husks (BH), which were free from impurities (dust and unwanted contaminants), and egg shells (ES) were dried and subsequently milled (Figure 1) using a 0.75 mm sieve on a mill (Retsch GmbH, Retsch SM 300, Haan, Germany).

Figure 1. SEM images of ground buckwheat husks (**left**) and egg shells (**right**).

2.1. Preparation of PLA Test Samples

The granulate of unfilled PLA matrix, ground egg shells and buckwheat husks were vacuum dried (Binder GmbH, VD53, Tuttlingen, Germany) for 24 h at 50 °C after moisture removal. The biocomposite materials were prepared by compounding on an extruder (Collin Lab & Pilot Solutions GmbH, Lab-Line ZK 25, Maitenbeth, Germany) with a granulator (Econ GmbH, Econ Ewa 10, Bergern, Austria) using the following process parameters:

extruder temperature profile 135 to 175 °C (extrusion head), extruder screw speed 150 rpm and granulator knife head speed 3000 rpm. The compounding produced granular PLA biocomposites with the addition of 20 wt% ground buckwheat husks (PLA/BH) and egg shells (PLA/ES). The composition of the material structures and biocomposite systems and their designation are shown in Table 1. Type 1A test bodies according to ISO 527-2 were produced from unfilled PLA granulate and biocomposite granulate (Arburg GmbH + Co KG, Arburg Allrounder 320 C, Lossburg, Germany) via injection moulding. The injection parameters were the following: melt temperature, 190 °C; tempering medium temperature, 20 °C; batch volume, 36 cm^3; injection pressure, 50 MPa; clamping force, 400 kN and total cycle time, 60 s. These parameters were evaluated as optimal for the production of the samples and were also based on the datasheet.

Table 1. Sample compositions.

Sample Marking	Matrix	Filler	Controlled Climatic Ageing
PLA	PLA	-	-
PLA/LA	PLA	-	Long-term
PLA/CA	PLA	-	Cyclical
PLA/BH	PLA	20 wt% BH	-
PLA/BH/LA	PLA	20 wt% BH	Long-term
PLA/BH/CA	PLA	20 wt% BH	Cyclical
PLA/ES	PLA	20 wt% ES	-
PLA/ES/LA	PLA	20 wt% ES	Long-term
PLA/ES/CA	PLA	20 wt% ES	Cyclical

2.2. *Ageing of Test Samples*

The test bodies were subjected to controlled climatic ageing according to DIN 75 220 in a solar chamber (Vötch, SUN 3600, Balingen, Germany) with MH 2 × 4 kW radiators under two different controlled climatic ageing conditions. In the first case, the test bodies were loaded with a long-term test under constant irradiation conditions (enclosure temperature of 42 ± 3 °C, relative humidity of 65 ± 5%, irradiation intensity of 1000 ± 100 $W{\cdot}m^{-2}$) with a total irradiation time of 240 h. The aim of these conditions was to simulate long-term solar irradiation. The second controlled climatic ageing condition was a cyclic test (cyclical) that was conducted according to DIN 75 220, where the first 15 cycles simulated a dry climate, and then 10 cycles simulated a humid climate. Each cycle lasted 24 h, and each cycle had different climatic parameters (humidity, temperature and irradiation intensity) for the two climates. The total cycling time was 600 h. The set parameters of the cycling test simulated outdoor climatic conditions that last 4 years under standard natural effects.

2.3. *Thermal Analysis (Differential Scanning Calorimetry)*

The thermal properties of the biocomposites were evaluated using a calorimeter (Mettler Toledo, DSC 1/700, Greifensee, Switzerland) according to ISO 11357 standard for the evaluation of non-isothermal properties of samples. The measurements were performed on 6 ± 0.5 mg samples heated from 0 to 200 °C at a heating rate of 10 °C/min. After the 1st heating isotherm, the samples were heated to 200 °C for 3 min to remove thermal history. Subsequently, the temperature was reduced to an initial 0 °C, and the 2nd heating started. DSC analysis was performed in the presence of nitrogen at a flow rate of 50 mL/min. The enthalpy of secondary (cold) crystallization (ΔH_{sc}), enthalpy of recrystallization (ΔH_{rc}), enthalpy of crystallite melting (ΔH_m), the temperature of secondary (cold) crystallization (T_{sc}), the temperature of recrystallization (T_{rc}) and the temperature of crystallite melting (T_m) were recorded from the 1st heating step.

The degree of crystallization (χ_c) was calculated according to Equation (1) as follows [28]:

$$\chi_c = (\Delta H_m - \Delta H_{rc} - \Delta H_{sc} - 100)/\Delta H^0{}_{hm}\ [\%] \quad (1)$$

where ΔH^0_{hm} denotes the enthalpy of the melting of perfectly crystalline PLA (93 J/g).

2.4. Study of the Structure on Electron Microscope

The study of surfaces and fracture surfaces of unfilled PLA matrix and biocomposite structures was performed using a scanning electron microscope (SEM) (Tescan, TESCAN MIRA 3, Brno, Czech Republic). Samples were taken from the test bodies and attached to the targets using carbon tape. Before scanning, the individual samples were coated with a platinum/palladium metal mixture with a coating thickness of 6 nm, using an instrument (Leica, Leica EM ACE200, Munich, Germany) to ensure surface conductivity. The deposition was carried out via physical vapor deposition (PVD) in a protective atmosphere using argon. The test bodies for studying the fracture surface microscopy were prepared and fractured after using liquid nitrogen before breaking the sample. This does not affect the fracture surface as in the case of samples after, for example, tensile or Charpy impact toughness testing.

2.5. Mechanical Properties

The test samples were conditioned according to ISO 291 at 35 °C with 62% relative humidity for 240 h before testing. The test pieces were then subjected to mechanical, tensile, bending and impact strength tests. These measurements were always taken on 10 test samples. The tensile properties were measured according to ISO 527 using an instrument (TIRA GmbH, TiraTest 2300, Schalkau, Germany) with a strain gauge (MF GmbH, MFL-300B, Velbert, Germany). The loading rate for the calculation of the tensile modulus was 1 mm/min until the failure of the body. The initial jaw spacing was 115 mm, with an initial gauge length of 50 mm and a preload of 2 N. The measured quantities were the tensile modulus (E_t) and the ultimate tensile strength (σ_m). The bending properties were measured according to ISO 178 using an instrument (Tinius Olsen, Hounsfield H10KT, Salfords, UK) with a sensing head of 500 N. The loading rate was 2 mm·min^{-1} at a preload of 2 N in three-point bending. The output of the measurements was the flexural modulus (E_f) and the flexural strength (σ_{fM}). The impact toughness of Charpy (a_{cU}) according to ISO 179 was measured using an instrument (Instron, Resil Ceast S.p.A, Norwood, MA, USA). The body was broken by striking the narrower side of the test body without indentation.

3. Results and Discussion

3.1. Differential Scanning Calorimetry (DSC)

The unfilled PLA material sample reached a degree of crystallinity of 1.9% at a secondary crystallization peak of 115.9 °C (Table 2). The morphological structure of the sample after processing showed a highly disordered structure. The addition of 20 wt% of ground waste natural fillers resulted in a significant increase in the degree of crystallinity; for the PLA/ES biocomposite, the degree of crystallinity was 7.7% (an increase of about 4 times), and for the PLA/BH biocomposite, the degree of crystallinity was 12.4% (increase of about 6.5 times). The addition of waste fillers led to a higher degree of crystallinity, as observed, for example, by Koutsomitopoulou et al. [29] when olive pith powder was added to the PLA polymer. Similarly, Sivagnanamani et al. [24] measured a higher degree of crystallinity in the PLA by adding ground egg shells.

The biocomposite materials PLA/BH and PLA/ES were secondary crystallized at a lower temperature than the unfilled PLA. Adding natural filler to the biopolymer system leads to a higher crystal structure formation and lowers the temperature at which the PLA secondary crystallizes by about 10 °C.

The exposure of the biocomposite samples to controlled climatic ageing significantly changed the degree of crystallinity, especially for the biocomposites filled with ground buckwheat husks (PLA/BH). After long-term (LA) and cyclic (CA) controlled ageing, the samples showed a higher degree of crystallinity by approximately four times than the PLA/BH composites that were not exposed to the climate chamber (higher degree of crystallinity was measured for long-term (LA) controlled climatic ageing). A similar increase (approx. four times) was also observed for the unfilled PLA polymer (higher degree of crystallinity was again measured for long-term (LA) controlled climatic ageing). In contrast, for the PLA/ES biocomposite, the effect of climate-controlled ageing on the degree of crystallinity was not statistically significant for long-term climate-controlled ageing, but a higher degree of crystallinity was measured for cyclic (CA) climate-controlled ageing, by approximately a 25% increase.

Table 2. Thermal analysis (DSC) data of PLA biocomposites exposed to long-term and cyclic controlled climatic ageing. Measured values are from the first heating. The symbols are explained as follows: the enthalpy of secondary (cold) crystallization (ΔH_{sc}); enthalpy of recrystallization (ΔH_{rc}); enthalpy of crystallite melting (ΔH_m); the temperature of secondary (cold) crystallization (T_{sc}); the temperature of recrystallization (T_{rc}); the temperature of crystallite melting (T_m); degree of crystallization (χ_c).

Material	ΔH_{sc} (J/g)	ΔH_{rc} (J/g)	ΔH_m (J/g)	X_c (%)	T_{sc} (°C)	T_{rc} (°C)	T_m (°C)
PLA	35.71	-	37.52	1.9	115.9	-	170.1
PLA/LA	30.06	0.48	37.33	7.3	103.3	155.7	168.6
PLA/CA	31.76	-	37.23	5.9	106.7	-	169.2
PLA/BH	20.95	0.64	30.78	12.4	104.5	155.7	169.0
PLA/BH/LA	-	1.60	38.67	49.8	-	152.7	168.3
PLA/BH/CA	-	1.90	36.15	46.0	-	150.1	167.7
PLA/ES	25.27	-	31.02	7.7	108.2	-	169.2
PLA/ES/LA	25.01	2.34	33.78	8.6	97.7	153.4	168.3
PLA/ES/CA	24.10	2.30	33.48	9.5	97.5	153.4	168.2

3.2. Study of the Structure on Electron Microscope

The surface texture of the unfilled PLA and unfilled PLA samples exposed to long-term (PLA/LA) and cyclic (PLA/CA) climate-controlled ageing (see Figure 2) exhibits a smooth surface without significant damage or cracks.

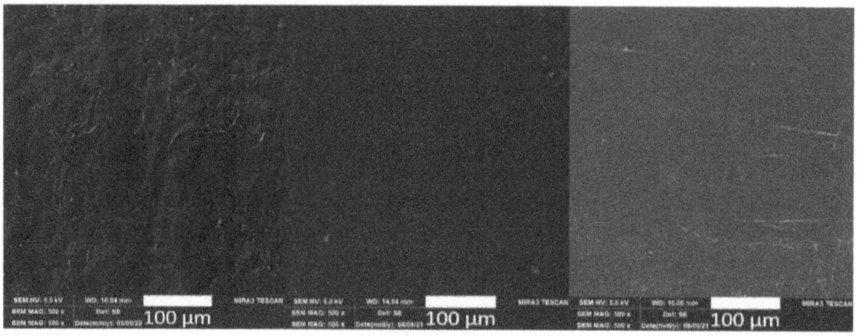

Figure 2. SEM images of the surface of unfilled PLA before exposure to controlled climatic ageing (**left**), after the effects of long-term climatic ageing (PLA/LA, 240 h, **middle**) and after the effects of cyclic controlled ageing (PLA/CA, 600 h, **right**).

The addition of ground natural waste fillers led to a visible effect on the surface of the PLA/BH and PLA/ES biocomposite structures (Figures 3 and 4) in terms of the surface tribology. Both the buckwheat husks and egg shell particles are free of clumps and have good filler distribution and dispersion.

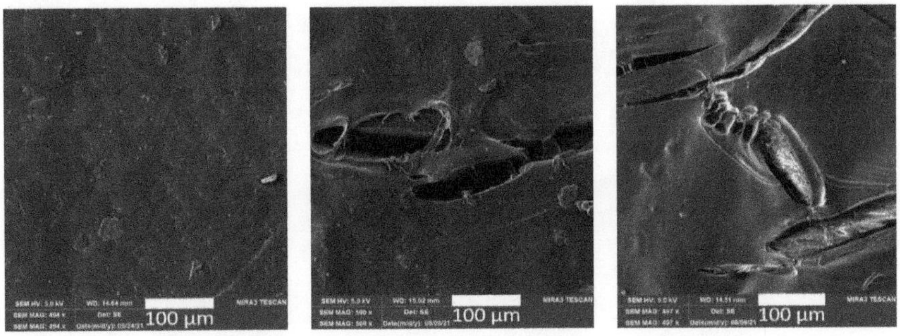

Figure 3. SEM images of the PLA/BH biocomposite surface before exposure to climate-controlled ageing (**left**), after the effects of long-term climate ageing (PLA/BH/LA, 240 h, **middle**) and after the effects of cyclic controlled ageing (PLA/BH/CA, 600 h, **right**).

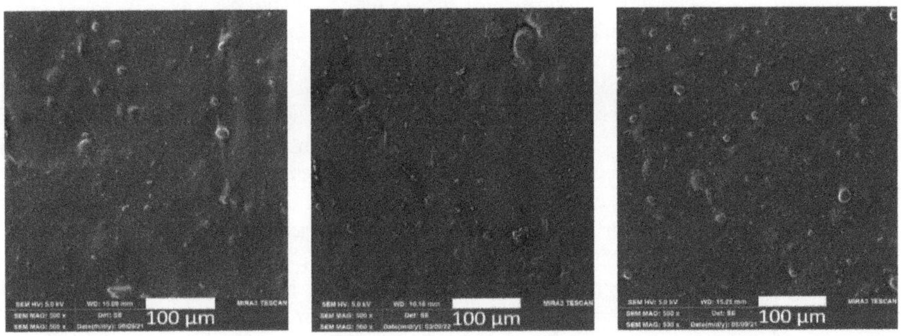

Figure 4. SEM images of the PLA/ES biocomposite surface before exposure to climate-controlled ageing (**left**), after the effects of long-term climate ageing (PLA/ES/LA, 240 h, **middle**) and after the effects of cyclic controlled ageing (PLA/ES/CA, 600 h, **right**).

After the effects of long-term and cyclic controlled ageing, the PLA/BH biocomposites show significant disruption of the surface structure, and cracks and fissures with visible fibrils on the surface. In contrast, the surface structures of the PLA/ES biocomposite do not show significant surface destruction after controlled climatic ageing (see Figure 3).

The fractured structure of the unfilled PLA (Figure 5) exhibits a brittle type of failure. The buckwheat husks in the biopolymer system do not have 100% adhesion to the PLA matrix, and the size and properties of these particles can lead to structural defects and affect the mechanical properties. Jalbrzykowski et al. [22] similarly observed a lack of adhesion between the filler and matrix in the fracture of the biopolymer structure for the PLA polymer with the addition of buckwheat husks, which was probably due to the sensing of the test body after tensile testing. The egg shells in the PLA matrix do not have good adhesion, as seen in Figure 5. The waste filler may act as a structural defect. This can be eliminated according to Sivagnanamani et al. [24]. They removed the impurities and inner membranes of the egg shells before grinding the egg shells. After grinding, they achieved a particle size of fewer than 25 microns, and subsequently observed strong interfacial adhesion in the biopolymer system. Cleaning the egg shells before milling and actual application could lead to a more favourable particle adhesion to the PLA matrix.

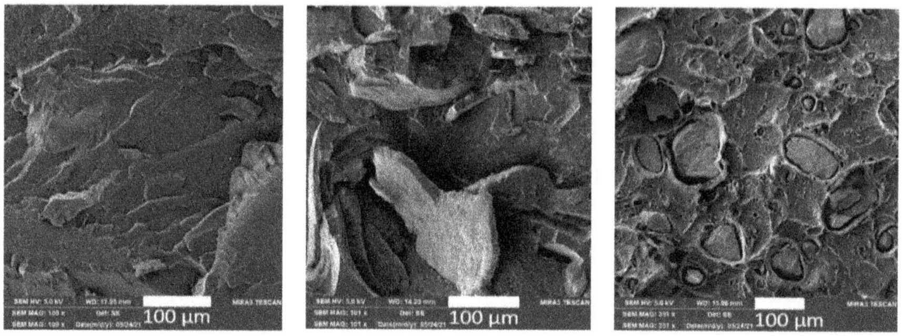

Figure 5. SEM images of fracture surfaces of unfilled PLA (**left**), PLA/BH biocomposite (**middle**) and PLA/ES (**right**) at the initial state.

3.3. Tensile Properties

The tensile modulus of the unfilled PLA polymer was measured to be 3644 MPa (Table 3). The addition of ground buckwheat husks to the PLA biopolymer matrix increased the tensile modulus by about 15%, and the ground egg shells increased the tensile modulus by about 20%. The exposure of the unfilled PLA and PLA/BH and the PLA/ES biocomposite samples to controlled climatic ageing decreased the tensile modulus. For the unfilled PLA, the effect of long-term climatic ageing (PLA/LA) led to a decrease of about 5% in the tensile modulus, and the effect of cyclic controlled ageing (PLA/CA) led to a decrease of about 12%. For the PLA/BH biocomposite, there was a similarly higher decrease in the tensile modulus for the cyclic controlled ageing parameters of about 11% (for PLA/BH/CA) and about 10% for the PLA/BH/LA biocomposite. In contrast, the PLA/ES biocomposite material had a higher decrease of about 12% in the modulus for long-term climate-controlled ageing (PLA/E/LA) and a decrease of about 5% for cyclic controlled ageing.

Table 3. Tensile properties of PLA biocomposites exposed to long-term and cyclic controlled climatic ageing.

Material	E_t (MPa)	σ_m (MPa)
PLA	3644 ± 32	63.6 ± 0.5
PLA/LA	3461 ± 154	70.0 ± 1.3
PLA/CA	3200 ± 75	64.2 ± 2.9
PLA/BH	4163 ± 29	47.8 ± 0.5
PLA/BH/LA	3727 ± 124	49.0 ± 2.0
PLA/BH/CA	3704 ± 554	47.4 ± 1.8
PLA/ES	4336 ± 58	48.9 ± 0.4
PLA/ES/LA	3798 ± 350	49.7 ± 0.8
PLA/ES/CA	4174 ± 159	47.7 ± 0.9

The ultimate tensile strength of the unfilled PLA was measured to be 63.6 MPa, and an increase in this value (about 10%) was measured for the samples after long-term ageing (PLA/LA). After cyclic ageing, the values are almost identical. The addition of the natural raw waste materials BH and ES led to a decrease in the ultimate strength value of about 25%. The addition of filler led to the forming of a heterogeneous structure with limited adhesion at the interfacial interface between the matrix and filler for the biocomposites PLA/BH and PLA/ES. Exposing the biocomposite samples to controlled long-term and cyclic climatic ageing did not significantly affect the ultimate tensile strength, and the change in the values is statistically insignificant.

3.4. Impact Toughness

The impact toughness of Charpy was significantly reduced by the addition of ground waste fillers to the PLA biopolymer matrix (Table 4). While the unfilled PLA biopolymer had a Charpy impact toughness value of 18.6 kJ/m^2, the PLA/BH biocomposite had a decrease of about 55%, and the PLA/ES biocomposite had a decrease of 40% due to structural defects at the interfacial interface of the particulate filler. Semicrystalline polymers or composites generally exhibit a coarser fracture structure, which is produced by crack propagation at the crystalline or matrix–filler interface. Exposing the unfilled PLA samples to controlled climatic ageing increased the Charpy impact toughness value by approximately 30% for both climate conditions. The increase was probably due to the higher degree of crystallinity after the effects of climate exposure.

Table 4. Charpy impact toughness of PLA biocomposites exposed to long-term and cyclic controlled climatic ageing.

Material	a_{cU} (k/m^2)
PLA	18.6 ± 1.5
PLA/LA	24.1 ± 1.1
PLA/CA	23.8 ± 1.0
PLA/BH	8.4 ± 1.5
PLA/BH/LA	9.8 ± 0.8
PLA/BH/CA	10.4 ± 1.0
PLA/ES	11.1 ± 1.4
PLA/ES/LA	14.0 ± 0.4
PLA/ES/CA	13.2 ± 1.6

The increases in the Charpy impact toughness values for the samples that were exposed to controlled climatic ageing were measured not only for the unfilled PLA, but also for the PLA with the addition of buckwheat husks and egg shells to the biopolymer matrix. The increase in the impact toughness value was about 20% for the PLA/BH biocomposite after the effects of long-term and cyclic controlled ageing. For the PLA biocomposite samples containing egg shells, the Charpy impact toughness values increased by about 25% after the effects of controlled climatic ageing. Thus, when compared with the buckwheat husks, the Charpy impact toughness was higher for the ground egg shells, but lower than that for the unfilled PLA matrix.

3.5. Flexural Properties

In addition to the uniaxial tensile loading of the test bodies, a three-point bending test was performed on prepared specimens of the unfilled PLA and biocomposites PLA/BH and PLA/ES, which resulted in the multi-axial loading of the test body.

The flexural modulus of the pure unfilled PLA was measured to be 3066 MPa (Table 5). The addition of ground buckwheat husks to the biopolymer matrix increased the flexural modulus by about 12%. In comparison, the addition of ground egg shells led to an increase of about 30% in the flexural modulus. Manshor et al. [30] achieved an increase in the flexural modulus for the PLA polymer samples by adding ground durian shells. Long-term ageing (PLA/LA) increased the flexural modulus value by about 10%, and cyclic ageing (PLA/CA) increased it by about 12%. Similar to the unfilled samples, the climate-controlled ageing of the PLA/BH samples increased the flexural modulus, with long-term ageing (PLA/BH/LA) resulting in an increase of 18%, and cyclic ageing (PLA/BH/CA) resulting in an increase of 13% in the flexural modulus. Thus, long-term climate-controlled ageing has a stronger effect than the unfilled PLA. The biopolymer composite systems containing egg shells (PLA/ES) showed an increase of about 8% in the flexural modulus after subjecting the samples to long-term ageing, and an increase of about 6% after cyclic ageing. The measured values show that controlled climatic ageing increases the flexural modulus of the biocomposite systems and unfilled PLA. Thus, in contrast to the tensile test, exposing the unfilled PLA and biocomposite PLA/BH and PLA/ES samples to controlled climatic ageing increased the flexural modulus.

Table 5. Flexural properties of PLA biocomposites exposed to long-term and cyclic controlled climatic ageing.

Material	E_f (MPa)	σ_{fM} (MPa)
PLA	3066 ± 77	101.3 ± 0.9
PLA/LA	3367 ± 45	112.7 ± 0.7
PLA/CA	3443 ± 79	112.5 ± 1.0
PLA/BH	3448 ± 171	84.5 ± 3.0
PLA/BH/LA	4084 ± 58	89.0 ± 7.7
PLA/BH/CA	3895 ± 79	84.3 ± 8.5
PLA/ES	3929 ± 56	89.2 ± 0.4
PLA/ES/LA	4266 ± 182	92.4 ± 2.2
PLA/ES/CA	4158 ± 213	89.1 ± 4.9

The flexural strength of the unfilled PLA was measured to be 101.3 MPa. Controlled climatic ageing increased the flexural strength value for the PLA/LA and PLA/CA samples by 11%. After processing during their experiments, Ramesh et al. [31] achieved similar flexural strengths for the injection moulded PLA samples. The addition of raw waste materials (BH and EC) led to a decrease in the flexural strength of the biopolymer composite systems. The buckwheat husks decreased the flexural strength by 17%, and the ground egg shells decreased the flexural strength by 12%. The PLA/BH biocomposite samples subjected to long-term and cyclic controlled ageing achieved more or less similar flexural strength values, with long-term ageing increasing the value by about 5%. The test bodies with egg-shell-blended biocomposites (PLA/ES) exposed to controlled climatic ageing showed a similar trend to the buckwheat husks. Thus, the flexural strength values after controlled climatic ageing were almost identical to those measured for the samples without ageing.

4. Conclusions

In recent years, biopolymer materials and biopolymer composites have become one of the most emerging research areas. PLA-based materials and biocomposites with natural fillers have great potential for a wide range of applications and are beginning to gain significant importance. The use of biopolymers and biocomposites is no longer only for packaging, film and disposable products, but also for technical products in view of the increasing demands of industrial practice. Furthermore, the indisputable advantage in terms of application potential is not only the knowledge of the properties of biocomposites, but especially the knowledge of the effects of climate and controlled climatic ageing on the final and performance properties of biocomposites in terms of environmental sustainability.

The experimental results obtained from the evaluation of the degradation effects of controlled climatic ageing on the properties of the unfilled PLA matrix and PLA biocomposite systems containing buckwheat husks and egg shells were related to the matrix used, the type of filler and the parameters of long-term and cyclic controlled ageing.

The degree of crystallinity was increased several times by adding natural waste fillers to the PLA matrix, confirming the results and conclusions known so far in this field as observed by Koutsomitopoulou et al. [29] and Sivagnanamani et al. [24]. At the same time, the degree of crystallinity increased significantly due to the effects of controlled climatic ageing for both the unfilled PLA matrix and the biocomposites filled with ground buckwheat husks. In contrast, for the PLA/ES biocomposite, the effect of controlled climatic ageing on the degree of crystallinity was statistically significant only for cyclic controlled climatic ageing.

When the mechanical properties were evaluated via tensile testing, it was found that adding ground buckwheat husks and egg shells to the PLA biopolymer matrix resulted in an increase of about 20% in the modulus compared to the unfilled PLA biopolymer.

The exposure of the unfilled PLA and biocomposite PLA/BH and PLA/ES samples to long-term and cyclic controlled climatic ageing decreased the tensile modulus for all material systems and structures evaluated.

The tensile strength of the unfilled PLA biopolymer increased after long-term climatic ageing. In contrast, cyclic ageing was statistically insignificant. The addition of natural raw waste materials, ground buckwheat husks and egg shells decreased the ultimate strength value by about 25%. Exposing the prepared biocomposite structure samples to controlled long-term and cyclic climatic ageing did not significantly affect the tensile strength, and the change in the values was statistically insignificant.

The flexural modulus increased significantly with the addition of ground waste fillers to the polymer matrix, as observed by Manshor et al. [30], especially after adding ground egg shells to the matrix. For the buckwheat husks, the increase was about half of that of the egg shells. The results confirm the findings of other publications and studies. The application of controlled climatic ageing increased the flexural modulus for both the unfilled PLA matrix and the biocomposite structures. The highest increase was measured when buckwheat husks were used. Long-term (LA) controlled climatic ageing has a higher effect on the flexural modulus for the biocomposite structures. In contrast to the tensile test, exposing the samples to controlled climatic ageing increased the flexural modulus.

The PLA matrix specimens subjected to controlled climatic ageing showed an increase in the flexural strength value, confirming the known findings from available publications. The addition of raw waste materials (BH and ES) led to a decrease in the flexural strength of the biopolymer composite systems by about 15% on average. Controlled climatic ageing has almost no effect on the flexural strength values of the biocomposites filled with ground buckwheat husks and egg shells.

The impact toughness value of Charpy decreased significantly by adding ground waste fillers to the PLA biopolymer matrix. Exposing the unfilled PLA samples to controlled climatic ageing increased the Charpy impact toughness value by approximately 30% for both climate conditions. The increase in the Charpy impact toughness values for the samples exposed to controlled climatic ageing was measured for both the unfilled PLA and the biopolymer matrix with the addition of both buckwheat husks and egg shells.

The surface texture of the unfilled PLA samples before and after the effects of controlled climatic ageing showed a smooth surface without significant damage and cracks. The addition of ground natural waste fillers resulted in a visible effect on the tribology of the surface of the biocomposite structures. However, the particles of the added waste materials have good distribution and dispersion in the biopolymer matrix. The PLA/BH biocomposites show significant surface structure distortion after the effects of long-term and cyclic controlled ageing. In contrast, the surface structures of the PLA/ES biocomposites do not show significant surface destruction after controlled climatic ageing. The fracture structure of the unfilled PLA shows a brittle type of failure. The ground egg shells and buckwheat hull particles in the biopolymer matrix did not show good adhesion, which could be because since they were not chemically cleaned, as confirmed by other studies and publications such as that by Sivagnanamani et al. [24].

The knowledge gained about biopolymers and biopolymer composites, including about the characterization of their properties after the effects of degradative climate-controlled ageing, is important in terms of potential applications and environmental and ecological aspects.

Author Contributions: Conceptualisation, P.L., L.B. and A.V.; methodology, P.L., L.B. and A.V.; validation, P.L., L.B. and A.V.; formal analysis, A.V. and P.L.; investigation, P.L., L.B. and A.V.; data curation, P.L., L.B. and A.V.; writing—original draft preparation, A.V. and P.L.; writing—review and editing, A.V. and P.L.; supervision, P.L.; project administration, P.L.; funding acquisition, P.L. All authors have read and agreed to the published version of the manuscript.

Funding: This research was funded by the Ministry of Education, Youth and Sports of the Czech Republic and the European Union—European Structural and Investment Funds in the frames of Operational Program Research, Development and Education—project Hybrid Materials for Hierarchical Structures (HyHi, Reg. No. CZ.02.1.01/0.0/0.0/16_019/0000843).

Institutional Review Board Statement: Not applicable.

Informed Consent Statement: Not applicable.

Data Availability Statement: The data presented in this study are available on request from the corresponding author.

Conflicts of Interest: The authors declare no conflict of interest.

References

1. Udayakumar, G.P.; Muthusamy, S.; Selvaganesh, B.; Sivarajasekar, N.; Rambabu, K.; Banat, F.; Sivamani, S.; Sivakumar, N.; Hosseini-Bandegharaei, A.; Show, P.L. Biopolymers and composites: Properties, characterization and their applications in food, medical and pharmaceutical industries. *J. Environ. Chem. Eng.* **2021**, *9*, 105322. [CrossRef]
2. Tănase, E.E.; Râpă, M.; Popa, O. Biopolymers based on renewable resources—A review. In Proceedings of the International Conference Agriculture for Life, Life for Agriculture, Bucharest, Romania, 5–7 June 2014; pp. 5–7.
3. *EUBIO_Admin EuBP_FS_What_Are_Bioplastics*; European Bioplastics e.V.: Berlin, Germany, 2020.
4. Current Figures on the Bioplastics Market: IfBB Presents New Edition of "Biopolymers-Facts and Statistics". IfBB-Institute for Bioplastics and Biocomposites. Available online: https://www.ifbb-hannover.de/en/press-release/current-figures-on-the-bioplastics-market-ifbb-presents-new-edition-of-biopolymers-facts-and-statistics.html (accessed on 27 June 2023).
5. Hamad, K.; Kaseem, M.; Yang, H.; Deri, F.; Ko, Y. Properties and medical applications of polylactic acid: A review. *Express Polym. Lett.* **2015**, *9*, 435–455. [CrossRef]
6. Inkinen, S.; Hakkarainen, M.; Albertsson, A.-C.; Södergård, A. From Lactic Acid to Poly(lactic acid) (PLA): Characterization and Analysis of PLA and Its Precursors. *Biomacromolecules* **2011**, *12*, 523–532. [CrossRef]
7. MacDonald, R.T.; McCarthy, S.P.; Gross, R.A. Enzymatic degradability of poly(lactide): Effects of chain stereochemistry and material crystallinity. *Macromolecules* **1996**, *29*, 7356–7361. [CrossRef]
8. Pang, X.; Zhuang, X.; Tang, Z.; Chen, X. Polylactic acid (PLA): Research, development and industrialization. *Biotechnol. J.* **2010**, *5*, 1125–1136. [CrossRef] [PubMed]
9. Jin, K.; Tang, Y.; Zhu, X.; Zhou, Y. Polylactic acid based biocomposite films reinforced with silanized nanocrystalline cellulose. *Int. J. Biol. Macromol.* **2020**, *162*, 1109–1117. [CrossRef] [PubMed]
10. Abdelrahman, M.S.; Nassar, S.H.; Mashaly, H.; Mahmoud, S.; Maamoun, D.; Khattab, T.A. Polymerization products of lactic acid as synthetic thickening agents for textile printing. *J. Mol. Struct.* **2020**, *1203*, 127421. [CrossRef]
11. Zhao, X.; Li, J.; Liu, J.; Zhou, W.; Peng, S. Recent progress of preparation of branched poly(lactic acid) and its application in the modification of polylactic acid materials. *Int. J. Biol. Macromol.* **2021**, *193*, 874–892. [CrossRef]
12. Murariu, M.; Dubois, P. PLA composites: From production to properties. *Adv. Drug Deliv. Rev.* **2016**, *107*, 17–46. [CrossRef]
13. Ögmundarson, Ó.; Sukumara, S.; Laurent, A.; Fantke, P. Environmental hotspots of lactic acid production systems. *GCB Bioenergy* **2020**, *12*, 19–38. [CrossRef]
14. Auras, R.A.; Lim, L.-T.; Selke, S.E.; Tsuji, H. *Poly (Lactic Acid): Synthesis, Structures, Properties, Processing, and Applications*; John Wiley & Sons: Hoboken, NJ, USA, 2011; ISBN 1-118-08813-1.
15. Boisseau, A.; Davies, P.; Choqueuse, D.; Nellissen, P.; Peters, L.; Nickel, R.; Adolphs, G.; Renaud, C.; OCV, O.C.; Thiebaud, F. Seawater Ageing of Composites for Ocean Energy Conversion Sytems. In Proceedings of the ICCM-17th International Conference on Composite Materials, Edinburgh, UK, 27–31 July 2009.
16. Vu, T.; Nikaeen, P.; Chirdon, W.; Khattab, A.; Depan, D. Improved Weathering Performance of Poly(Lactic Acid) through Carbon Nanotubes Addition: Thermal, Microstructural, and Nanomechanical Analyses. *Biomimetics* **2020**, *5*, 61. [CrossRef] [PubMed]
17. White, C.; White, K.M.; Pickett, J. *Service Life Prediction of Polymers and Plastics Exposed to Outdoor Weathering*; William Andrew: Norwich, NY, USA, 2017; ISBN 0-323-49777-2.
18. Litauszki, K.; Kovács, Z.; Mészáros, L.; Kmetty, Á. Accelerated photodegradation of poly(lactic acid) with weathering test chamber and laser exposure—A comparative study. *Polym. Test.* **2019**, *76*, 411–419. [CrossRef]
19. Ammala, A.; Bateman, S.; Dean, K.; Petinakis, E.; Sangwan, P.; Wong, S.; Yuan, Q.; Yu, L.; Patrick, C.; Leong, K.H. An overview of degradable and biodegradable polyolefins. *Prog. Polym. Sci.* **2011**, *36*, 1015–1049. [CrossRef]
20. Battegazzore, D.; Noori, A.; Frache, A. Natural Wastes as Particle Filler for Poly(Lactic Acid)-Based Composites. *J. Compos. Mater.* **2019**, *53*, 783–797. [CrossRef]
21. Kumar, B.A.; Saminathan, R.; Tharwan, M.; Vigneshwaran, M.; Babu, P.S.; Ram, S.; Kumar, P.M. Study on the mechanical properties of a hybrid polymer composite using egg shell powder based bio-filler. *Mater. Today Proc.* **2022**, *69*, 679–683. [CrossRef]
22. Jałbrzykowski, M.; Oksiuta, Z.; Obidziński, S.; Czyżewska, U.; Osiecki, T.; Kroll, L.; Yildiz, M.J. Assessment of innovative PLA biopolymer compositions with plant waste fillers. *Eng. Fail. Anal.* **2022**, *139*, 106496. [CrossRef]

23. Zemnukhova, L.A.; Shkorina, E.D.; Fedorishcheva, G.A. Composition of Inorganic Components of Buckwheat Husk and Straw. *Russ. J. Appl. Chem.* **2005**, *78*, 324–328. [CrossRef]
24. Sivagnanamani, G.S.; Begum, S.R.; Siva, R.; Kumar, M.S. Experimental Investigation on Influence of Waste Egg Shell Particles on Polylactic Acid Matrix for Additive Manufacturing Application. *J. Mater. Eng. Perform.* **2022**, *31*, 3471–3480. [CrossRef]
25. Mittal, A.; Teotia, M.; Soni, R.K.; Mittal, J. Applications of egg shell and egg shell membrane as adsorbents: A review. *J. Mol. Liq.* **2016**, *223*, 376–387. [CrossRef]
26. Daengprok, W.; Garnjanagoonchorn, W.; Mine, Y. Fermented pork sausage fortified with commercial or hen eggshell calcium lactate. *Meat Sci.* **2002**, *62*, 199–204. [CrossRef] [PubMed]
27. NatureWorks Resources. Available online: https://www.natureworksllc.com/resources?tags=fbf5e27a42144a05aaced0d2fdc23667 (accessed on 26 June 2023).
28. Brdlík, P.; Borůvka, M.; Běhálek, L.; Lenfeld, P. Biodegradation of Poly(Lactic Acid) Biocomposites under Controlled Composting Conditions and Freshwater Biotope. *Polymers* **2021**, *13*, 594. [CrossRef] [PubMed]
29. Koutsomitopoulou, A.F.; Bénézet, J.C.; Bergeret, A.; Papanicolaou, G.C. Preparation and characterization of olive pit powder as a filler to PLA-matrix bio-composites. *Powder Technol.* **2014**, *255*, 10–16. [CrossRef]
30. Manshor, M.R.; Anuar, H.; Aimi, M.N.N.; Fitrie, M.I.A.; Nazri, W.B.W.; Sapuan, S.M.; El-Shekeil, Y.A.; Wahit, M.U. Mechanical, thermal and morphological properties of durian skin fibre reinforced PLA biocomposites. *Mater. Des.* **2014**, *59*, 279–286. [CrossRef]
31. Ramesh, P.; Prasad, B.D.; Narayana, K.L. Effect of MMT Clay on Mechanical, Thermal and Barrier Properties of Treated Aloevera Fiber/PLA-Hybrid Biocomposites. *Silicon* **2020**, *12*, 1751–1760. [CrossRef]

polymers

Article

Effect of Drug Encapsulation and Hydrothermal Exposure on the Structure and Molecular Dynamics of the Binary System Poly(3-hydroxybutyrate)-chitosan

S. G. Karpova [1], A. A. Olkhov [1,2], I. A. Varyan [1,2,*], A. A. Popov [1,2] and A. L. Iordanskii [3,*]

[1] Department of Biological and Chemical Physics of Polymers, Emanuel Institute of Biochemical Physics, Russian Academy of Sciences, 4 Kosygina Street, 119334 Moscow, Russia; karpova@sky.chph.ras.ru (S.G.K.); anatoly.popov@mail.ru (A.A.P.)
[2] Academic Department of Innovational Materials and Technologies Chemistry, Plekhanov Russian University of Economics, 36 Stremyanny Lane, 117997 Moscow, Russia
[3] N. N. Semenov Federal Research Center for Chemical Physics Academy of Science, 119991 Moscow, Russia
* Correspondence: ivetta.varyan@yandex.ru (I.A.V.); aljordan08@gmail.com (A.L.I.)

Abstract: In this work, film materials based on binary compositions of poly-(3-hydroxybutyrate) (PHB) and chitosan with different ratios of polymer components in the range from 0/100 to 100/0 wt. % were studied. Using a combination of thermal (DSC) and relaxation (EPR) measurements, the influence of the encapsulation temperature of the drug substance (DS) of dipyridamole (DPD) and moderately hot water (at 70 °C) on the characteristics of the PHB crystal structure and the diffusion rotational mobility of the stable TEMPO radical in the amorphous regions of the PHB/chitosan compositions is shown. The low-temperature extended maximum on the DSC endotherms made it possible to obtain additional information about the state of the chitosan hydrogen bond network. This allowed us to determine the enthalpies of thermal destruction of these bonds. In addition, it is shown that when PHB and chitosan are mixed, significant changes are observed in the degree of crystallinity of PHB, degree of destruction of hydrogen bonds in chitosan, segmental mobility, sorption capacity of the radical, and the activation energy of rotational diffusion in the amorphous regions of the PHB/chitosan composition. The characteristic point of polymer compositions was found to correspond to the ratio of the components of the mixture 50/50%, for which the inversion transition of PHB from dispersed material to dispersion medium is assumed. Encapsulation of DPD in the composition leads to higher crystallinity and to a decrease in the enthalpy of hydrogen bond breaking, and it also slows down segmental mobility. Exposure to an aqueous medium at 70 °C is also accompanied by sharp changes in the concentration of hydrogen bonds in chitosan, the degree of PHB crystallinity, and molecular dynamics. The conducted research made it possible for the first time to conduct a comprehensive analysis of the mechanism of action of a number of aggressive external factors (such as temperature, water, and the introduced additive in the form of a drug) on the structural and dynamic characteristics of the PHB/chitosan film material at the molecular level. These film materials have the potential to serve as a therapeutic system for controlled drug delivery.

Keywords: poly-(3-hydroxybutyrate); TEMPO; stable radical; correlation times; amorphous phase

Citation: Karpova, S.G.; Olkhov, A.A.; Varyan, I.A.; Popov, A.A.; Iordanskii, A.L. Effect of Drug Encapsulation and Hydrothermal Exposure on the Structure and Molecular Dynamics of the Binary System Poly(3-hydroxybutyrate)-chitosan. *Polymers* **2023**, *15*, 2260. https://doi.org/10.3390/polym15102260

Academic Editor: Andrea Sorrentino

Received: 8 April 2023
Revised: 3 May 2023
Accepted: 9 May 2023
Published: 10 May 2023

1. Introduction

Currently, a significant amount of scientific and practical developments is devoted to the use of biodegradable composites/blends in biomedicine, the packaging industry, and tailoring environmental problems [1–3]. In the framework of the circular economy [4–6], for biomedicine platforms and food packaging, there is a direct tendency to move from conventional synthetic plastics to sustainable and biodegradable ones [5,6]. One of the key reasons for the intensive use of such materials is the fact that polymers' blending could lead to not only a significant improvement in performance, but also the appearance of new physicochemical parameters of the composites that were not inherent in the original components. The triad made of biodegradable polymers such as poly(3-hydroxybutyrate) (PHB), polylactide (PLA), and chitosan has practically inexhaustible sources of (bio)chemical synthesis and covers a wide range of hydrophilicity, and its constituents are characterized by high ecological and cell/tissue compatibility. In addition, these bio-based polymers are often used in combination with other polymers as constructional materials, innovative active packaging, and materials for environmental remediation [7–9].

The wide range of implementations enables the above biopolymers to dominate the world trade markets of bio-materials and bio-articles [10–12].

PHB is the main representative of natural polyesters of the polyhydroxyalkanoate family [11–13]. Along with useful properties, this polyester has a number of undesirable characteristics: high cost and fragility. To overcome these drawbacks, its copolymers with the units of the PHA family such as 3-hydroxyvalerate, 3-hydroxyoctanoate, and 4-hydroxybutyrate are used [12–14]. The second way to improve PHB characteristics for the exploitation object is the development of blends and composites that modify its behavior in the implementation areas from biomedicine to environmental safety [15,16].

Chitosan is the second component of the studied binary bio-based system, and it is also biocompatible and capable of biodegradation. Similar to PHB, it is widely used in biomedical, packaging, and eco-friendly applications such as products of tissue engineering, drug delivery therapeutic systems, active barriers with inherent functionality, and adsorbents for environment remediation [17–20]. Unlike PHB, it is characterized by high hydrophilicity, which is determined by the presence of a large number of functional groups, such as amine and hydroxyl groups. Amine groups form strong intramolecular and intermolecular hydrogen bonds, which transform chitosan into a glassy state. This material is able to be resorbed in the environment into environmentally friendly degradation products for a sufficiently long time—from a month to a year. An important feature of both polymers, as well as the products made from them, is the biodegradation of the macromolecular chains by the mechanism of hydrolytic or enzymatic destruction. The end products of the decomposition process are products that are safe for the body: for PHB, these are carbon dioxide and water. However, the high sensitivity of chitosan to moisture limits its application. This disadvantage can be overcome by mixing chitosan with moisture-resistant polymers such as PHB [21–23], polylactides [23], and polycaprolactone [24] provided that the biodegradability of the blends' constituents must be saved.

The bio-based polymers' disintegration mechanisms, namely biodegradation, oxidation, and hydrolysis, are significantly affected by the structural organization of the amorphous and crystalline phases formed in the bulk of the composite [25–27]. The rate of these processes increases if the structure of the sample is not homogeneous, but partly or fully heterogeneous. The semicrystalline biopolymers in the blends have a mutual effect on their crystallization behavior and glass transition, leading to a change in diffusion-transport characteristics, drug delivery, kinetic biodegradation, oxidation/ozonolysis, hydrolysis, and thermal stability of composite films and fibers [28–30].

At mixing, hydrophobic PHB and extremely hydrophilic chitosan, which are located at the opposite ends of the polarity scale, can generate composites in a wide range of hydrophilicity. By varying the composition of the PHB/chitosan blends and thereby influencing their morphology and crystallinity, it is possible to obtain versatile materials with different physical and chemical characteristics and accordingly diverse functionalities, such as water diffusivity, appropriate thermal and mechanical behavior, and controlled degradation rate [9,21,31,32].

An effective way to assess the state of the amorphous and crystalline phases of both initial biopolymers (PHB and chitosan) and their blends is a combination of dynamic and structural methods. In this work, EPR spectroscopy (probe analysis) and differential scanning calorimetry (DSC) were used. Such a combination of research methods allowed the authors to obtain a more complete assessment of the structural evolution of a PHB/chitosan blend in an aqueous medium in the short time interval (minutes and hours), which precedes the hydrolytic decomposition of the polymer system.

The main purpose of this work was to study the effect of the composition of PHB/chitosan blends and the encapsulation of the drug substance (dipridamole) on the crystal structure, the state of hydrogen bonds of polymer molecules, and the segmental mobility of the polymers. In addition, the EPR method was used to study the influence of external aggressive factors, temperature, and the aqueous medium on the process of breaking the hydrogen bond network, as well as the evolution of PHB crystallinity and the molecular dynamics of a microprobe (TEMPO) in composites. The presented study will allow a deeper understanding of the effect of the loaded drug on the state of the composite system as well as contribute to the improvement of existing and the development of new therapeutic platforms and scaffolds with the function of controlled drug release.

2. Materials and Methods

In the work, mixed compositions based on biodegradable polymers hydroxybutyrate and chitosan were studied. To obtain films, we used a natural biodegradable polymer poly(3-hydroxybutyrate) (PHB) of the 16F series (BIOMER, Krailling, Germany), obtained by microbiological synthesis. The original polymer was a white fine powder. The molecular weight of PHB was $M_w = 2.06 \cdot 10^5$ g/mol (206 kDa), density d = 1.248 g/cm^3, T_m = 177 °C. Chitosan (Bioprogress, Shchelkovo, Russia), an infusible polysaccharide, was a fine powder. The molecular weight of this polymer was $M_w = 4.4 \cdot 10^5$ g/mol; degree of deacetylation—82.3%. Solvents were used in the production of films by casting: for PHB—$CHCl_3$ and dioxane grade ChDA (ZAO Ecos1, Russia); for chitosan—CH_3COOH grade ChDA. Films were prepared by mixing a solution of chitosan in aqueous acid media and a solution of PHB in dioxane. Chitosan solution was obtained by dissolving the powder in acetic acid. Dipyridamole was used as a model drug for controlled release (2,2′,2′′,2′′′-((4,8- Di-1-piperidinylpyrimido (5,4-d)pyrimidine- 2,6-diyl)dinitrilo]tetrakis(ethanol)). The pharmacological group to which dipyridamole (DPD) belongs is antiplatelet agents, angioprotectors and microcirculation correctors, and adenosinergic agents. DPD is a yellow crystalline powder. The molecular weight of DPD is 504.53 g/mol. Ultrathin PHB fibers were obtained by electroforming (ES). To obtain fibers, molding solutions were prepared: PHB and PHB with DPD in chloroform. The concentration of PHB in the solution was 7% by weight. The content of dipyridamole in the polymer composition was 5% by weight, relative to the mass of PHB. PHB molding solutions with DPD were prepared using an automatic magnetic stirrer with heating and an ultrasonic bath. The following solvents were used to obtain films by casting: for PHB—$CHCl_3$ and dioxane of the ChDA brand (JSC Ekos-1, Moscow, Russia); for chitosan—CH_3COOH of the ChDA brand. Chitosan solution was obtained by dissolving the powder in acetic acid. Films were prepared by mixing a solution of chitosan in aqueous acid media and a solution of PHB in dioxane.

X-band electron paramagnetic resonance (EPR) spectra were recorded on an EPR-V automated spectrometer (Federal Research Center for Chemical Physics, Russian Academy of Sciences, Moscow, Russia). The value of the microwave power to avoid saturation effects did not exceed 1 mW. The modulation amplitude was always significantly less than the width of the resonance line and did not exceed 0.5 G. The stable nitroxide radical TEMPO was used as a spin probe. The radical was introduced into the fibers from the gas phase at a temperature of 60 °C for an hour. The concentration of the radical in the polymer was determined by integrating the EPR spectra. The reference was an evacuated TEMPO solution in CCl_4 with a radical concentration of $\sim 1 \cdot 10^{-3}$ mol/L.

The probe rotation correlation time τ was found from the EPR spectra using the formula given earlier [33]:

$$\tau = \Delta H^+ \times [(I^+/I^-)^{0.5} - 1] \times 6.65 \times 10^{-10} \quad (1)$$

where ΔH^+—width of the spectrum component located in a weak field, I^+/I^-—intensity ratio of the components in a weak and strong field. The measurement error for τ was ± 5%.

The equilibrium concentration of the radical adsorbed in samples of the studied compositions of the same mass was calculated using Brucker (winer) software. In the process of identifying the spectra, the identified, weighed samples were recorded, so the calculation of the radicality in each sample was performed using the Origin program. The weight of the polymer was 0.02–0.03 g.

The samples were studied by DSC using a DSC Q-20 instrument from TA Instruments (New Castle, DE, USA) in a nitrogen atmosphere at a heating rate of 10 K/min. The average statistical error in measuring thermal effects was ±3%.

Water and thermal impacts on the polymer were determined as follows: Samples were placed in a bottle with distilled water and then placed in an oven and kept at T = 70 °C for 2 h. Then these samples were dried for 7 days, and their weight was checked. After 5 days, the weight of the samples did not change.

3. Results and Discussion

3.1. *Thermal Characteristics of PHB/Chitosan Compositions with Encapsulated DPD Exposed in an Aqueous Medium*

3.1.1. Binary Composition of PHB/Chitosan

When mixing two polymer components, PHB and chitosan, one should expect changes in their thermal, mechanical, and diffusion characteristics along with variability in the crystallinity and morphology of the mixture composition. In this work, the DSC method was used to study the crystalline phase of PHB in its mixed compositions with chitosan. For example, Figure 1 shows the heating endotherms of the composition 40/60% (PHB/chitosan) for binary (a) and triple compositions of PHB/chitosan/DPD (b) at 5% wt. The DSC curves of polymer samples of various compositions with component ratios from 80/20 to 30/70% showed the presence of three characteristic endothermic peaks in all thermograms.

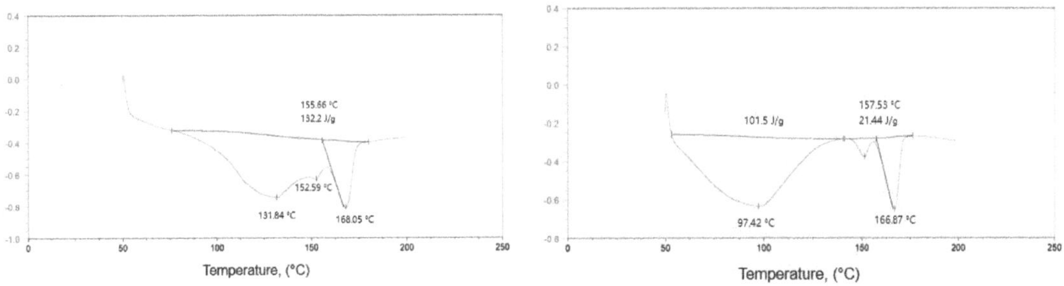

(a) (b)

Figure 1. Heating endotherms of 40/60% PHB/chitosan mixed compositions: (**a**) initial binary composition and (**b**) PHB/chitosan/DPD ternary composition.

With a rise in temperature, e.g., under heating of the chitosan samples, as provided by the DSC protocol, the partial and then full disrupture of the cross-linking formed by their hydrogen bonds transpires. On the thermograms, this process is accompanied by an endothermic gently sloping maximum that characterizes the intensity of H-bond destruction [23,34,35]. The second and third peaks correspond to different modes of PHB melting. The bimodal nature of the melting of PHB manifests itself in a wide range of compositions (see Table 1). The exception is samples with extremely low PHB/chitosan ratios, namely 0/100 and 20/80%, for which, as a result of extremely low crystallinity, melting endotherms are not observed. The melting region of the PHB homopolymer (Tm) has two maxima at 157 and 173 °C, which indicates the existence of two populations of the biopolyester crystal structure, differing both in size and degree of their perfection with a total specific enthalpy of melting ΔH = 74.4 J/g and a degree of crystallinity χ = 52% (see Table 1). The high-temperature peak in the region of 173 °C belongs to the melting of crystals with a more perfect structure, and the low-temperature peak in the region of 157 °C belongs to the melting of a less perfect PHB crystal structure. The bimodal nature of the phase transition of PHB was observed earlier for its ultrathin fibers and films [36,37]. With an increase in the content of chitosan, a slight shift in the T_m values to the low-temperature region is due to the influence of its molecules on the process of biopolyester crystallization.

Table 1. Degree of crystallinity (χ), melting enthalpy (ΔH, J/g), and melting point of PHB (T_m, °C) in a mixture with chitosan, and the maximum temperature of chitosan hydrogen bond destruction (T_D). All characteristics were obtained by DSC.

		PHB 100%	PHB/chit 80/20%	PHB/chit 70/30%	PHB/chit 50/50%	PHB/chit 40/60%	PHB/chit 30/70%	PHB/chit 20/80%	Chitosan 100%
Initial Composition of PHB/Chitosan									
PHB	χ, %	52	49	47	43	35	33	-	-
	T_m	157;173	155;170	155;170	156;171	152;168	153	-	-
Chitosan	Δ*H*	-	-	128	134	186	205	209	285
	T_D	-	-	85	95	132	98	117	126
PHB/chitosan with DPD									
PHB	χ, %	54	55	54	45	38	-	63	-
	T_m	153;170	153;168	153;168	151;167	151;167	-	163	-
Chitosan	Δ*H*	-	-	87	110	101	248	250	333
	T_D	-	-	85	90	97	96	97	103
PHB/chitosan in the Aquatic Environment									
PHB	χ, %	58	45	43	41	40	39	39	-
	T_m	155;170	155;170	155;170	155;170	155;170	155;170	171	-
Chitosan	Δ*H*	-	115	130	159	172	172	157	156
	T_D	-	87	88	89	90	95	95	95
PHB/chitosan with DPD, in the Aquatic Environment									
PHB	χ, %	80	76	70	66	47	23	25	
	T_m	153;170	152;169	152;169	152;168	152;168	172	172	-
Chitosan	Δ*H*	-	-	130	154	163	170	109	110
	T_D	-	-	85	87	97	92	95	95

Figure 2 and Table 1 present data on PHB crystallinity (χ), enthalpy (ΔH), and hydrogen bond breaking temperature (T_D) of chitosan, as well as melting temperatures of the PHB crystalline fraction (T_m) on the composition of the system. As noted above, in mixtures with a low concentration of biopolyester (<30%), its crystalline phase is not formed; therefore, all curves reflecting the thermal characteristics of PHB do not start from the zero composition of the mixture, but from the ratio of 30/70% (see Figure 2a). With this composition, the crystallinity is ~33%, which is almost 2 times lower than the crystallinity of the PHB homopolymer, and then, as its content in the system increases, the values of χ increase rapidly. The "crystallinity–composition" curve includes a break in the range of compositions at 50/50% PHB/chitosan, and with a higher content of PHB in the mixture, the crystallinity of PHB changes monotonously and rather weakly. In the opposite way, the enthalpy and cleavage temperature of H-bonds increase as the concentration of chitosan increases. The characteristic point of compositions (50/50%) appears most clearly for curve 1 in Figure 2b. It is also clearly seen here that in the region of high PHB concentrations (>50%), the dependence of ΔH on the composition is extremely weak. Similarly, the position of the flat maximum of the DSC curve, which belongs to chitosan and characterizes the temperature of the destruction of its hydrogen bonds (T_D), shifts from 126 to 85 °C. The DSC curves also show that the destruction of hydrogen bonds in native chitosan occurs at 150–104 °C. In mixed compositions, this process is realized at lower temperatures: 130–50 °C. The enthalpy of destruction of these bonds for chitosan in PHB/chitosan mixtures occupies a wide range of values. At a lower concentration of chitosan in the mixture (<30%), the maximum belonging to the heat of rupture of hydrogen bonds in the system was not observed.

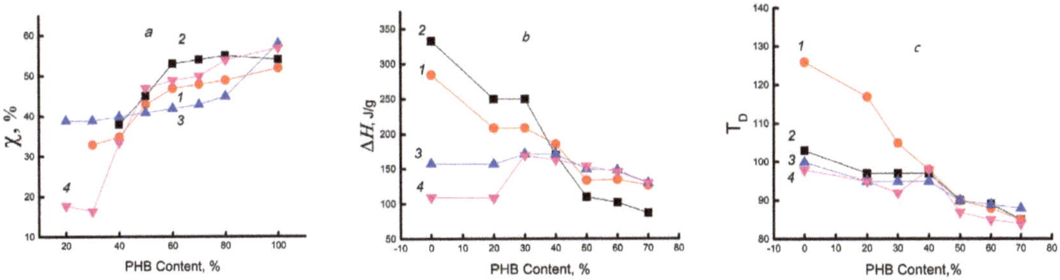

Figure 2. Dependence of the thermal characteristics on the composition of the PHB/chitosan composition: (**a**) the degree of crystallinity of PHB (χ), (**b**) the enthalpy of cleavage of hydrogen bonds of chitosan (ΔH), and (**c**) the maximum temperature of decomposition of hydrogen bonds of chitosan (T_D). The numbers on the curves: 1—binary composition of PHB/chitosan; 2—the same, but with encapsulated DPD; 3—binary composition of PHB/chitosan after exposure in the aquatic environment at 70 °C; 4—the same, but with encapsulated DPD after exposure in the aquatic environment at 70 °C.

A decrease in the degree of crystallinity and a decrease in the melting point of PHB with an increase in the content of chitosan indicate that the crystalline phase of PHB in the binary mixture under study becomes less perfect. In the process of crystallization, polysaccharide molecules exert a steric hindrance to the perfect packing of polyester molecules in crystallites and lamellae to a greater extent, the higher the content of polysaccharide.

The nature of the change in the thermal characteristics of the binary PHB/chitosan system depending on the composition of the polymer components allows us to make a preliminary conclusion that a specific concentration region is observed, 50/50%, the transition through which is accompanied by phase inversion, i.e., when the continuous (dispersed) phase of chitosan is transformed into dispersed. If a strong and dense network of hydrogen bonds is formed in the initial chitosan, which retains water molecules well, then as PHB molecules are introduced, a sharp drop in ΔH occurs, reflecting the reduction of such bonds and, consequently, a rapid decrease in the integral heat of their rupture. On the contrary, the increase in the crystallinity of the second component, PHB, shows that its macromolecules encounter fewer and fewer obstacles in the formation of crystalline regions. Along with steric limitations of crystallization, the role of intermolecular hydrogen bonds formed during the interaction between chitosan molecules containing amine groups and PHB molecules, including both ester groups of the main chain and terminal acid and hydroxyl fragments, cannot be excluded. Although such an interaction seems to be weaker than the energy of chitosan hydrogen bonds, it can also lead to the decompression of the network of chitosan hydrogen bonds, i.e., to an increase in the free volume in the system. The latter effect should enhance the segmental mobility of polymers, which will be analyzed further using the EPR method.

3.1.2. The Ternary System of PHB/Chitosan/Drug

In the transition from the binary to the ternary composition of PHB/chitosan/DPD, let us consider the effect of the encapsulated drug on the crystal structure of PHB and the strength of the hydrogen bonds of chitosan. In therapeutic systems and active packaging, the impermeable crystalline phase creates steric hindrances to DPD diffusion transport.

Therefore, its introduction into the system is of interest both for the analysis of the structural evolution of a mixture as a matrix that provides control of drug delivery, and for the assessment of segmental mobility, which determines such key processes in polymers as diffusion, barrier properties, sorption capacity, and controlled release.

Figure 2a (curve 2) shows the dependences of the degree of crystallinity of PHB in the ternary system PHB/chitosan/DPD. A comparison of these curves with similar results for the binary system shows that for all compositions of the ternary system, the degree of crystallinity of PHB with encapsulated DPD is higher than that in the initial binary systems. Similar to the binary system, the enthalpy of hydrogen bond breaking in the ternary system tends to decrease, and most importantly, they both intersect in the composition range of approximately 50/50%, i.e., at the same inversion point (see Figure 2b). The excess of crystallinity in the presence of DPD can be explained either by the additional nucleation of the crystalline phase, where the low-molecular-weight drug acts as a nucleation agent, or by the already mentioned plasticizing effect of DPD. In the range of compositions above the inversion point, the concentration of hydrogen bonds in the ternary system becomes noticeably lower than that in the binary system and also changes slightly with increasing PHB content. The data presented earlier in the work [38] for ultrathin PHB fibers also showed that the encapsulated drug in ultrathin PHB fibers affects the crystallization that occurs during electrospinning.

3.1.3. Hydrated Binary System PHB/Chitosan

At the next stage of the analysis of the properties of the PHB-chitosan system, we will consider the effect of water exposure on its structure at a standard elevated temperature of 70 °C. It is well known that when polymers come into contact with water, their structure and crystallinity change [39,40], and during water evaporation (drying), the polymer crystallinity is increased [41]. In each individual case, it is necessary to establish the degree of structural changes. It is especially important to do this for an amphiphilic blended composition formed by hydrophobic PHB and hydrophilic chitosan, which are often used in humid conditions in combination with elevated temperatures.

In general, water molecules can interact with polymer molecules in two different ways. On the one hand, they have a plasticizing effect on hydrophilic and moderately hydrophilic polymers, which, as a result of an increase in segmental mobility, can lead to an additional increase in the crystalline phase, i.e., to their recrystallization [42]. On the other hand, after desorption of water from the polymer matrix, water complexes remain in the form of clusters, which are typically characteristic of hydrophobic polymers, in particular in Zimm's clustering integral modeling [43,44], or hydrated functional groups, which is more typical for polysaccharides, proteins, polyalcohols, and other hydrophilic polymers [45].

The data on the degree of crystallinity of samples of the PHB/chitosan binary mixture exposed to distilled water at 70 °C for 120 min are shown in Figure 2a (curve 3). It can be seen from the figure that, after exposure to an aqueous medium, PHB/chitosan films with a composition of <50/50% crystallinity have higher values compared to the initial samples. This indicates the predominance of the plasticizing effect of water and, as a result, pre-crystallization in the participation of maximally oriented PHB feed-through circuits. Beyond the inversion point, i.e., in the range of values of the composition >50/50%, when its continuous phase is formed due to the hydrophobization of the system, the sorption of water decreases and the plasticizing effect is less noticeable. As a result, the increase in crystallinity is quite small.

Chitosan is a hydrophilic polymer. Exposure to an aqueous medium leads to the hydration of its functional groups and then leads to swelling and, as a result, to a sharp increase in the free volume of the polysaccharide [46]. A comparison of curve 3 and the previously considered curves 1 and 2 (see Figure 2b) shows that as a result of exposure to water of mixtures of composition <50/50%, the enthalpy of hydrogen bond rupture noticeably decreases. Additionally, the destruction of the hydrogen bond network can be indicated by a drop in the temperature of hydrogen bond destruction, shown in Figure 2c.

A characteristic feature of all three curves in Figure 2b is the preservation of the inversion point at ~50% ratio of polymer components; they all intersect with each other near this point.

3.1.4. Hydrated Ternary System PHB/Chitosan/DPD

In the conclusion of this section, let us consider the simultaneous effect of encapsulated DPD and exposure in an aqueous medium on the thermal characteristics of the initial binary composition of PHB/chitosan. Curves 4 on the corresponding fragments a, b, and c in Figure 2 reflect this effect. The crystal structure for this system, as well as for the non-aqueous ternary PHB/chitosan/DPD system (curve 3), begins to appear somewhat earlier on the concentration scale, namely already in samples containing 20% PHB containing 15–17% of the crystalline phase. A further sharp increase in the values of χ from ~15% to ~65% demonstrates the key effect of the drug and plasticizer on the crystallinity of PHB.

Up to the inversion point, i.e., at a high content of chitosan, its contact with water leads to intense swelling and an increase in the segmental mobility of polysaccharide chains. At the same time, the plasticization of the polyester, PHB, appears in an aqueous medium at a moderately high temperature. Both of these effects affect the crystallization ability of the latter in different ways. If chitosan in the absence of a plasticizer (water), as noted above, prevents the packing of PHB chains into crystallites and lamellae, then an increase in its segmental dynamics upon contact with hot water leads to (a) an increase in the crystallinity of the biopolyester and (b) an expansion of the range of PHB concentrations, where crystallization occurs. If the formation of a crystalline phase is observed only at a composition of 30/70% or more for non-watered systems PHB/chitosan and PHB/chitosan/DPD, then in similar hydrated systems, the presence of crystals is recorded at a lower composition value (at 20/80%). If we compare the crystallinity of hydrated systems that do not contain and contain DPD, curves 3 and 4 in Figure 2a, then the increase in crystallinity is more than 240%, namely 15% and 37%, respectively. Based on this difference, it can be concluded that, other things being equal, the encapsulation of DPD in a hydrated polymer medium does not contribute to the pre-crystallization of PHB. A similar negative effect of the presence of DPD is also observed when comparing the thermal effects of hydrogen bond breaking in chitosan (Figure 2b). Indeed, up to the inversion point, the hydrated PHB/chitosan composition (curve 3) has a stronger formation of hydrogen bonds than the same PHB/chitosan/DPD composition (curve 4). In general, the treatment of samples of mixed compositions with water at 70 °C has a stronger effect on their structural and thermal properties than the introduction of drugs. This is clearly seen when comparing the location of curves 1 and 2 in Figure 2b for the original systems with the location of similar curves 3 and 4 for watered systems. With this comparison, the difference in ΔH values is very clear, which shows the great role of water sorption as a specific aggressive agent.

3.2. *Dynamic Characteristics of the Amorphous Phase of Mixed Compositions of PHB/Chitosan with Encapsulated DPD after Exposure to an Aqueous Medium*

In partially crystalline polymers, the structure of amorphous regions is largely determined by the influence of their crystalline phase. Therefore, mixing highly crystalline PHB and practically non-crystallizing chitosan changes not only the degree of crystallinity of the biopolyester, but also the molecular dynamics in the amorphous regions formed mainly by the polysaccharide. To study molecular/segmental mobility, the EPR method was used using the stable TEMPO nitroxide radical, which acts as a molecular probe.

Let us consider the effect of the composition of the PHB/chitosan mixture on the dynamics of polymer molecules. Previously, it was shown that the EPR spectra of the radical in the matrices of the PHB homopolymer and the PHB/chitosan binary system represent a superposition of two spectra corresponding to two populations of radicals with their own characteristic correlation times τ_1 and τ_2 [47,48].

Time τ_1 reflects molecular mobility in denser amorphous regions with low free volume (slow component of the spectrum), while τ_2 reflects mobility in less dense regions with higher free volume (fast component of the same spectrum). The transition of the speed of rotation on the TEMPO radical from fast to slow is shown in Figure 3. At the same time, the heterogeneous nature of amorphous regions is due to the difference in the packing density of polymer molecules and, consequently, the difference in molecular dynamics. However, as shown by Chumakova [49], the characteristic mobility in a certain range of probe rotation speeds can be represented by a single integral value—the characteristic relaxation time (τ).

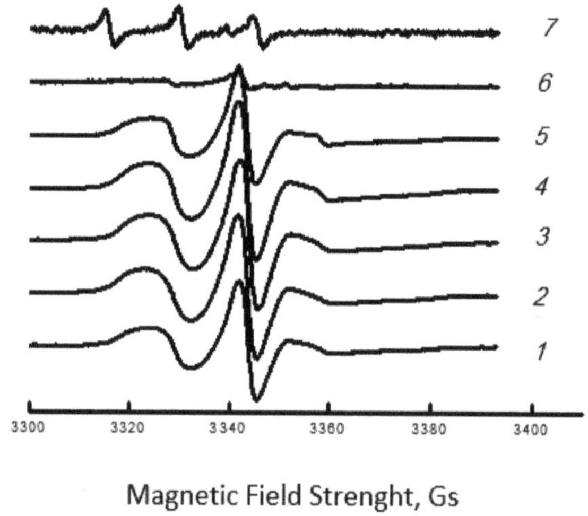

Figure 3. ESR spectra of nitroxyl radical TEMPO for the blends of PHB-chitosan with the content of the polysaccharide: 1—0, 2—30, 3—50, 4—60, 5—70, 6—80, 7—100%.

Because of their high potential barrier to internal rotation of chain links, polysaccharide molecules such as chitosan are highly rigid. This potential (energy barrier) is due to steric hindrances in the movement of the elementary link, as well as the formation of a network of hydrogen bonds. Depending on the history of the chitosan sample, this network may include immobilized water molecules. The glassy dense network of this polysaccharide prevents the effective penetration of the radical, which is confirmed by its low concentration in samples with a low PHB/chitosan ratio (up to 30/70%), as shown in Figure 4a. With a further increase in the concentration of biopolyester in the mixture, the concentration of the radical increases sharply and reaches a maximum in the region of ~60% PHB. As already shown in Section 3.1, the range of 50–60% belongs to the inversion point, where practically all thermal characteristics change the nature of the dependence on the composition. Due to the low compatibility of the components and the phase transition, the dispersed phase of PHB passes into the dispersion phase. In this case, the loosest structure is formed. At a high content of PHB (~50–100%), the concentration of the sorbed probe also significantly decreases due to the growth of crystallinity in the polymer mixture, since the radical practically does not penetrate into the densely packed crystalline regions of PHB. Figure 4b shows the dependence of the characteristic correlation time τ on the composition of the mixture, which is symbatic with the previous dependence 4a. A maximum is also observed here, located in the same concentration range (~50% PHB). The closeness of the shape of the concentration curves, reflecting the structural (a) and dynamic (b) nature of the polymer mixture, as in the previous Section 3.1, is determined by its phase transformation, namely the transition of PHB from dispersed material to dispersion medium.

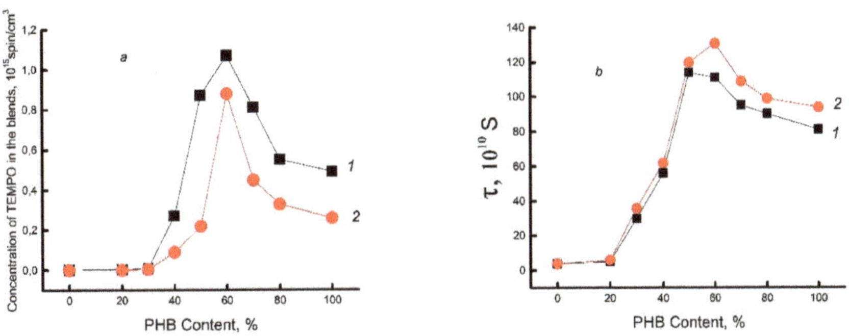

Figure 4. Concentration of TEMPO radical (**a**) and its time of correlation (**b**) in the PHB–chitosan blends. Binary system (1) and the same system with embedded DPD (2).

For chitosan, in addition to intra- and intermolecular hydrogen bonds, when mixed with PHB, another type of interaction arises, which is determined by the formation of hydrogen bonds between the ester groups of PHB and the amine groups of chitosan formed as a result of deacetylation. It is reasonable to assume that the introduction of PHB molecules into the composition and their interaction with chitosan molecules can partially destroy the previously formed hydrogen network of the polysaccharide, which becomes less compact as a result. This fact is confirmed by the course of the descending branch of the concentration dependence of the integral correlation time (τ) (see curve 1 in Figure 4b). The intensive growth of this characteristic in the range of low and moderate PHB concentrations (20–50%) up to the concentration inversion point of 50/50% is determined by the screening effect of polyester molecules, which hinder the contact of chitosan molecules with water. Hydrophobic PHB molecules partially prevent the penetration of water into chitosan and prevent the plasticizing and loosening effects that usually accompany the sorption of solvent molecules by chitosan samples.

Let us consider the effect of DPD on the molecular dynamics of PHB/chitosan compositions (see curve 2 in Figure 4a). It can be seen that the nature of this dependence differs little from the analogous dependence for the PHB/chitosan system, which does not contain drugs. However, in the range of high biopolyester concentrations (>50%), DPD encapsulation in the system leads to a slight decrease in the sorption capacity of the stable radical and an insignificant drop in molecular mobility.

It is well known that water, penetrating into a binary polymer matrix, can affect both its physicochemical and mechanical properties and the diffusion–sorption characteristics of the third component (drug, electrolyte, modifying agent, etc.). As previously shown [21], for monolithic film samples, mixing moderately hydrophobic PHB polymer with highly hydrophilic chitosan at room and physiological temperatures (25 and 37 °C, respectively) leads to partial hydrophilization of the system, which manifests itself in an increase in its sorption capacity and in a sharp increase in drug diffusion coefficients.

Figure 5 shows the dependences of the correlation time on the composition of the PHB/chitosan mixture for different exposure times in the aquatic environment. All curves, except for the original curve 1, have the character of a curve with an extremum and have fairly close values. Consequently, structural changes that determine the segmental mobility of the composition occur already in the first 30 min of exposure. Further heating in water to a temperature of 70 °C has little effect on the dynamics of polymer molecules. The encapsulation of the drug in the polymer system slightly shifts the inversion point corresponding to the maxima on the curves to the 60/40% PHB/chitosan composition. Here it can also be assumed that the appearance of DPD in the system contributes to the preliminary crystallization of PHB; therefore, a comparison of Figure 5a,b shows a small contribution of the drug to the dynamics of the molecular chains of the composition.

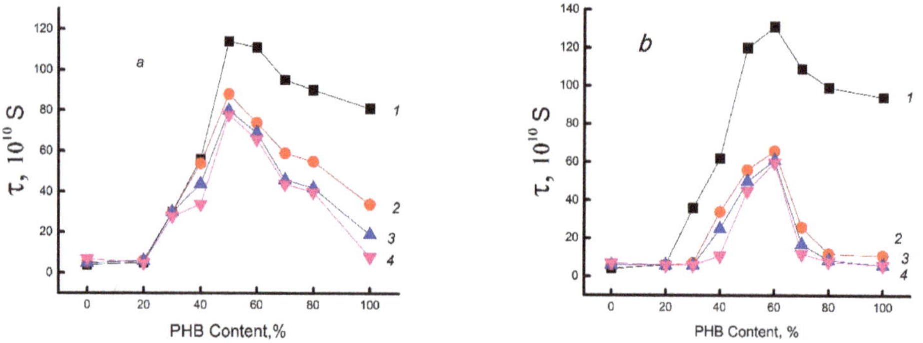

Figure 5. Characteristic time correlation of TEMPO as the blend characterization of hydrothermal action at 70 °C. Time of sample contact with the aqueous medium. (**a**) Initial binary system; (**b**) the same system with encapsulated DPD. Time of exposure: 1—0, 2—30, 3—60, 4—120 min.

Previously, it was suggested that the transition of the phase state of PHB from dispersed material to dispersion medium shields finely dispersed regions of chitosan from the action of water molecules. With this in mind, the overall picture of the dynamic behavior of molecules is as follows: (a) Ascending branches of curves 1–4 in Figure 5, which reflect a slowdown in segmental dynamics, are due to the binding of water by the functional groups of chitosan. In the same range of compositions, a noticeable pre-crystallization of PHB is observed, which prevails over the plasticizing effect. (b) After passing the maximum at the inversion point of the compositions, the segmental mobility increases; since here the biopolyester is not shielded by chitosan molecules, it forms a dispersed phase and, in direct contact with water, undergoes its plasticizing effect with a rather weak pre-crystallization (see Figure 2a).

It should be noted that the values of τ noticeably decrease during the exposure time in the aquatic environment, which is very likely due to the diffusion of water into the volume of PHB [50]. Encapsulation of DPD in the PHB/chitosan binary system in combination with the thermal process of water exposure (Figure 5b) reduces the time of its exposure to biopolyester molecules. If, in the absence of a drug, the change in segmental dynamics occurred within 120 min and possibly a little longer, then in its presence, the process of changing the dynamics is completed much faster, in 20 min. As seen in this figure, curves 1 and 2 diverge significantly, while subsequent curves 3 and 4 practically coincide with curve 2. As already reported, the drug distributed in the polymer matrix makes the polymer structure more disordered and, due to the presence of functional groups, creates additional osmotic forces. Both of these effects increase the diffusion flow of water in the system and shorten the time for structural and dynamic changes.

Additional information on the dynamic behavior of the PHB/chitosan system of various compositions was obtained by studying the temperature dependence of the molecular probe rotation speed and determining the corresponding effective activation energy Eτ values (see Figure 6). A characteristic feature of Eτ as a function of the content of chitosan in PHB is the presence of its maximum in the region of the inversion point, which is in accordance with all previous thermal and dynamic results.

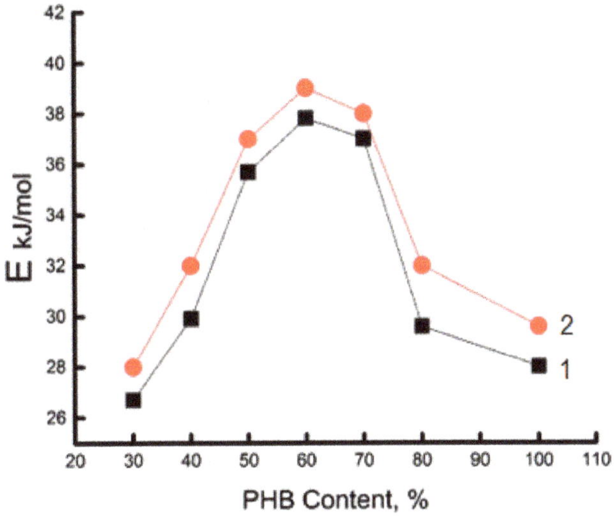

Figure 6. Effect of the composition of the PHB/chitosan mixture on the effective activation energy of the rotational mobility of the TEMPO molecular probe. 1—binary system, PHB-chitosan; 2—the same system with encapsulated DPD.

4. Conclusions

The study of the structural and dynamic state of the composite made it possible for the first time to analyze the effect of external factors, such as temperature, drug encapsulation, and an aqueous medium at an elevated temperature (70 °C). A decrease in crystallinity and melting temperature of PHB with an increase in the content of chitosan in the binary mixture indicates that chitosan molecules prevent the formation of the PHB crystalline phase, making it less perfect. During the crystallization of the biopolyester, the polysaccharide molecules sterically hinder the perfect packing of PHB molecules in crystallites. The change in the thermal characteristics of the PHB/chitosan system depending on its composition made it possible to identify a specific concentration region, with a composition of 50/50%, where phase inversion occurs, namely the process where the continuous (dispersion medium) phase of chitosan is converted into a dispersed material. As PHB molecules are introduced into the system, the values of ΔH decrease sharply, which reflects a decrease in the number of H-bonds. The transition from the PHB/chitosan binary system to the PHB/chitosan/drug ternary system showed that the degree of PHB crystallinity increases due to additional nucleation of the crystalline phase. As a result of exposure to water at 70 °C for the compositions with a ratio of components <50/50%, the enthalpy of hydrogen bond breaking sharply decreases. Additionally, the destruction of the network of hydrogen bonds may be indicated by a drop in the temperature of destruction of hydrogen bonds. All characteristics' dependences, namely χ, ΔH, and T_D, for hydrated and non-hydrated compositions intersect each other near a 50/50% concentration point for both binary and ternary systems. Exposure to an aqueous medium at 70 °C is also accompanied by sharp changes in the concentration of hydrogen bonds in chitosan and molecular dynamics. The results presented herein could serve potentially for the development of drug delivery therapeutic systems.

In this work, crystallinity measurements were performed by DSC using a DSC Q-20 instrument from TA Instruments (New Castle, DE, USA) in a nitrogen atmosphere at a heating rate of 10 K/min. The average statistical error in measuring thermal effects was ±3%. (Center for Shared Use of the Institute of Biochemical Physics "New Materials and Technologies", Russian Academy of Sciences).

Author Contributions: Conceptualization, S.G.K. and A.L.I.; investigation, I.A.V., S.G.K. and A.A.O.; resources, I.A.V.; data curation, S.G.K.; writing—original draft preparation, S.G.K. and A.A.O.; writing—review and editing, A.A.P. and I.A.V.; visualization, S.G.K.; supervision, S.G.K.; project administration, I.A.V.; funding acquisition, A.L.I. All authors have read and agreed to the published version of the manuscript.

Funding: This research received no external funding.

Institutional Review Board Statement: Not applicable.

Data Availability Statement: Not applicable.

Acknowledgments: The authors are thankful to Anna Zhulkina for supplying the FTIR data and presenting the majority of the spectra for further interpretation and treatment. With great pleasure, we want to thank Urs Haenggi and Biomer GMbH (Krailling, Germany) for providing the PHB for the presented study. The study was carried out using scientific equipment of the Center of Shared Usage "New Materials and Technologies" of Emanuel Institute of Biochemical Physics.

Conflicts of Interest: The authors declare no conflict of interest.

References

1. Luzi, F.; Puglia, D.; Torre, L. Natural fiber biodegradable composites and nanocomposites: A biomedical application. In *Biomass, Biopolymer-Based Materials, and Bioenergy Construction, Biomedical, and Other Industrial Applications*; Woodhead Publishing Series in Composites Science and Engineering: Kidlington, UK, 2019; Chapter 10; pp. 179–201. [CrossRef]
2. Wu, F.; Misra, M.; Mohanty, A.K. Challenges and new opportunities on barrier performance of biodegradable polymers for sustainable packaging. *Prog. Polym. Sci.* **2021**, *117*, 101395. [CrossRef]
3. Corsi, I.; Venditti, I.; Trotta, F.; Punta, C. Environmental safety of nanotechnologies: The eco-design of manufactured nanomaterials for environmental remediation. *Sci. Total Environ.* **2023**, *864*, 161181. [CrossRef] [PubMed]
4. Kalita, N.K.; Hakkarainen, M. Integrating biodegradable polyesters in a circular economy. *Curr. Opin. Green Sustain. Chem.* **2023**, *40*, 100751. [CrossRef]
5. Morsada, Z.; Hossain, M.; Islam, M.T.; Mobin, M.A.; Saha, S. Recent progress in biodegradable and bioresorbable materials: From passive implants to active electronics. *Appl. Mater.* **2021**, *25*, 101257. [CrossRef]
6. Cheng, J.; Jacquin, B.J.; Pujo-Pay, M.; Meistertzheim, A.L. Biodegradability under marine conditions of bio-based and petroleum-based polymers as substitutes of conventional microparticles. *Polym. Degrad. Stab.* **2022**, *206*, 110159. [CrossRef]
7. Briassoulis, D.; Tserotas, P.; Athanasoulia, L.G. Alternative optimization routes for improving the performance of poly(3-hydroxybutyrate) (PHB) based plastics. *J. Clean. Prod.* **2021**, *318*, 128555. [CrossRef]
8. Swetha, T.A.; Bora, A.; Mohanrasu, K.; Balaji, P.; Raja, R.; Ponnuchamy, K.; Muthusamy, G.; Arun, A. A comprehensive review on polylactic acid (PLA)—Synthesis, processing and application in food packaging. *Int. J. Biol. Macromol.* **2023**, *234*, 123715. [CrossRef]
9. Chen, S.; Tian, H.; Mao, J.; Ma, F.; Zhang, M.; Chen, F.; Yang, P. Preparation and application of chitosan-based medical electrospun nanofibers. *Int. J. Biol. Macromol.* **2023**, *226*, 410–422. [CrossRef]
10. Price, S.; Kuzhiumparambil, U.; Pernice, M.; Ralph, P. Techno-economic analysis of cyanobacterial PHB bioplastic production. *J. Environ. Chem. Eng.* **2022**, *10*, 107502. [CrossRef]
11. Palmeiro-Sanchez, T.; O'Flaherty, V.; Lens, P.N.L. Polyhydroxyalkanoate bio-production and its rise as biomaterial of the future. *J. Biotechnol.* **2022**, *348*, 10–25. [CrossRef]
12. Jost, V.; Langowski, H.-C. Effect of different plasticisers on the mechanical and barrier properties of extruded cast PHBV films. *Eur. Polym. J.* **2015**, *68*, 302–312. [CrossRef]
13. Eraslan, K.; Aversa, C.; Nofar, M.; Barletta, M.; Gisario, A.; Salehiyan, R.; Goksu, Y.A. Poly(3-hydroxybutyrate-co-3-hydroxyhexanoate) (PHBH): Synthesis, properties, and applications—A review. *Eur. Polym. J.* **2022**, *167*, 111044. [CrossRef]
14. Martin, D.P.; Williams, S.F. Medical applications of poly-4-hydroxybutyrate: A strong flexible absorbable biomaterial. *Biochem. Eng. J.* **2003**, *16*, 97–105. [CrossRef]
15. Mirmusavi, M.H.; Zadehnajar, P.; Semnani, D.; Karbasi, S.; Fekrat, F.; Heidari, F. Evaluation of physical, mechanical and biological properties of poly 3-hydroxybutyrate-chitosan-multiwalled carbon nanotube/silk nanomicro composite scaffold for cartilage tissue engineering applications. *Int. J. Biol. Macromol.* **2019**, *132*, 822–835. [CrossRef] [PubMed]
16. Boisseaux, P.; Hopkinson, P.; Santillo, D.; Smith, C.; Garmulewicz, A.; Powel, Z.; Galloway., T. Environmental safety of second and third generation bioplastics in the context of the circular economy. *Ecotoxicol. Environ. Saf.* **2023**, *256*, 114835. [CrossRef] [PubMed]
17. Rezaei, F.S.; Sharifianjazi, F.; Esmaeilkhanian, A.; Salehi, E. Chitosan films and scaffolds for regenerative medicine applications: A review. *Carbohydr. Polym.* **2021**, *273*, 118631. [CrossRef]

18. Ahmad, A.; Mubarak, N.M.; Naseem, K.; Tabassum, H.; Rizwan, M.; Najda, A.; Kashif, M.; Bin-Jumah, M.; Hussain, A.; Shaheen, A.; et al. Recent advancement and development of chitin and chitosan-based nanocomposite for drug delivery: Critical approach to clinical research. *Arab. J. Chem.* **2020**, *13*, 8935–8964. [CrossRef]
19. Florez, M.; Guerra-Rodríguez, E.; Cazon, P.; Vazquez, M. Chitosan for food packaging: Recent advances in active and intelligent films. *Food Hydrocoll.* **2022**, *124*, 107328. [CrossRef]
20. Zhang, Y.; Zhao, M.; Cheng, Q.; Li, Z. Research progress of adsorption and removal of heavy metals by chitosan and its derivatives: A review. *Chemosphere* **2021**, *279*, 130927. [CrossRef]
21. Ivantsova, E.L.; Iordanskii, A.L.; Kosenko, R.Y.; Rogovina, S.Z.; Grachev, A.V.; Prut, E.V. Poly(3-hydroxybutyrate)-chitosan: A new biodegradable composition for prolonged delivery of biologically active substances. *Pharm. Chem. J.* **2011**, *45*, 51–55. [CrossRef]
22. Zhuikova, Y.V.; Zhuikov, V.; Varlamov, V. Biocomposite Materials Based on Poly(3-hydroxybutyrate) and Chitosan: A Review. *Polymers* **2022**, *14*, 5549. [CrossRef] [PubMed]
23. Rogovina, S.; Zhorina, L.; Yakhina, A.; Shapagin, A.V.; Iordanskii, A.L.; Berlin, A.A. Hydrolysis, Biodegradation and Ion Sorption in Binary Biocomposites of Chitosan with Polyesters: Polylactide and Poly(3-Hydroxybutyrate). *Polymers* **2023**, *15*, 645. [CrossRef]
24. Ma, L.; Shi, X.; Zhang, X.; Li, L. Electrospinning of polycaprolacton/chitosan core-shell nanofibers by a stable emulsion system. *Colloids Surf. A* **2019**, *583*, 123956. [CrossRef]
25. Bule Možar, K.; Miloloža, M.; Martinjak, V.; Cvetnić, M.; Kušić, H.; Bolanča, T.; Kučić Grgić, D.; Ukić, Š. Potential of Advanced Oxidation as Pretreatment for Microplastics Biodegradation. *Separations* **2023**, *10*, 132. [CrossRef]
26. Olkhov, A.A.; Karpova, S.G.; Tyubaeva, P.M.; Zhulkina, A.; Zernova, Y.N.; Iordanskii, A.L. Effect of Ozone and Ultraviolet Radiation on Structure of Fibrous Materials Based on Poly(3-hydroxybutyrate) and Polylactide. *Inorg. Mater. Appl. Res.* **2020**, *11*, 1130–1136. [CrossRef]
27. Kucera, F.; Petrusa, J.; Jancara, J. The structure-hydrolysis relationship of poly(3-hydroxybutyrate). *Polym. Test.* **2019**, *80*, 106095. [CrossRef]
28. Karpova, S.G.; Ol'khov, A.A.; Popov, A.A.; Zhulkina, A.L.; Kosenko, R.Y.; Iordanskii, A.L. Study of the Effect of External Factors on the Structural and Dynamic Parameters of Film Materials Based on Poly(3-Hydroxybutyrate) and Polyactide. *Nanobiotechnol. Rep.* **2021**, *16*, 211–221. [CrossRef]
29. Aversa, C.; Barletta, V.; Cappiello, G.; Gisario, A. Compatibilization strategies and analysis of morphological features of poly(butylene adipate-co-terephthalate) (PBAT)/poly(lactic acid) PLA blends: A state-of-art review. *Eur. Polym. J.* **2022**, *173*, 111304. [CrossRef]
30. Kervran, M.; Vagner, C.; Cochez, M.; Ponçot, M.; Reza Saeb, M.; Vahabi, H. Thermal degradation of polylactic acid (PLA)/polyhydroxybutyrate (PHB) blends: A systematic review. *Polym. Degrad. Stab.* **2022**, *201*, 109995. [CrossRef]
31. Medvecky, L.; Giretova, M.; Stulajterova, R. Properties and in vitro characterization of polyhydroxybutyrate–chitosan scaffolds prepared by modified precipitation method. *J. Mater. Sci. Mater. Med.* **2014**, *25*, 777–789. [CrossRef]
32. Toloue, E.B.; Karbasia, S.; Salehi, H.; Rafienia, M. Potential of an electrospun composite scaffold of poly (3-hydroxybutyrate)-chitosan/alumina nanowires in bone tissue engineering applications. *Mater. Sci. Eng.* **2019**, *99*, 1075–1091. [CrossRef] [PubMed]
33. Zhang, K.; Monteiro, M.J.; Jia, Z. Stable organic radical polymers: Synthesis and applications. *Polym. Chem.* **2016**, *7*, 5589–5614. [CrossRef]
34. Khalida, M.N.; Agnelya, F.; Yagoubib, N.; Grossiorda, J.L.; Couar, G. Water state characterization, swelling behavior, thermal and mechanical properties of chitosan based networks. *Eur. J. Pharm. Sci.* **2002**, *15*, 425–432. [CrossRef] [PubMed]
35. Dresvyanina, E.N.; Grebennikov, S.F.; Elokhovskii, V.Y.; Dobrovolskaya, I.P.; Ivan'Kova, E.M.; Yudin, V.E.; Heppe, K.; Morganti, P. Thermodynamics of interaction between water and the composite films based on chitosan and chitin nanofibrils. *Carbohydr. Polym.* **2020**, *245*, 116552. [CrossRef] [PubMed]
36. Karpova, S.G.; Ol'Khov, A.A.; Iordanskii, A.L.; Lomakin, S.M.; Shilkina, N.S.; Popov, A.A.; Berlin, A.A. Nonwoven blend composites based on poly(3-hydroxybutyrate)–chitosan ultrathin fibers prepared via electrospinning. *Polym. Sci.-Ser. A Focus Phys.* **2016**, *58*, 76. [CrossRef]
37. Gunaratne, L.M.W.K.; Shanks, R.A.; Amarasinghe, G. Thermal history effects on crystallisation and melting of poly(3-hydroxybutyrate). *Thermochim. Acta* **2004**, *423*, 127–135. [CrossRef]
38. Iordanskii, A.L.; Ol'Khov, A.A.; Karpova, S.G.; Chalykh, A.E.; Berlin, A.A. Influence of the structure and morphology of ultrathin poly(3-hydroxybutyrate) fibers on the diffusion kinetics and transport of drugs. *Polym. Sci.-Ser. A* **2017**, *59*, 352–362. [CrossRef]
39. Jeck, S.; Scharfer, P.; Schabel, W.; Kind, M. Water sorption in semicrystalline poly(vinyl alcohol) membranes: In situ characterisation of solvent-induced structural rearrangements. *J. Membr. Sci.* **2012**, *389*, 162–172. [CrossRef]
40. Iordanskii, A.; Karpova, S.; Olkhov, A.; Borovikov, P.; Kildeeva, N.; Liu, Y. Structure-Morphology Impact upon Segmental Dynamics and Diffusion in the Biodegradable Ultrafine Fibers of Polyhydroxy-butyrate-Polylactide blends. *Eur. Polym. J.* **2019**, *19*, S0014. [CrossRef]
41. Shin, S.Y.; Jang, M.; Cheon, H.J.; Go, S.; Yoon, H.; Chang., M. Nanostructure-assisted solvent vapor annealing of conjugated polymer thin films for enhanced performance in volatile organic compound sensing. *Sens. Actuators B Chem.* **2022**, *351*, 130951. [CrossRef]
42. Xiao, H.; Lu, W.; Yeh, J.-T. Effect of plasticizer on the crystallization behavior of poly(lactic acid). *J. Appl. Polym.* **2009**, *113*, 112–121. [CrossRef]
43. Jayakumar, A.; Radoor, S.; Siengchin, S.; Shin, G.H.; Kim, J.T. Recent progress of bioplastics in their properties, standards, certifications and regulations: A review. *Sci. Total Environ.* **2023**, *878*, 163156. [CrossRef] [PubMed]

44. Kocherbitov, V.; Argatov, I. A thermodynamic theory of sorption in glassy polymers. *Polymers* **2021**, *233*, 124195. [CrossRef]
45. Maurer, M.; Oostenbrink, C. Water in protein hydration and ligand recognition. *J. Molec. Recognit.* **2019**, *32*, 2810. [CrossRef]
46. Jiang, B.; Kasapis, S.; Kontogiorgosc, V. Combined use of the free volume and coupling theories in the glass transition of polysaccharide/co-solute systems. *Carbohydr. Polym.* **2011**, *83*, 926–933. [CrossRef]
47. Klonos, P.; Terzopoulou, Z.; Koutsoumpis, S.; Pissis, P. Rigid amorphous fraction and segmental dynamics in nanocomposites based on poly(L–lactic acid) and nano-inclusions of 1-3D geometry. *Eur. Polym. J.* **2016**, *82*, 16–34. [CrossRef]
48. Kamaev, P.P.; Aliev, I.I.; Iordanskii, A.L.; Wasserman, A.M. Molecular dynamics of the spin probes in dry and wet poly(3-hydroxybutyrate) films with different morphology. *Polymer* **2001**, *42*, 515–520. [CrossRef]
49. Vorobiev, A.K.; Bogdanov, A.V.; Yankova, T.S.; Chumakova, N.A. Spin Probe Determination of Molecular Orientation Distribution and Rotational Mobility in Liquid Crystals: Model-Free Approach. *J. Phys. Chem. B* **2019**, *123*, 5875–5891. [CrossRef]
50. Pankova, Y.N.; Shchegolikhin, A.N.; Iordanskii, A.L.; Olkhov, A.A.; Zaikov, G.E. The characterization of novel biodegradable blends based on polyhydroxybutyrate: The role of water transport. *J. Mol. Liq.* **2010**, *156*, 65–69. [CrossRef]

polymers

MDPI

Article

Assessing the Biodegradability of PHB-Based Materials with Different Surface Areas: A Comparative Study on Soil Exposure of Films and Electrospun Materials

Kristina G. Gasparyan [1], Polina M. Tyubaeva [1,2], Ivetta A. Varyan [1,2,*], Alexandre A. Vetcher [3,4,*] and Anatoly A. Popov [1,2]

1 Department of Physical Chemistry of Synthetic and Natural Polymer Compositions, Emanuel Institute of Biochemical Physics, Russian Academy of Sciences, 4 Kosygina Street, 119334 Moscow, Russia; anatoly.popov@mail.ru (A.A.P.)
2 Academic Department of Innovational Materials and Technologies Chemistry, Plekhanov Russian University of Economics, 36 Stremyanny Per., 117997 Moscow, Russia
3 Complementary and Integrative Health Clinic of Dr. Shishonin, 5 Yasnogorskaya Str., 117588 Moscow, Russia
4 Institute of Biochemical Technology and Nanotechnology, Peoples' Friendship University of Russia (RUDN), 6 Miklukho-Maklaya St., 117198 Moscow, Russia
* Correspondence: ivetta.varyan@yandex.ru (I.A.V.); vecher_aa@pfur.ru (A.A.V.)

Abstract: Due to the current environmental situation, biopolymers are replacing the usual synthetic polymers, and special attention is being paid to poly-3-hydroxybutyrate (PHB), which is a biodegradable polymer of natural origin. In this paper, the rate of biodegradation of films and fibers based on PHB was compared. The influence of exposure to soil on the structure and properties of materials was evaluated using methods of mechanical analysis, the DSC method and FTIR spectroscopy. The results showed rapid decomposition of the fibrous material and also showed how the surface of the material affects the rate of biodegradation and the mechanical properties of the material. It was found that maximum strength decreased by 91% in the fibrous material and by 49% in the film. Additionally, the DSC method showed that the crystallinity of the fiber after exposure to the soil decreased. It was established that the rate of degradation is influenced by different factors, including the surface area of the material and its susceptibility to soil microorganisms. The results obtained are of great importance for planning the structure of features in the manufacture of biopolymer consumer products in areas such as medicine, packaging, filters, protective layers and coatings, etc. Therefore, an understanding of the biodegradation mechanisms of PHB could lead to the development of effective medical devices, packaging materials and different objects with a short working lifespan.

Keywords: biodegradable polymers; poly-3-hydroxybutyrate; electrospinning; fiber; film; decomposition; biodegradation in soil

Citation: Gasparyan, K.G.; Tyubaeva, P.M.; Varyan, I.A.; Vetcher, A.A.; Popov, A.A. Assessing the Biodegradability of PHB-Based Materials with Different Surface Areas: A Comparative Study on Soil Exposure of Films and Electrospun Materials. *Polymers* **2023**, *15*, 2042. https://doi.org/10.3390/polym15092042

Academic Editor: Cornelia Vasile

Received: 30 March 2023
Revised: 21 April 2023
Accepted: 23 April 2023
Published: 25 April 2023

1. Introduction

In recent years, people have become more aware of environmental pollution and in this regard have begun to actively develop the study and application of bioplastics as an excellent alternative to disposable products which can be produced and decomposed naturally by microorganisms [1]. Currently, many bioplastics are used, including polylactide (PLA), polycaprolactone (PCL) and poly-3-hydroxybutyrate (PHB), which can replace the usual petroleum-based plastics [2].

Of special interest is a natural biodegradable polyester PHB which is synthesized from renewable natural sources. This polymer has good service properties (for example, high thermal stability over a broad temperature interval up to 150 °C), is biodegradable, biocompatible with the human body and can be sterilized by various physical and chemical methods [3,4]. Moreover, PHB can serve as a perfect polymer host matrix for the development of efficient materials for different areas including medical devices, packaging materials and different objects with a short working lifespan. However, its main advantage is considered to be its good biodegradability, whereby the problem of further utilization of PHB-based materials can be easily solved [5,6].

PHB (Figure 1) is the first bioplastic and was discovered in 1926 by the French researcher Maurice Lamoine. His discovery was ignored for many decades because coal and oil were inexpensive, but in the mid-1970s the world economy experienced an oil crisis and began to explore alternatives. In nature, this polymer is synthesized by some types of microorganisms that belong to the genres *Alcaligenes*, *Azobacter*, *Bacillus* and *Pseudomonas* [3], it is found in the cytoplasm of microbial cells in the form of granules and usually makes up more than 40% of their dry weight [4].

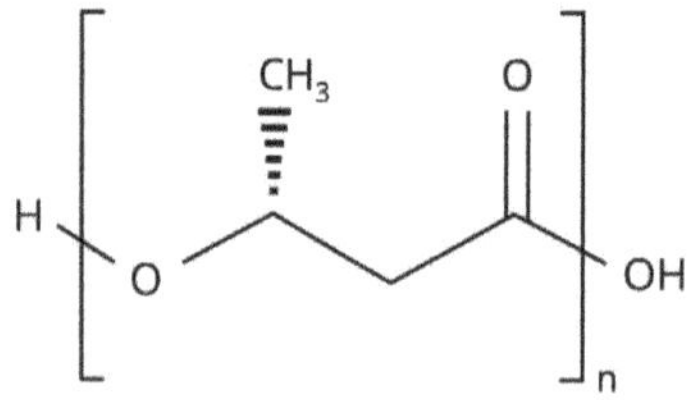

Figure 1. Structural formula of PHB.

So far, PHB is not particularly common because of its fragility: it does not stretch well and breaks easily compared to other types of bioplastics [5,6]. To improve its mechanical and other desired properties, various substances are added: polymers, nanoparticles, metals, plasticizers and the other additives given in Table 1. PHB is used in many fields: the textile industry, fabric engineering, biomedical products, food industry (disposable trays, cups, plates) [7,8]. Moreover, it is of great importance that the modification of PHB does not affect its high biodegradability [9].

Table 1. Various additives.

Additives	Effect
Glycerol-based additives: glycerol triacetate (GTA), glycerol tributyrate (GTB) [10]	Act as nucleating agents, thermal stability
Hydroxyvalerate and hydroxyhexanoate [11]	Increased toughness
Triethyl citrate [12]	Improves mechanical, dynamic-mechanical and thermal properties
Acetyl o-tributyl citrate (ATBC) and tributyl citrate (TBC) [13]	Improvement of thermal, mechanical and barrier properties
Sugarcane bagasse [14]	cheaper materials, good fertilizers

Despite the popularity of PHB, its biodegradation mechanisms are still an area of a great interest [15]. The hydrolytic and enzymatic degradation of PHB have been well studied separately [16,17]. The essence of the PHB biodegradation mechanism is the activity of the depolymerase enzymes, which provide three key stages of the decomposition process: transition from a high-molecular compound to monomers and oligomers; transition from monomers and oligomers to biomass; transition from biomass to CO_2 and H_2O [18]. It is the enzymes that catalyze the hydrolytic erosion of the surface that triggers the process of biodegradation [19,20]. Of course, the biodegradation of the material proceeds as a complex process that is influenced by many factors, especially: surface roughness, available surface area, hydrophobicity, charge, chemical and supramolecular structure, etc. [21]. However, during composting in controlled conditions the process of biodegradation will be largely determined by the specifics of the structural organization of the material. Two key ways in which material are destroyed should be distinguished (Figure 2): surface biodegradation and bulk biodegradation. In the main part, porous material with a highly developed surface area undergoes bulk destruction which is accompanied by autocatalytic hydrolysis, and dense monolithic bodies such as films undergo surface destruction [22].

Figure 2. The mechanism of biodegradation: (**1**) surface, (**2**) bulk.

Considering that over the past 5 years PHB production has increased four-fold [23], this study of PHB biodegradation was carried out in various environments, such as: soil, drinking water, the marine environment, etc. [8,24]. In the main part, in the studies conducted the loss of mass of the composite based on PHB was observed. Among the key environmental factors determining the rate of biodegradation, the following were particularly noted: type of microorganism, temperature, pH, humidity of the medium and availability of oxygen and nutrients in the medium [25]. Furthermore, it was found that degradation proceeded faster in an alkaline medium [26] and at a higher temperature [27]. The mechanism of the PHB decomposition in the soil is almost aerobic, accompanied by hydrolysis and oxidation by soil components. Additionally, it is known that the most active microorganisms consume polymers of monocarboxylic acids, with polyhydroxyalkanoates occupy a leading position among these polymers [28]. At the moment, different groups of bacteria and fungi are particularly capable of destroying PHB molecular chains: gram-positive and gram-negative bacteria of the genus *Streptomyces, classes of Firmicutes, Pseudomonas lemoigne, Comamonas* sp., *Acidovorax faecalis, Aspergillus fumigatus* and *Variovorax paradoxus* [28].

Considering the designated issues, the main purpose of this study was to find out how the surface area of PHB materials affects the rate of the biodegradation process. Electrospun fibrous material was chosen as the model of the highly developed porous material and a film was chosen as the comparison object.

This work offers a simple and versatile approach to studying, under natural conditions, the surface area of PHB-based nonwoven fibrous materials in comparison to film materials. This comprehensive approach may provide a more thorough understanding of the effects exposure to soil has on the structure and properties of the biomaterials for different properties, since this article compares, for the first time, the factors characterizing the contribution made by the surface to the speed and type of the mechanism of initiation of the biodegradation process.

2. Materials and Methods

2.1. Materials

A biopolymer of natural origin—poly-3-hydroxybutyrate (PHB)—was used in this work. Commercial PHB was obtained by microbial synthesis (16F series, production by BIOMER, Frankfurt, Germany), and was characterized by 60% of crystalline phase, 206 kDa of molecular weight, 1.248 g/cm^3 of density, melt flow index (MFI) = 10 g/10 min (180 °C, 5 kg).

Materials were obtained from PHB using two methods: electrospinning and pressing.

PHB films were obtained by pressing on the laboratory hydraulic press (Moscow, Russia). The advantages of the pressing method are the simple design of the molds and the relatively low cost of the equipment [29]. This method made it possible to obtain uniform thin films comparable to nonwoven analogues. The dosage of the PHB for pressing was 1.1 g, the temperature on the heating plates was 180–190 °C and pressure was 50–60 MPa. The duration of pressing was 2 min. The obtained samples of PHB films were cooled in water at a temperature of about 20 °C for 5 min. The thickness of the PHB films was 130 ± 0.098 μm.

PHB fibers were obtained by using the electrospinning [30] method on the single-capillary laboratory unit EFV-1 (Moscow, Russia). The scheme and the main structural components, including the solution reservoir, a stainless-steel needle, a high-voltage power supply and a vertically grounded collector, are shown in Figure 3 [31]. A homogeneous polymer solution was prepared by dissolving PHB in chloroform at 7% (w/v). The electrical conductivity of the solution was 10 μS/cm and its viscosity was 1.0 Pa s. The layers of the fibrous material were obtained from 25 mL of PHB solution. The conditions of the electrospinning process were voltage 17 kV, the distance between the electrodes—200 mm and the gas pressure on the solution—14 kg (f) cm^{-2}. The electrospun materials were dried at 24 °C for 48 h to remove residual solvents and moisture [32]. The thickness of the PHB fibrous materials was 130 ± 0.047 μm.

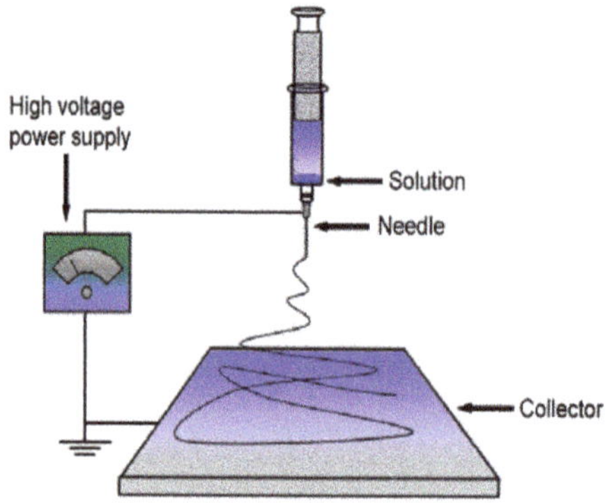

Figure 3. The scheme of the single-capillary laboratory unit for electrospinning [31].

2.2. Methods

The main methods used in this work to study the structure and properties of materials in the process of biodegradation were scanning electron and optical microscopy, determination of morphology characteristics (surface density, thickness and porosity), mechanical analysis, differential scanning calorimetry, infrared spectroscopy and indoor exposure by composting in the soil.

2.2.1. Scanning Electron Microscopy (SEM)

The determination of the morphology of the surface area of the PHB materials was carried out by SEM observations using a Tescan VEGA3 microscope (Wurttemberg, Czech Republic). The SEM microphotographs were obtained at an accelerating voltage of 20 kV. Before scanning, the samples were vacuumized and covered with a layer of patina.

2.2.2. Optical Microscopy

The determination of structural changes, such as the diameter of fibers, changes in morphology etc., were carried out using an Olympus BX43 (Olympus, Japan, Tokyo). The main morphological properties of the fibers were measured by the micrograph using Olympus Stream Basic software (Tokyo, Japan).

2.2.3. Surface Density, Thickness and Porosity Estimation

The determination of the surface density was carried out using the analytical weighing-machine Balance XPR106DUHQ/A (Mettler Toledo, Columbus, OH, USA). Surface density, δ, g/cm^3, was calculated as:

$$\delta = \frac{m}{l \times B \times b} \quad (1)$$

where m is the weight of the sample; l is the length; B is the width; b is the thickness. The average value was obtained from 10 measurements taken at different parts of the sample.

The determination of the porosity was carried out using the analytical weighing-machine Balance XPR106DUHQ/A (Mettler Toledo, Columbus, OH, USA). Porosity, W, %, was calculated as:

$$W = \left(1 - \frac{V_{PHB}}{V_s}\right) \times 100\% \tag{2}$$

where V_f is the volume of the polymer; V_s is the volume of the the sample of the material. The average value was obtained from 10 measurements taken at different parts of the sample.

The determination of the thickness was carried out using the analytical weighing-machine Digital micrometer 0–25 mm, NMD-165D (Norgau, Germany) The value was obtained as an average of five measurements made at three points of the sample.

2.2.4. Mechanical Analysis

Mechanical properties, including maximum strength and breaking strain, were ascertained using compression testing machine DVT GP UG 5 (Devotrans, Istanbul, Turkey). The stretching speed was 25 mm/min without preload pressure according to ASTM D5035-11. The working area of the samples was 10 × 40 mm.

Maximum strength, Fmax, N, was registered automatically. The average value was obtained from 5 measurements.

Elongation at break, ε, %, was calculated as:

$$\varepsilon = \frac{\Delta l}{l_0} \times 100\% \tag{3}$$

where Δl—the difference between the final and initial length of the sample; l_0—the initial length of the sample. The average statistical error in measuring was ±0.2%.

2.2.5. Differential Scanning Calorimetry (DSC)

Thermal characteristics including the melting enthalpy, melting temperature, degree of crystallinity were ascertained using the DSC 214 Polyma (Netzsch, Selb, Germany). The DSC temperature program included 2 heating cycles (from 20 °C to 220 °C) and 2 cooling cycles (from 220 °C to 20 °C). Samples were tested in an argon atmosphere, with a heating rate of 10 K/min and with a cooling rate of 10 K/min with sample weights of 6–7 mg.

Enthalpy of melting, ΔH, J/g, was calculated by NETZSCH Proteus software according to the standard technique.

Temperature of melting, T_m, °C, was calculated by NETZSCH Proteus software according to the standard technique

Crystallinity degree, χ, %, was calculated as:

$$\chi = \frac{\Delta H}{H_{PHB}} \times 100\% \tag{4}$$

where ΔH—melting enthalpy; H_{PHB}—melting enthalpy of the initial ideal crystalline of the PHB, 146 J/g [33].

2.2.6. Infrared Spectroscopy (FTIR)

The chemical composition of the materials was studied using a FTIR Lumos spectrometer (BRUKER, Berlin, Germany) at a temperature of 24 °C in the range of wavenumbers of $4000 \leq \nu \leq 600$ cm^{-1} in the mode of reflected light FTIR on the diamond crystal. The intensity of several absorption peaks related to polymer degradation (functional groups) was estimated by OPUS software (BRUKER, Berlin, Germany). The intensity of several absorption peaks related to polymer degradation (functional groups) was estimated. The resolution was 2 cm^{-1}.

2.2.7. Composting in Soil

The study of biodegradation in soil was carried out using reconstituted soil simulating real soil with a moisture capacity of 60%, at temperature 22 ± 3 °C, and a soil pH of 6. The soil was prepared by the standard technique [34,35].

The rate of the mass lost during biodegradation in soil, R, %, was calculated as:

$$R = \frac{m_i - m_s}{m_i} \times 100\% \quad (5)$$

where m_i—mass of the initial sample; m_s—mass of the sample after the soil.

3. Results

The method of production has a high impact on the appearance and structure of the surface of the material. Figure 4 shows a significant difference between the surface structures of nonwoven and pressed materials. It can clearly be seen that the film based on PHB has its own specific relief and low pores, but it cannot be comparable with the proportion of open pores in the inter-fiber space formed as a result of electrospinning.

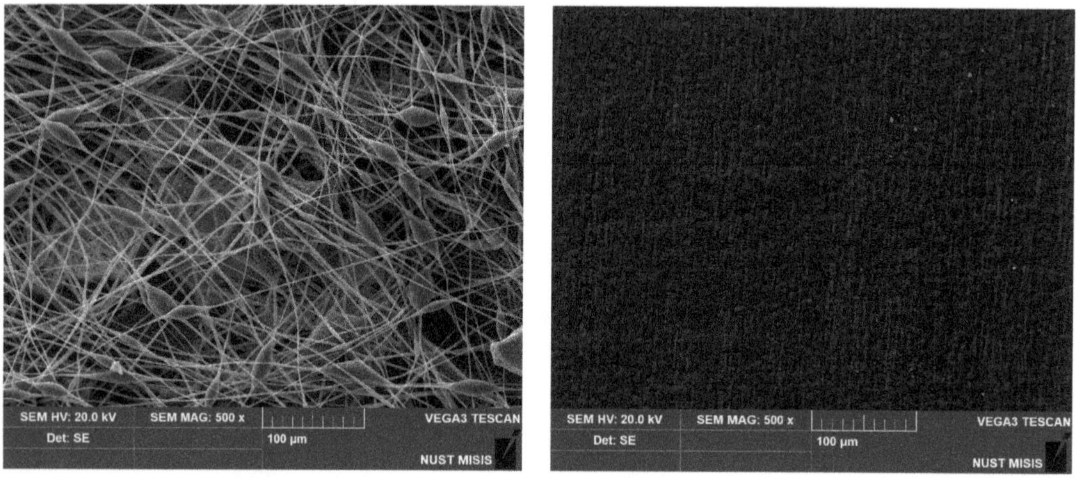

(a) (b)

Figure 4. SEM images of the materials based on PHB obtained by different methods: (**a**) electrospinning, (**b**) pressing.

For quantitative comparison of the morphology of the surface of the two materials, the following indicators were used: surface density and porosity (Table 2). It should be mentioned that electrospun material and film of the same thickness 130 μm displayed a great difference in the friability of the materials' volume and surface.

Table 2. Morphology of the materials based on PHB.

Material	Surface Density, (±SD, n = 10) δ, g/m^3	Porosity, (±SD, n = 10) W,%
Fibrous material	0.098 ± 0.01	80 ± 2.0
Film	0.155 ± 0.03	6 ± 1.0

The degree of development of the surface certainly played an important role in the rate of biodegradation, given that all the conditions of the process were the same. So the first changes for nonwovens were recorded as early as on the 3rd day of composting, while for films noticeable changes were only recorded on the 30th day. Such observations indicate an increase in the rate of biodegradation depending on surface structure.

As previously reported, it should be noted that there are three stages of PHB degradation in soil: transition from a high-molecular compound to monomers and oligomers; transition from monomers and oligomers to biomass; transition from biomass to CO_2 and H_2O [17]. Figure 5 shows the changes in the material's structure at the first stage of degradation.

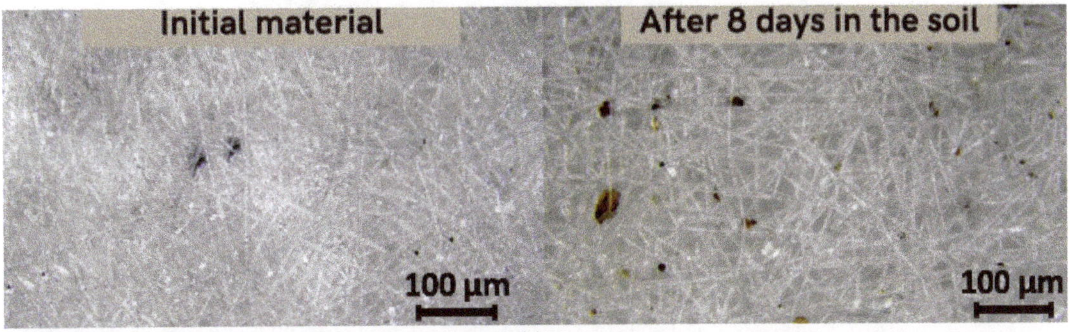

(a)

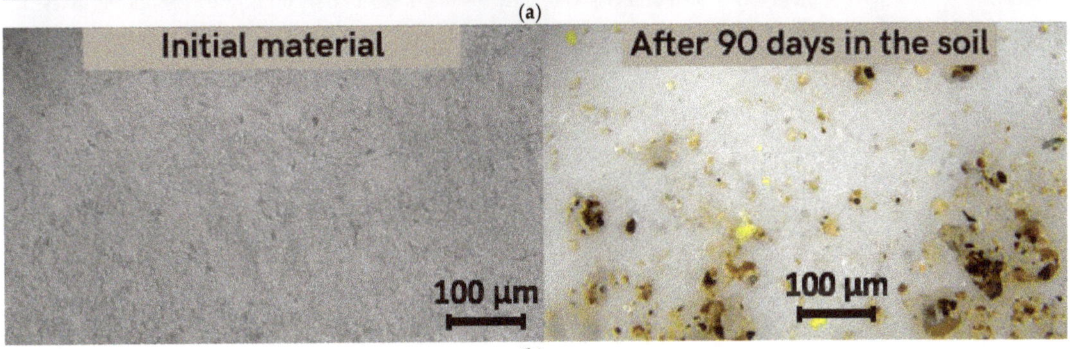

(b)

Figure 5. Microscopy of the original material and the material after placement in the soil (magnification 200): (**a**) fiber, (**b**) film.

This stage is mainly represented by the mechanical action of soil adhesion and the activation of microbiota, and the beginning of oxidative and hydrolytic action under the influence of enzymes and soil components [27]. After 8 days of exposure to laboratory soil, the microstructure of the fibers changed significantly (Figure 5a). These changes were quantified by the morphological characteristics of the material. The average diameter of the fibers decreased from 5.6 microns to 5 microns and surface density decreased from 0.098 g/m^3 to 0.047 g/m^3. Moreover, inclusion of soil particles in the material's structure and in pores was observed. However, for the film, the exposure time of 8 days had no effect on either the structure or the properties. In 90 days, however, the film had undergone noticeable changes. Surface density increased from 0.155 g/m^3 to 0.192 g/m^3. This increase could be explained by the fact that the soil particles are added to the material's surface and the film could accumulate excess moisture or microbiotic component, something which is typical for the first stage of degradation in soil.

Figure 6 shows the duration of the biodegradation of the highly porous electrospun material based on PHB.

Figure 6. Biodegradation of fibrous material from PHB.

It should be noted that significant fragmentation and embrittlement of the sample were observed. In this process, it is difficult to assess the degree of mass loss in a nonwoven material sample. However, for the sake of the clarity of the biodegradation process, Figure 7c shows an averaged curve of the mass loss for three experiments. All electrospun materials based on PHB were fully disintegrated under composting conditions in less than 20 days.

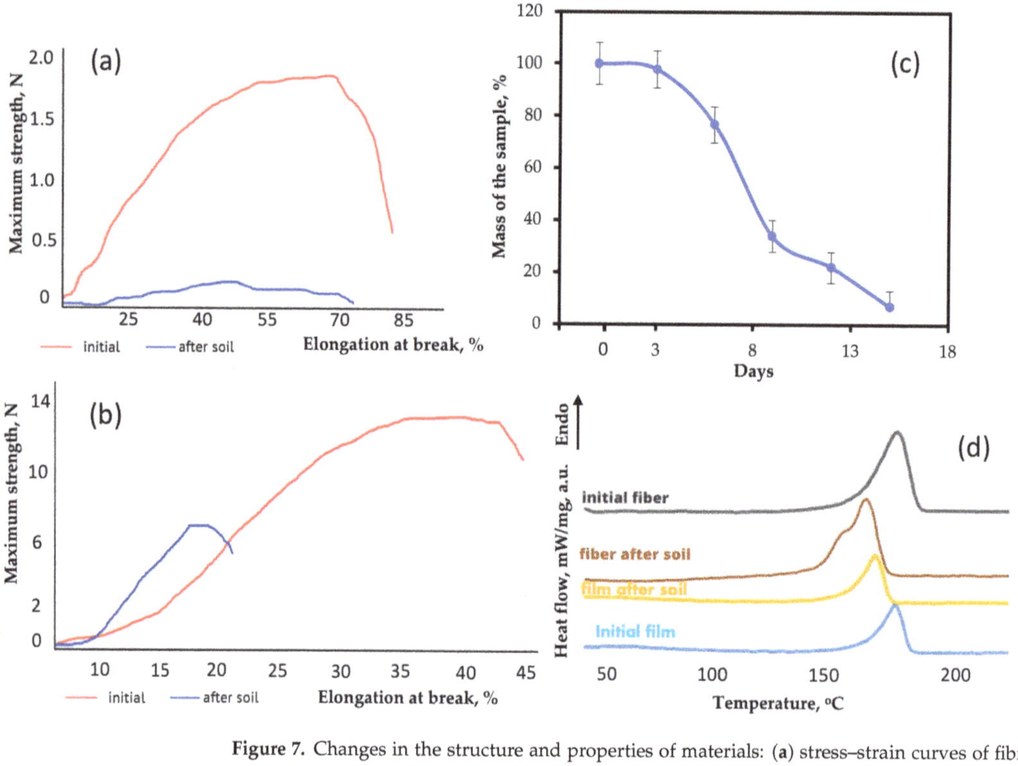

Figure 7. Changes in the structure and properties of materials: (**a**) stress–strain curves of fibrous material before and after the soil; (**b**) stress–strain curves of film before and after the soil; (**c**) the rate of the mass of the fibrous material based on PHB lost during biodegradation in the soil; (**d**) DSC curves of the material before and after the soil.

Thus, not only was surface hydrolytic erosion observed [19], but disintegration of nonwoven samples with high porosity due to the space between the fibers (Figure 6) was also seen. Thus, there is a fragmentation stage corresponding to the beginning of the biodegradation process, which includes, at the first stage, the formation of monomers and oligomers as a result of chemical and mechanical actions; at the second stage, microbiological transformation to biomass; and at the last stage, chemical transformation to CO_2 and H_2O [18].

To establish the difference in the rates of biodegradation of film and electrospun fibrous material, various characteristics of the material were evaluated.

Mechanical properties are an important indicator of structural changes in the material. Changes in mechanical properties could also be triggered by the activity of microorganisms and by the influence of water and other factors. The effect exposure to the soil had on the mechanical properties of the polymer material are shown in Table 3.

Table 3. Mechanical properties of the materials.

Material	Maximum Strength, Fmax, N	Elongation at Break, ε, %
Fibrous material	1.91	67.53
Fibrous material after soil (8 days)	0.16	49.93
Initial film	12.34	45.75
Film after soil (90 days)	6.31	19.61

It should be noted that nonwoven fabric and film with the same thicknesses have fundamentally different physical and mechanical properties. Materials based on PHB are generally considered to be quite fragile [36]. The features of the crystallization process often negatively affect the strength of the material, which is significantly inferior to synthetic polyesters [36]. The maximum strength of thin films based on PHB was 12.34 N. The strength of electrospun samples with the same thickness was decreased by a factor of nearly six. The main reason for this decrease is the fragility of individual fibers, which are easily torn, concentrating the residual stress of the undocrystallized areas when plasticizing additives are removed or in the case of post-treatment [37]. However, the decrease in strength is compensated for by an increase in maximum elongation. Table 3 shows that elongation at break of fibrous material is higher by almost 50%. Despite the fact that individual fibers tear more easily than a monolithic film, fibers provide higher flexibility in a whole system of nonwoven material [38]. Fibers have a certain degree of freedom, the restriction of which depends on gluing, thickening and other morphological defects [39]. In general, nonwoven fabric based on PHB is able to withstand larger elongations than a film of the same thickness due to the structural organization of the fibrous layer, as can be seen from the results.

Particular attention should be paid to the contribution of composting to the change in mechanical properties. Figure 7a shows changes in the stress–strain curves of the electrospun material after 8 days of exposure and Figure 7b shows changes the in stress–strain curves films after 90 days of exposure. These time periods were chosen due to the need to note changes in the first stage of biodegradation.

After exposure to the soil, the strength of the fibrous material decreased by 90% and the strength of the film decreased by 49%. The elongation at break also decreased by 26% for the fiber and by 60% for the film. The monolithic structure of the film was found to be to be more resistant to exposure and its biodegradation is slower. However, a significant embrittlement of the film, which was confirmed by a more intense decrease in elongation, indicates a chemical process occurring in the material. Although biodegradation in film samples proceeds according to the mechanism of surface destruction (Figure 2), the hydrolytic process could appear in whole system through the pores. The proportion of open pores in nonwoven fibrous material is many times higher than that in film, and the mechanism is approaching bulk degradation (Figure 2).

These data indicate a significant decrease in the strength and strain of polymer materials [40]. It is clearly seen that the method of polymer production makes a significant contribution to the rate of biodegradation. It should be mentioned that the key role in this process may be played by the surface area and the degree of accessibility of the polymer to soil microorganisms and chemical substances. The surface area of the fibrous material is many times larger than the surface area of the film.

Confirmation of this proposal follows from the analysis of the supramolecular structure using the DSC method. This method makes it possible to determine and investigate melting and crystallization temperatures [41]. Figure 7d shows the DSC curves of the material after biodegradation in soil. It is important to note that the method by which the PHB is obtained makes a significant contribution to the crystallization process and the perfection of the crystal structure of PHB [36]. It can clearly be seen that the shape of the melting peaks of the film and the fibrous material is quite similar, but the degree of crystallinity of the fibers is slightly higher. The crystallinity degree for the film and the fibrous material was 56.3% and 58.5%, respectively. The melting temperature of the samples was 174.4 °C and 174.5 °C, respectively.

It was found that the crystallinity degree of the film decreased by almost a factor of two from 58.5% to 28% during exposure to the soil, while the crystallinity degree of the fibrous material decreased from 56.3% to 45%. A decrease in crystallinity is an important marker of the destruction of the crystalline phase of the polymer [9]. As it can be seen from Figure 8, the shape of the melting peak of the fibrous material changed a great deal. Despite a slight decrease in the degree of crystallinity of the fibers, it can clearly be seen that separation of the crystalline fraction occurred. After exposure to the soil, the crystallites began to melt at a lower temperature and some of them formed a low-temperature shoulder in the range of 140–160 °C. This indicates the availability of the crystalline phase for the effects of biodegradation factors. The melting point of the fibrous material decreased from 174.5 °C to 168.5 °C. which confirms the fragmentation of crystallites in the process of biodegradation. Smaller and defective crystalline fractions melt at lower temperatures [36]. At the same time, the melting point of the film decreased from 174.4 °C to 170.6 °C. Thus, a decrease in the melting point indicates a restructuring of the crystalline phase. Additionally, the decrease in the proportion of the crystal fraction was more noticeable. The results of the DSC are consistent with changes in physical and mechanical properties, since the structure and perfection of the crystalline phase largely affect the elongation and strength of the whole material. This indicates a noticeable and rapid effect from the soil microbiota, which is able to attack both the amorphous and crystalline phases more actively [42].

In addition, the FTIR method was able to confirm biodegradation [4]. The FTIR spectra are shown on Figure 8.

The general decrease in the intensity of the peaks after biodegradation is primarily due to the warping of the surface and the change in its relief under the action of the soil microbiota [43]. The most pronounced changes for the peaks at 1721 cm^{-1} (C=O group) and 1052 cm^{-1} (C-O-C group) for PHB at the first stages of biodegradation should be noted. Finally, a decrease in intensity of 1278 cm^{-1} (CH_3 group) is expected for a more chemically stable part of the PHB chain [36]. In other works, the carboxyl index is used to characterize biodegradation, represented by the ratio of intensities of 1721 cm^{-1} and 1278 cm^{-1} [44]. The carbonyl index of PHB generally remains stable during the first stage of degradation (about 20 days) [45]. So, for nonwoven fabric it was 1.15 before degradation and 1.10 after 8 days of degradation (Figure 8a). Two processes may take place on the surface of the PHB in soil: accumulation on the surface of carbonyl groups resulting from the ester linkage hydrolysis [45] and the overall decrease of C=O groups as a consequence of their transition to -OH groups. These processes probably compensate for each other. At the same time, the peak of the ester linkage (1052 cm^{-1}) has changed its shape and is difficult to identify. These observations indicate a break in the PHB chain and chemical changes that are usual for enzymatic hydrolysis [18].

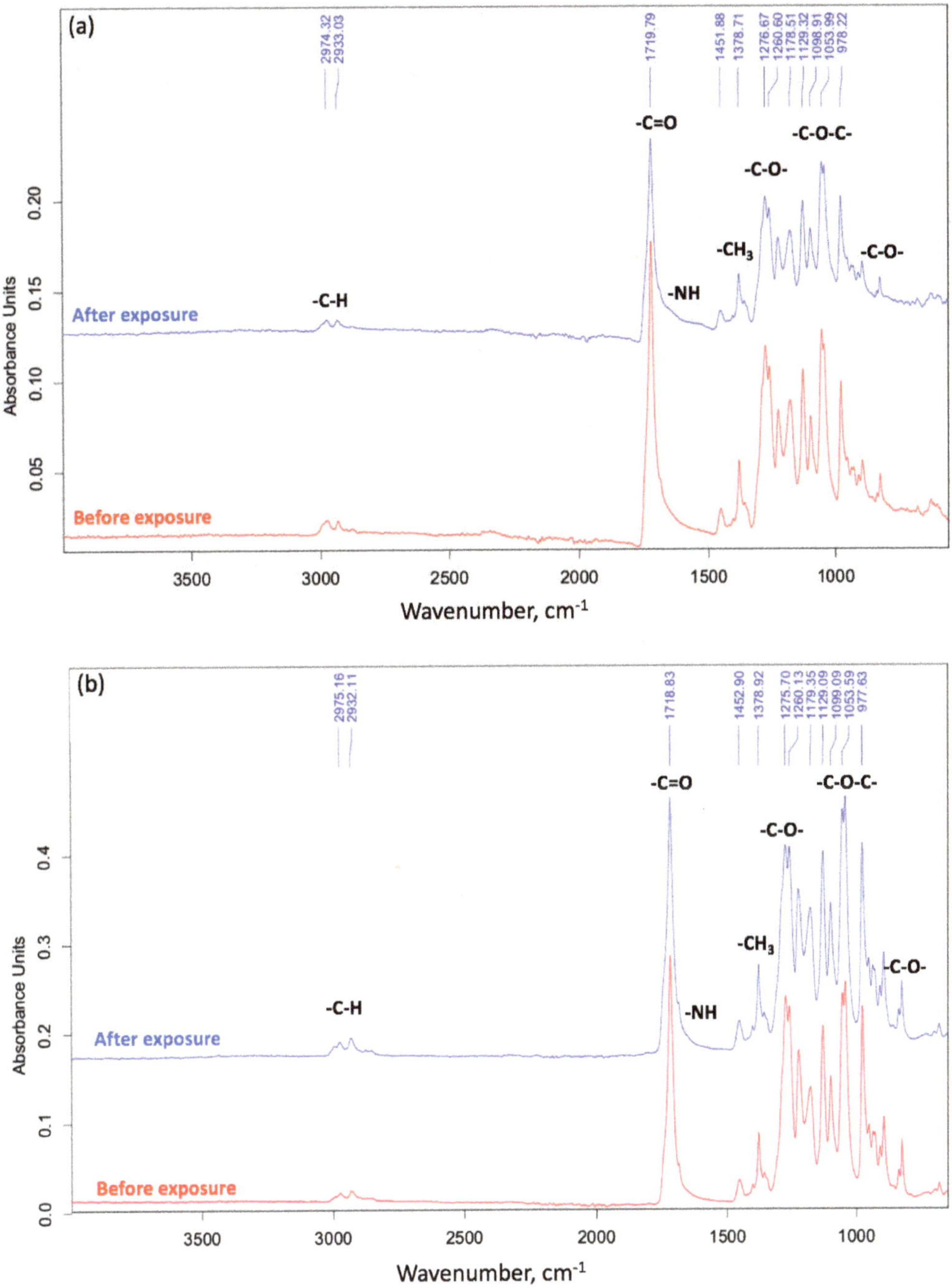

Figure 8. The FTIR spectra: (**a**) fiber, (**b**) film.

For the film, the carbonyl index was 1.73 before degradation and 1.44 after 90 days of degradation (Figure 8b). These observations are already consistent with ideas about the process of biodegradation. This is consistent with the fact that after the first stage of biodegradation (about 20 days), the oligomers obtained as a result of the enzymatic hydrolysis of ester bonds leave the surface of the film [45]. Moreover, in the region of 900–700 and 1700–1500 cm^{-1}, noticeable changes are observed for fibrous material (Figure 8a). This fact suggests the breakage of C-C bonds and the accumulation of OH groups, processes which are able to proceed faster when there is a greater availability of PHB macromolecules in the entire volume of the sample. This also confirms our ideas about the mechanism of biodegradation, something which proceeds in accordance with bulk degradation time.

Therefore, it is possible to generically characterize the biodegradation of nonwoven electrospun material as bulk degradation which occurs in the whole volume of the sample, while the degradation of the film proceeds according to a well-studied mechanism of surface degradation. In both cases, degradation is initiated from the polymer surface, having a significant effect on the amorphous and crystalline phases of the semi-crystalline PHB. But for a fibrous material, the contribution of the surface is so great that it accelerates degradation several times, leading to greater changes in the supramolecular and chemical structures than in the case of a film.

This theory is confirmed by the results obtained and by the differences in the degree of crystallinity. The crystallinity degree of the film decreased from 58.5% to 28% during 90 days in the soil, and the material was still comprehensive but fragile. At the same time, the crystallinity degree of the fibrous material decreased from 56.3% to 45% during 8 days, but the material had already disintegrated into small fragments. Such dynamics fully correspond to inversely proportional changes in mechanical properties. The elongation at break of the film decreased from 45.8% to 19.6% during 90 days in the soil. At the same time, the elongation at break of the fibrous material decreased from 67.5% to 49.9% during 8 days. At the same time, the high fragmentation of the nonwoven sample after 8 days in the soil is caused by a catastrophic decrease in strength to 0.16 N. The strength of the film was more than 6 N, which is almost three times higher than that of the fibrous material before the start of testing in the soil, with the samples of the same thickness.

4. Conclusions

Poly-3-hydroxybutyrate is a perfectly biocompatible polyester which does not cause environmental damage since its biodegradation products are water and CO_2 [46,47]. Despite its low mechanical properties and low elasticity, in comparison with large-tonnage polymers, and taking into account its biodegradation abilities, it could be used in biomedicine, food eco-packaging, etc. [48].

This work established consistent patterns of changes in the supramolecular structure of the PHB, the chemical structure of the surface, and the physical and mechanical properties of the material during biodegradation in soil. Particular attention was paid to the contribution of the surface structure to the rate of biodegradation.

It was shown that a high degree of surface development, as in a nonwoven fibrous material, with a porosity of more than 80% and a high proportion of open pores, is able to disintegrate in the soil several times faster than a monolithic film material with a pore fraction of less than 10% with samples of the same thickness.

Thus, it is necessary to note the high contribution of the surface to the rate of biodegradation of the material. This leads to prospects for the creation of new materials with a wide range of applications and with controlled stability and speed of biodegradation.

Author Contributions: Conceptualization, P.M.T.; methodology, P.M.T.; software, K.G.G.; validation, I.A.V.; formal analysis, P.M.T.; investigation, K.G.G.; resources, A.A.V.; data curation, P.M.T. and A.A.V.; writing—original draft preparation, A.A.V. and K.G.G.; writing—review and editing, A.A.V. and A.A.P.; visualization, P.M.T.; supervision, P.M.T.; project administration, I.A.V. All authors have read and agreed to the published version of the manuscript.

Funding: There is no funding for P.M.T., A.A.P., K.G.G. and I.A.V., but the paper has been supported by the RUDN University Strategic Academic Leadership Program (recipient A.A.V.).

Institutional Review Board Statement: Not applicable.

Data Availability Statement: Not applicable.

Acknowledgments: The study was carried out using scientific equipment of the Center of Shared Usage New Materials and Technologies of the Emanuel Institute of Biochemical Physics and the Common Use Centre of Plekhanov Russian University of Economics.

Conflicts of Interest: The authors declare no conflict of interest.

References

1. Bharti, S.N.; Swetha, G. Need for Bioplastics and Role of Biopolymer PHB: A Short Review. *J. Pet. Environ. Biotechnol.* **2016**, *7*, 5–8.
2. Mohd Ishak, Z.A.; Ahmad Thirmizir, M.Z. Editorial corner-a personal view producing green composites via polymer blending. *Express Polym.* **2021**, *15*, 910–911. [CrossRef]
3. Ugur, A.; Sahin, N.; Beyatli, Y. Accumulation of poly-hydroxybutyrate in Streptomyces species during growth with different nitrogen sources. *Turk. J. Biol.* **2002**, *26*, 171.
4. Krutko, E.T.; Prokopchuk, N.R.; Globa, A.I. Natural biodegradable poly-measures. In *Book Technology of Biodegradable Polymer Materials*, 1st ed.; Krul, L.P., Kudina, E.F., Eds.; Belarusian State Technological University: Minsk, Belarus, 2014; Volume 1, p. 105.
5. Raza, Z.A.; Khalil, S.; Abid, S. Recent progress in development and chemical modification of poly(hydroxybutyrate)-based blends for potential medical applications. *Int. J. Biol. Macromol.* **2020**, *160*, 77–100. [CrossRef] [PubMed]
6. Ansari, S.; Fatma, T. Polyhydroxybutyrate-a Biodegradable Plastic and its Various Formulations. *Int. J. Innov. Res. Sci. Eng. Technol.* **2014**, *3*, 5.
7. Sreedevi, S.; Unni, K.N.; Sajith, S.; Priji, P.; Josh, M.S.; Benjamin, S. Bioplastics: Advances in poly-3-hydroxybutyrate research. In *Advances in Polymer Science*; Springer: Berlin/Heidelberg, Germany, 2014; pp. 1–30.
8. Woolnough, C.A.; Yee, L.H.; Charltonb, T.; Fostera, L.G.R. Environmental degradation and biofouling of 'green' plastics including short and medium chain length polyhydroxyalkanoates. *Polym. Int.* **2009**, *59*, 658–667. [CrossRef]
9. Weng, Y.-X.; Wang, X.-L.; Wang, Y.-Z. Biodegradation behavior of PHAs with different chemical structures under controlled composting conditions. *Polym. Test.* **2011**, *30*, 372–380. [CrossRef]
10. Quispe, M.M.; Lopez, O.V.; Boina, D.A.; Stumb, J.F.; Villar, M.A. Glycerol-based additives of poly(3-hydroxybutyrate) films. *Polym. Test.* **2021**, *93*, 15. [CrossRef]
11. Bonartsev, A.P.; Bonartseva, G.A.; Reshetov, I.V.; Kirpichnikov, M.P.; Shaitan, K.V. Application of polyhydroxyalkanoates in medicine and the biological activity of natural poly(3-Hydroxybutyrate). *Acta Nat.* **2019**, *11*, 4–16. [CrossRef]
12. Rodrigo, T.U.; Felisbert, M.I. Plasticization of poly(3-hydroxybutyrate) with triethyl citrate: Thermal and mechanical properties, morphology, and kinetics of crystallization. *J. Appl. Polym. Sci.* **2020**, *138*, 14.
13. Râpă, M. Raluca Nicoleta Darie-Nita Effect of plasticizers on melt processability and properties of PHB. *J. Optoelectron. Adv. Mater.* **2015**, *17*, 1778–1784.
14. Yu, J.; Stahl, H. Microbial utilization and biopolyester synthesis of bagasse hydrolysates. *Bioresour. Technol.* **2008**, *99*, 7. [CrossRef] [PubMed]
15. Reinerte, S.; Kirpluks, M.; Cabulis, U. Thermal degradation of highly crosslinked rigid PU-PIR foams based on high functionality tall oil polyol. *Polym. Degrad. Stab.* **2019**, *167*, 50–57. [CrossRef]
16. Jeepery, F.I.; Sudesh, K.; Abe, H. Miscibility and enzymatic degradability of poly(3-hydroxybutyrate-co-3-hydroxyhexanoate)-based polyester blends by PHB depolymerase and lipase. *Polym. Degrad. Stab.* **2021**, *192*, 12. [CrossRef]
17. Tosin, M.; Pischedda, A.; Degli-Innocenti, F. Biodegradation kinetics in soil of a multi-constituent biodegradable plastic. *Polym. Degrad. Stab.* **2019**, *166*, 213–218. [CrossRef]
18. Poltronieri, P.; Mezzolla, V.; D'Urso, O.F. PHB Production in Biofermentors Assisted through Biosensor Applications. *Proceedings* **2016**, *1*, 4. [CrossRef]
19. Rutkowska, M.; Krasowska, K.; Heimowska, A.; Adamus, G.; Sobota, M.; Musioł, M.; Janeczek, H.; Sikorska, W.; Krzan, A.; Žagar, E.; et al. Environmental degradation of blends of atactic poly[(R,S)-3-hydroxybutyrate] with natural PHBV in Baltic sea water and compost with activated sludge. *J. Polym. Environ.* **2008**, *16*, 183–191. [CrossRef]
20. Tsuji, H.; Suzuyoshi, K. Environmental degradation of biodegradable polyesters 2 Poly(ε-caprolactone), poly[(R)-3-hydroxybutyrate], and poly(L-lactide) films in natural dynamic seawater. *Polym. Degrad.* **2002**, *75*, 357–365. [CrossRef]
21. Meereboer, K.W.; Misra, M.; Mohanty, A.K. Review of recent advances in the biodegradability of polyhydroxyalkanoate (PHA) bioplastics and their composites. *RSoC* **2020**, *22*, 40. [CrossRef]
22. Marina, P.A.; Perdiguero, M.; Fiori, S.; Kenny, J.M.; Peponi, L. Biodegradable electrospun PLA-PHB fibers plasticized with oligomeric lactic acid. *Polym. Degrad. Stab.* **2020**, *179*, 7.
23. European Bioplastics. Available online: https://www.european-bioplastics.org/new-market-data-the-positive-trend-for-the-bioplastics-industry-remains-stable/ (accessed on 10 February 2023).
24. Deroiné, M.; César, G.; Le Duigou, A.; Davies, P.; Bruzaud, S. Natural degradation and biodegradation of poly(3-hydroxybutyrate-co-3-hydroxyvalerate) in liquid and solid marine environments. *J. Polym. Environ.* **2015**, *23*, 493–505. [CrossRef]
25. Altaee, N.; El-Hiti, G.A.; Fahdil, A.; Sudesh, K.; Yousif, E. Biodegradation of different formulations of polyhydroxybutyrate films in soil. *Springerplus* **2016**, *5*, 762. [CrossRef] [PubMed]

26. Muhamad, I.I.; Joon, L.K.; Azemi, M.; Noor, M. Comparing the degradation of poly-β-(hydroxybutyrate), poly-β-(hydroxybutyrate-co-valerate)(PHBV) and PHBV/cellulose triacetate blend. *Polym. J.* **2006**, *1*, 39–46.
27. Mouhoubi, R.; Lasschuijt, M.; Carrasco, S.; Gojzewski, H.; Wurm, F. End-of-life biodegradation? how to assess the composting of polyesters in the lab and the field. *Waste Manag.* **2022**, *154*, 36–48. [CrossRef]
28. Prudnikova, S.V.; Korobikhina, K.I.; Boyandin, A.N.; Volkova, T.G. Regularities of biodegradation of polyhydroxyalkanoates on the territory of Vietnam and Central Siberia. *J. Sib. Fed. Univ. Biol.* **2012**, *3*, 311–321.
29. Fang, L.; David, W. Grainger Fluorinated Biomaterials. In *Biomaterials Science*; Academic Press: Cambridge, MA, USA, 2013; Volume 5.
30. Filatov, Y.N. *Electrospinning of Fibrous Materials*, 1st ed.; Krivchenko, V.N.: Moscow, Russia, 2001; p. 231.
31. Alsvik, I.L.; Hägg, M.-B. Pressure Retarded Osmosis and Forward Osmosis Membranes: Materials and Methods. *Polymers* **2013**, *5*, 303–327. [CrossRef]
32. Sun, L.; Songa, L.; Zhang, X.; Zhou, R.; Yin, J.; Luan, S. Poly(γ-glutamic acid)-based electrospun nanofibrous mats with photodynamic therapy for effectively combating wound infection. *Mater. Sci. Eng. A* **2020**, *113*, 10. [CrossRef]
33. Zhu, G.; Kremenakova, D.; Wang, Y.; Militky, J. Air permeability of polyester nonwoven fabrics. *Autex Res. J.* **2015**, *15*, 8–12. [CrossRef]
34. Varyan, I.; Tyubaeva, P.; Kolesnikova, N.; Popov, A. Biodegradable Polymer Materials Based on Polyethylene and Natural Rubber: Acquiring, Investigation, Properties. *Polymers* **2022**, *14*, 2457. [CrossRef]
35. *GOST R 53381-2009*; Soils and Soils Are Nutritious Soils. Technical Conditions: Date of Introduction 2010-07-01. Standartinform: Moscow, Russia, 2009.
36. Chuan, J.; Yeo; Muiruri, J.; Thitsartarn, W.; Li, Z.; He, C. Recent advances in the development of biodegradable PHB-based toughening materials: Approaches, advantages and applications. *Mater. Sci. Eng. C* **2018**, *92*, 4.
37. Zykova, A.; Morokov, E.; Tyubaeva, P. Influence of Processing Methods on the Mechanical Behavior of Poly-3-hydroxybutyrate Nonwoven Scaffolds. *Macromol. Symp.* **2022**, *404*, 1092–1116. [CrossRef]
38. Khramtsova, E.; Morokov, E.; Antipova, C.; Krasheninnikov, C.; Lukanina, K.; Grigoriev, T. How the Nonwoven Polymer Volume Microstructure Is Transformed under Tension in an Aqueous Environment. *Polymers* **2022**, *14*, 13. [CrossRef] [PubMed]
39. Yano, T.; Higaki, Y.; Tao, D.; Murakami, D.; Kobayashi, M.; Ohta, N.; Koike, J.; Horigome, M.; Masunaga, H.; Ogawa, H.; et al. Orientation of poly(vinyl alcohol) nanofiber and crystallites in non-woven electrospun nanofiber mats under uniaxial stretching. *Polymer* **2012**, *53*, 4702–4708. [CrossRef]
40. *GOST 28840-90*; Machines for Testing Materials for Tension, Compression and Bending: Date of Introduction 1993-01-01. Resolution of the State Standard: Moscow, Russia, 29 December 1990; 8p.
41. Spivak, L.V.; Simonov, Y.N.; Dyshlyuk, M.A. Differential scanning calorimetry: New experimental possibilities. *Physics* **2019**, *3*, 52–57.
42. Volova, T.G.; Boyandin, A.N.; Prudnikov, S.V. Biodegradation of Polyhydroxyalkanoates in Natural Soils. *J. Sib. Fed. Univ.* **2015**, *8*, 152–167. [CrossRef]
43. Redkin, N.A. *IR-Fourier Spectrometry and Mass Spectrometry in the Identification of Organic Compounds*, 1st ed.; Samara University: Samara, Russia, 2019; p. 92.
44. Kuptsov, A.H.; Zhizhin, G.N. *Fourier-KR and Fourier-IR Spectra of Polymers*, 1st ed.; M. Fizmatlit: Moscow, Russia, 2013; p. 657.
45. Salomez, M.; George, M.; Fabre, P.; Touchaleaume, F.; Cesar, G.; Lajarrige, A.; Gastaldi, E. A comparative study of degradation mechanisms of PHBV and PBSA under laboratory-scale composting conditions. *Polym. Deg. Stab.* **2019**, *167*, 102–113. [CrossRef]
46. Manna, A.; Paul, A.K. Degradation of microbial polyester poly(3-hydroxybutyrate) in environmental samples and in culture. *Biodegradation* **2000**, *11*, 323–329. [CrossRef]
47. Guerin, P.; Renard, E.; Langlois, V. Degradation of natural and artificial poly[(R)-3-hydroxyalkanoate]s: From biodegradation to hydrolysis. *Microbiology* **2010**, *14*, 284–313.
48. Sudesh, K.; Abe, H.; Doi, Y. Synthesis, structure and properties of polyhydroxyalkanoates: Biological polyesters. *Prog. Polym.* **2000**, *25*, 1503–1555. [CrossRef]

polymers

MDPI

Article

Structural Rearrangements of Polylactide/Natural Rubber Composites during Hydro- and Biotic Degradation

Yulia V. Tertyshnaya [1,2,*], Maria V. Podzorova [1,2], Anastasia V. Khramkova [3], Vasily A. Ovchinnikov [2] and Aleksey V. Krivandin [1,*]

[1] Department of Biological and Chemical Physics of Polymers, Emanuel Institute of Biochemical Physics, Russian Academy of Sciences, 4 Kosygina Str., Moscow 119334, Russia
[2] Department of Chemistry of Innovative Materials and Technologies, Plekhanov Russian University of Economics, 36 Stremyanny per., Moscow 117997, Russia
[3] Politecnico di Milano, 32 Piazza Leonardo da Vinci, 20133 Milan, Italy
* Correspondence: terj@rambler.ru (Y.V.T.); avk955@mail.ru (A.V.K.); Tel.: +7-495-939-7186 (A.V.K.)

Abstract: In the work, the impact of the biological medium and water on structural rearrangements in pure polylactide and polylactide/natural rubber film composites was studied. Polylactide/natural rubber films with a rubber content of 5, 10, and 15 wt.% were obtained by the solution method. Biotic degradation was carried out according to the Sturm method at a temperature of 22 ± 2 °C. Hydrolytic degradation was studied at the same temperature in distilled water. The structural characteristics were controlled by thermophysical, optical, spectral, and diffraction methods. Optical microscopy revealed the surface erosion of all samples after exposure to microbiota and water. Differential scanning calorimetry showed a decrease in the degree of crystallinity of polylactide by 2–4% after the Sturm test, and a tendency to an increase in the degree of crystallinity after the action of water was noted. Changes in the chemical structure were shown in the spectra recorded by infrared spectroscopy. Due to degradation, significant changes in the intensities of the bands in the regions of 3500–2900 and 1700–1500 cm^{-1} were shown. The X-ray diffraction method established differences in diffraction patterns in very defective and less damaged regions of polylactide composites. It was determined that pure polylactide hydrolyzed more readily under the action of distilled water than polylactide/natural rubber composites. Film composites were more rapidly subjected to biotic degradation. The degree of biodegradation of polylactide/natural rubber composites increased with the rise in the content of natural rubber in the compositions.

Keywords: polylactide; hydrolytic degradation; biodegradation; polymer composites; crystallinity; X-ray; Sturm test

Citation: Tertyshnaya, Y.V.; Podzorova, M.V.; Khramkova, A.V.; Ovchinnikov, V.A.; Krivandin, A.V. Structural Rearrangements of Polylactide/Natural Rubber Composites during Hydro- and Biotic Degradation. *Polymers* **2023**, *15*, 1930. https://doi.org/10.3390/polym15081930

Academic Editor: Agustín Lara-Sánchez

Received: 27 February 2023
Revised: 14 April 2023
Accepted: 17 April 2023
Published: 19 April 2023

1. Introduction

The degradation of polymers in aqueous media is a multifactorial process studied in physics, chemistry, and biology. Polylactide (PLA) and its degradation have attracted the attention of researchers for decades [1–4]. The use of new fillers and the creation of new polymer composites always left the question of biotic and abiotic degradation open.

Considering the process of biodegradation of polymers, it must be taken into account that this is not a monoprocess but a complex system of chemical and biological interactions, which may include hydrolysis and the action of enzymes secreted by microorganisms. There are also certain characteristics of the polymers themselves that affect the biodegradation process such as molecular weight and crystallinity. The high molecular weight and high crystallinity of PLA results in slower degradation [5,6].

The biodegradability and characteristics of the environment where the biodegradation process takes place are also of great importance. Unlike other biodegradable polymers, which degrade primarily in a single step by microbial attack, PLA degradation follows a sequential mechanism [7,8]. The first step involves reducing the molecular weight of PLA through hydrolysis, which is considered the rate-limiting step and is the main depolymerization mechanism. The second step involves microbial assimilation and metabolism of low molecular weight PLA oligomers and monomers [9]. The degree of degradation of aliphatic polyesters, to which PLA belongs, depends on the type and nature of enzymes produced by microorganisms [10]. The enzymes such as protease, esterase, and lipase produced by microbes show the ability to degrade PLA [11]. Proteinase and protease degrade polymer foils more rapidly than esterase and lipase [11,12].

However, studies [12–15] indicate that PLA-degrading bacteria are not as common in the environment. It was shown in the works [12,13] that PLA biodegradation mainly occurs due to microorganisms such as *Chryseobacterium* sp., *Sphingobacterium* sp., *Stenotrophomonas pavanii, Pseudomonas geniculata,* and *Thermopolyspora flexuosa.* In this case, the process inducers contain L-alanine units, which are similar to PLA L-lactic acid units in the stereochemical position of the chiral carbon [11]. Subsequently, the depolymerase acts on the intracellular ester bonds of PLA, resulting in the formation of oligomers, dimers and, consequently, monomers. Then, these low molecular weight compounds are absorbed by microorganisms, and they are converted into carbon dioxide, water, and methane as a result of the activity of intercellular enzymes [11,16–18]. However, as mentioned above, an important role for the biodegradation of PLA is played not only by the microorganisms themselves and hydrolytic enzymes but also by the chemical process of hydrolysis itself.

The course of the process of hydrolytic degradation occurs in several stages [19]:

- the diffusion of water into the material;
- the hydrolysis of chains in the amorphous region due to the lower resistance to water;
- the reducing of molecular weight due to the hydrolytic cleavage of side bonds and the formation of water-soluble compounds;
- the hydrolysis of the crystalline phase, which can occur by an autocatalytic mechanism due to acidic decomposition products as well as due to an increase in the concentration of carboxylic acid at the end of the chain.

The hydrolysis process is influenced by both external and internal factors. Important external factors are the temperature and pH of the medium.

It is known that an increase in temperature can accelerate the process of hydrolytic degradation. The temperature below which the PLA hydrolytic degradation rate decreases is 60 °C [19]. At a temperature of 37 °C, the rate of degradation is reduced by about two orders of magnitude compared with the process taking place at a temperature equal to or higher than 60 °C. This effect is associated with the transition of polylactide from a glassy to a highly elastic state, and the glass transition temperature of PLA is in the range of 58–63 °C.

In addition, such changes in the rates of hydrolytic degradation also occur in systems of mixtures of PLA with another component. For example, in a mixture of polylactide and polylactic-glycolic acid (in a ratio of 50:50), the rate of hydrolytic degradation under neutral conditions increases significantly at a process temperature above 60 °C [19].

The pH value is also important. It significantly affects the destruction of PLA since it affects not only the reaction mechanism but also the kinetics of the process. Destruction in an alkaline medium begins with a nucleophilic attack on the terminal hydroxyl of another carbonyl group. As a result of this process, oligomers of lactic acid and lactide are formed, which are degraded to lactic acid [20]. In an acidic environment, the destruction of polylactide is initiated by the protonation of the terminal ester group, after which an intramolecular hydrogen bridge is formed. The products of hydrolytic degradation by this mechanism are lactic acid and lactic acid oligomers [20]. As pH values increase from 1 to 10, the observed constant first decreases, reaching a minimum around pH 4, and then increases at higher pH values. At the same time, the growth of the constant is very significant—by about four orders of magnitude. The presence of a minimum in the dependence of the kinetic constant on pH confirms that hydrolysis is catalyzed either by an acidic or alkaline medium [19].

One of the most important internal factors is the crystallinity of the polymer [2,4]. The crystalline regions of PLA are resistant to hydrolysis due to the limited access of water molecules to the chains within the solid crystalline region. This causes selective hydrolytic chain cleavage in amorphous regions and the removal of water-soluble oligomers and monomers [21]. An increase in the concentration of D-units in PLA reduces optical purity and regularity, which leads to greater diffusion of water through the polymer matrix in amorphous regions and accelerated hydrolysis [21,22].

In a sample of complete amorphous PLA, a higher degree of destruction is observed in comparison with semicrystalline PLA under the same conditions [19,21]. A decrease in the molecular weight of amorphous PLA by 14% is observed after 18 weeks of being in a phosphate-buffered solution at pH = 4 and at a temperature of 37 °C. At the same time, it will take about 20 months to reduce the molecular weight of semi-crystalline PLA by the same value and under the same conditions [19].

Plasticizers and fillers also affect the process of PLA hydrolytic degradation. The reasons for the acceleration of the process may be the presence of a phase boundary, an increase in the proportion of the amorphous phase in the composite, and the hydrophilicity of the filler.

Rubbers are often used to increase the elasticity of brittle polymers [23–25]. Rubbers increase elongation at break and toughness of composites [24,26]. Natural rubber (NR) is an elastomer obtained from the sap of rubber trees. The application of NR makes it possible to improve some of the mechanical properties of PLA composites and preserve the bioavailability of the resulting material [27].

Most of the works are devoted to the study of morphological features and mechanical properties of PLA/NR. The structure changes of these composites during a degradation have not been sufficiently studied. It is expected that the natural rubber used in the current work will increase the rate of PLA hydro- and biotic degradation due to an increase in the proportion of the amorphous phase in the composites and the bioavailability of NR. The DSC method was used to determine the phase transition temperatures and the degree of crystallinity. Morphology was studied by optical microscopy. Changes in the chemical and physical structure were recorded by spectral methods and X-ray diffraction, respectively.

2. Materials and Methods

2.1. Sample Preparation

Thermoplastic poly(lactic acid) (PLA) 4032D (with about 2% of D-lactide) with a molecular weight (M_w) of 1.9×10^5 g/mol was procured from Nature Works (Minnetonka, MN, USA). Natural rubber (NR), SVR-3L with a Mooney viscosity of 50 ± 5 (100 °C) and poly(cis-1,4-isoprene) content 91–96 by wt.% was kindly supplied by Vietnam Rubber Group (Ho Chi Minh City, Vietnam). The polylactide was previously dried at 40 °C for 3 h to remove moisture.

Film samples were obtained from a solution. The polymer solutions were prepared by dissolving PLA and PLA/NR in the right ratio in 100 mL of chloroform (Component-Reactiv, Moscow, Russia). The sample weight was 9 g per 100 mL, and the ratio of the components (PLA:NR, wt.%) was 100:0, 95:5, 90:10, and 85:15. The resulting samples were dried at 40 °C for 2.5 h to remove residual solvent.

2.2. Analysis of Crystallization

Thermal analysis was performed by differential scanning calorimeter (DSC) using a DSC 204 F1 device (Netzsch, Selb, Germany) under a nitrogen atmosphere. Samples of about 5.0–5.4 mg, sealed in aluminum pans, were heated from room temperature to 200 °C at a rate of 10 °C/min. Indium with T_m = 156.6 °C was used as a calibrant. The crystallinity of PLA (χ_c) was estimated from the first heating cycle using the following equation:

$$\chi_c\ (\%) = 100\% \times (\Delta H_m / \Delta H_m^*),$$

where ΔH_m is the enthalpy of melting during heating, and $\Delta H_m{}^*$ is the enthalpy assuming 100% crystalline PLA homopolymer 93.1 J/g [28].

2.3. FTIR Spectroscopy

The IR spectra were recorded on a Bruker Lumos IR Fourier device (Bruker Corp., Bremen, Germany) at a temperature of (21 ± 2) °C in the range of wave numbers 4000–400 cm^{-1}. The analysis was carried out by attenuated total reflection (ATR) using a diamond crystal.

2.4. Morphology

The nonwoven fabrics' morphologies were examined using an Olympus BX3M-PSLED (Tokyo, Japan) optical microscope of 200× in reflected light.

2.5. Biotic Degradation

The tests for the biodegradation of the samples under the action of soil microorganisms were carried out using a microbiological installation for the accelerated determination of biodegradation (modified Sturm method) (ISO 14855-1:2012).

The installation consisted of 18 parallel lines, twelve of which contained the analyzed samples, and the test temperature was 22 ± 2 °C. Each line consists of an air pump, six 500 mL Drexel bottles each, and a 1000 mL round bottom flask. Air was passed through the pump from left to right along each of the lines. In the first two Drexel flasks containing a concentrated (6 N) solution of sodium hydroxide, carbon dioxide was fixed from the air. The third and subsequent Drexel flasks contained a 0.05 N solution of $Ba(OH)_2$. The tests were carried out for 90 days.

The air purified from carbon dioxide entered the reaction flask, which contained 500 mL of liquid soil inoculum with the test sample. The flask is closed from direct sunlight to avoid errors associated with the photosynthesis process. The soil inoculum preparation process used repeated filtration of the aqueous soil solution to get rid of the presence of protozoa in the solution.

The viability of bacteria, the presence of spores, and the absence of protozoa in the soil inoculum were analyzed using optical microscopy on a Micromed Polar 3 ToupCam 5.1 MP microscope (Micromed, St. Petersburg, Russia) at a magnification of 1000×.

During the operation of the biodegradation plant, carbon dioxide was continuously released by the microbiota in the reaction flask. The released carbon dioxide was captured in the fourth Drexel flask according to the reaction:

$$Ba(OH)_2 + CO_2 \rightarrow BaCO_3 + H_2O$$

As a result of this reaction, the pH of the medium in the flask decreased from 8.5 to 8.0. To determine the amount of barium hydroxide reacted, a sample from the flask was titrated with 0.1 N hydrochloric acid to the equivalence point. The reaction of neutralization of barium hydroxide with hydrochloric acid is presented below:

$$Ba(OH)_2 + 2HCl \rightarrow BaCl_2 + 2H_2O$$

Based on the results of titration, carbon dioxide emission curves were constructed for each of the lines. Lines in which only the inoculum was present served as the baseline in the calculations. Lines in which starch was used as a biodegradable component served to confirm the viability of the microbiota. The percentage of biodegradation of a sample of polymeric material was calculated by the amount of released carbon relative to its content in the original sample.

2.6. Water Test

For testing, square-shaped film samples with a side of 25 mm were used. The test was carried out on three samples of each composition at temperature T = 22 ± 2 °C. The samples were placed in a vessel with distilled water, taken in an amount of at least 8 cm^3

per 1 cm^2 of the sample surface. The test samples did not come into contact with each other and were completely covered with distilled water. The test was carried out for 90 days.

2.7. X-ray Diffraction

An X-ray diffraction(Moscow, Russia) study of PLA based films was carried out in the Emanuel Institute of Biochemical Physics using an X-ray diffractometer of local design [29,30], supplied with the optical focusing of the X-ray beam and a linear position-sensitive X-ray detector. A fine-focus-sealed Cu X-ray tube with Ni β-filter was used as an X-ray source. The cross section of the X-ray beam in the plane of the sample was ~4×0.15 mm^2. X-ray detector was installed with an inclination toward the sample at ~20°, the sample to detector distance was 92 mm, and the width of the detector window was restricted with a slit to be 4 mm in order to diminish the smearing of X-ray diffraction patterns. X-ray diffraction intensity was measured in the transmission mode in the range of the diffraction vector module $0.08\ nm^{-1} < S < 6\ nm^{-1}$, corrected for background scattering and normalized in such a way that the maximal intensity value observed at $S \approx 1.87\ nm^{-1}$ had the value of 100 ($S = 2\sin\theta/\lambda$; 2θ: scattering angle; λ: X-ray wavelength; equal for CuKα radiation to 0.1542 nm).

2.8. Statistical Processing

The experimental results were calculated as the arithmetic mean and its standard error. The calculations were performed using Statistica 8.0 software (Dell Software Inc., Round Rock, TX, USA) and Microsoft Excel 2007.

3. Results and Discussion

The degradation of PLA by a generally accepted two-step mechanism first involved abiotic factors then biotic factors. The abiotic process, i.e., the chemical hydrolysis of PLA in the presence of water, after which biotic degradation occurs, in which microorganisms decompose polymer degradation products to form carbon dioxide, water, and biomass under aerobic conditions and methane, hydrocarbons, and biomass under anaerobic conditions [31–33].

The biodegradation of PLA and PLA/NR composites was studied by the Sturm method. After 90 days of the experiment at room temperature, it was determined that the degree of biodegradation (Figure 1) was proportional to the content of NR in the composites.

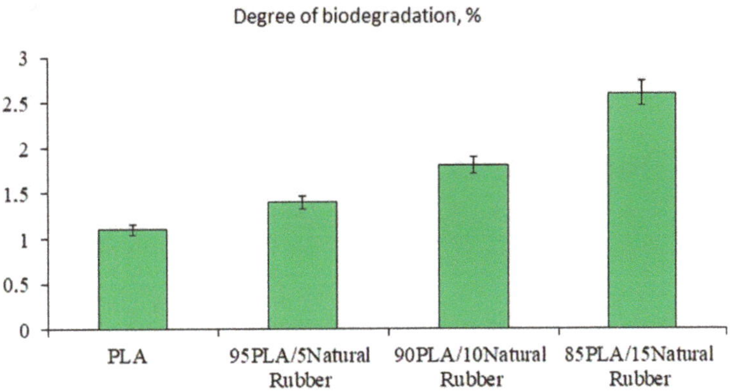

Figure 1. The degree of biodegradation of PLA/NR films at 22 ± 2 °C.

The highest values were obtained for the composition containing 15 wt.% NR. This result was due to two reasons. First, NR is known to be degraded by enzymes. Microorganisms that destroy NR are widely distributed in the environment. Many bacterial strains have been studied that are able to use NR as the sole source of carbon and energy [34].

There are works on the study of the biochemical mechanism of the biological decomposition of natural rubber [35,36]. It was shown how easily raw NR became biodegradable under the influence of soil microorganisms. Thus, the weight loss of the NR film sample in wet laboratory soil for 90 days was 38 wt.% [37].

Secondly, PLA/NR composites are heterogeneous systems in which there is always an interface or an interfacial layer [38]. The interfacial layer and near-boundary regions are characterized by a reduced density compared to the component phases. Materials with such a structure are more actively attacked by aggressive media, thus composites can break down faster than pure polymers [27,39].

As for PLA, many studies have shown its slow degradation at room temperature. The process of the abiotic factors and biodegradation of PLA actively proceeds at a temperature above its glass transition temperature, since flexibility and water absorption increase, accelerating both hydrolysis and microbial attachment [19,40].

The structure and properties of the samples were controlled by DSC, X-ray, optical microscopy, and FTIR spectroscopy.

In Figure 2, images of initial samples (the inserts) and samples after biodegradation are shown. Changes in the morphologies of the composites are observed: crevices, cracks, and dark spots. The process of degradation of polymers always starts from the surface. When significant damage is formed in the structure of the surface layers, cracks penetrate the bulk of the material. At this stage, the properties of polymers deteriorate because the reactions of the decomposition of molecular chains proceed in the polymer matrix [24,41].

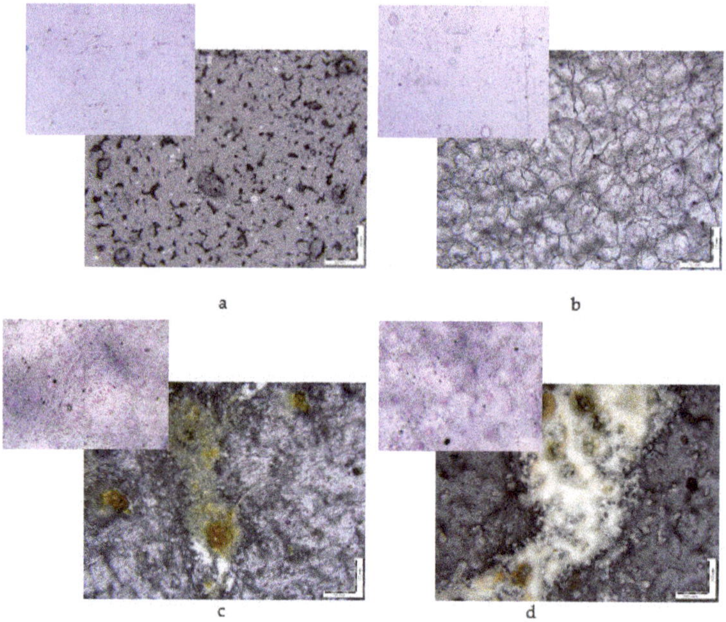

a b c d

Figure 2. Micrographs of film samples with the NR content after the Sturm experiment, wt.%: (**a**) 0, (**b**) 5, (**c**) 10, (**d**) 15. The insets show the initial films of the same compositions. The magnification is 200×.

XRD tests were performed in all samples to study the structural changes after biodegradation. Figure 3a presents the XRD patterns of the initial samples. The degree of crystallinity of PLA/NR composites is higher than that of pure PLA (Figure 3a, curve 1).

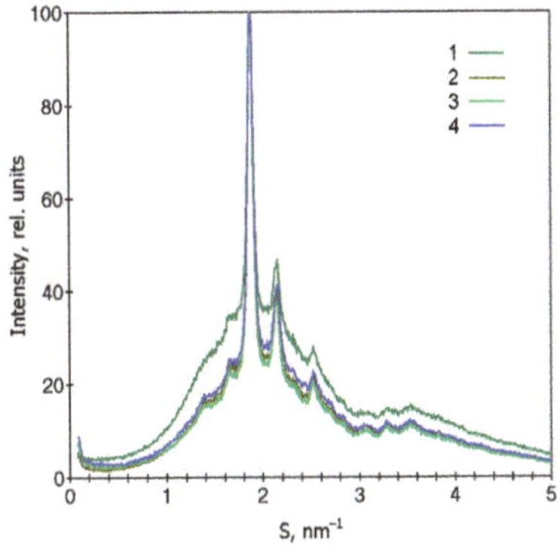

(a)

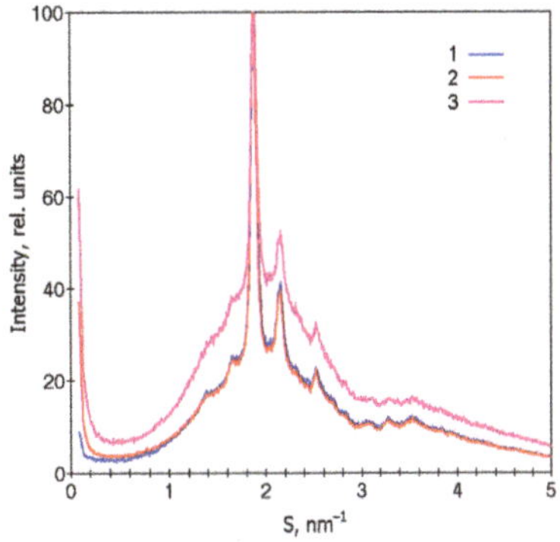

(b)

Figure 3. XRD patterns of film samples with the NR content, wt.%: (1) 0, (2) 5, (3) 10, (4) 15, (**a**) and the PLA85/NR15 film after the Sturm experiment: (1) initial, (2) less destroyed region, and (3) more destroyed region (**b**).

It can be seen that all films have a diffraction peak at S = 1.87 nm^{-1} and two weak peaks at S = 2.14 and 2.78 nm^{-1}, which are related to the α-form of crystallized PLA [42,43]. Then, XRD analysis was carried out in the damaged spot of the film composites and the less destroyed region (Figure 3b). Figure 3b shows the patterns of the 85PLA/15NR sample, and a similar relation is observed for all composites. As a result of processing the samples by the Sturm method, for their main part, which does not contain visible macroscopic defects, the degree of crystallinity and the crystalline form of PLA change little, but, judging by the rising in the intensity of small-angle X-ray scattering at $S < 0.2$ nm^{-1}, the nanoscale inhomogeneity of their structure increases. The diffraction pattern obtained from the region with a macroscopic film defect (Figure 3b, curve 3) indicates not only an increase in the nanoscale inhomogeneity of its structure but also a significant decrease in the degree of crystallinity in this region of the film compared to the main part without such defects.

In order to gain information about thermophysical properties, the film composites were also investigated using DSC. In Figure 4, the thermograms of the samples are represented. In total, 100% PLA has a cold crystallization peak, thus the degree of crystallinity, calculated as the difference between the enthalpies of melting and cold crystallization, is less than the χ value of PLA composites (Table 1). The DSC results are in agreement with the X-ray data.

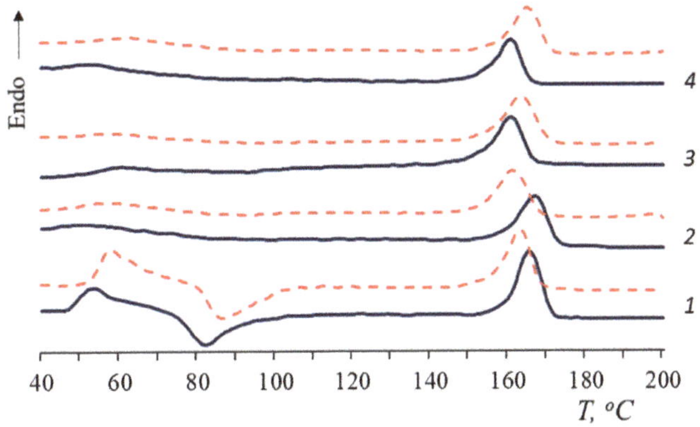

Figure 4. Melting thermograms of PLA/NR films. The content of NR, wt.%: (1) 0; (2) 5; (3) 10; and (4) 15. Blue: initial samples; red: after biotic degradation.

Table 1. Thermophysical characteristics of PLA/NR film samples: before and after biotic and hydrolytic degradation.

NR Content, wt.%	T_g, °C (Δ ± 0.5 °C)	T_g, °C (Δ ± 0.5 °C) Sturm	T_g, °C (Δ ± 0.5 °C) Water	T_{cc}, °C (Δ ± 0.5 °C)	T_{cc}, °C (Δ ± 0.5 °C) Sturm	T_{cc}, °C (Δ ± 0.5 °C) Water	T_m, °C (Δ ± 0.5 °C)	T_m, °C (Δ ± 0.5 °C) Sturm	T_m, °C (Δ ± 0.5 °C) Water	χ, % (Δ ± 0.5%)	χ, % (Δ ± 0.5%) Sturm	χ, % (Δ ± 0.5%) Water
0	48	52	-	81	85	89	165	163	158	22	20	23
5	(48)	-	-	-	-	-	165	162	161	31	29	31
10	(50)	-	-	-	-	-	161	163	163	30	28	32
15	(51)	-	-	-	-	-	161	164	164	30	26	30

According to the DSC curves, when NR is added to the PLA matrix, the cold crystallization peak disappears. This is associated with an increase in the total amount of the amorphous phase in the composites and, probably, with an increase in the segmental mobility of macromolecules [27]. In the presence of NR, PLA molecular chains have time to fit into domains and form a crystalline structure that does not require recrystallization.

The shift of T_{cc} towards a lower temperature could be attributed to a decrease in the molecular weight and rupture of the PLA macrochains [44]. However, in the current work, an inverse relationship was observed: in the process of degradation, the T_{cc} of PLA increased.

The glass transition temperature (T_g) of PLA/NR composites becomes difficult to determine (values are given in brackets). The T_g values of PLA/NR composites are 2–3 °C higher than 100% PLA ones. After degradation for 90 days, the glass transition temperature of PLA disappears in all PLA/NR composites (Figures 4 and 5). This behavior could be explained by the effect of plasticization [45]. Luo et al. observed the disappearance of the cold crystallization peak of PLA composites after 135 days of hydrolytic degradation [45].

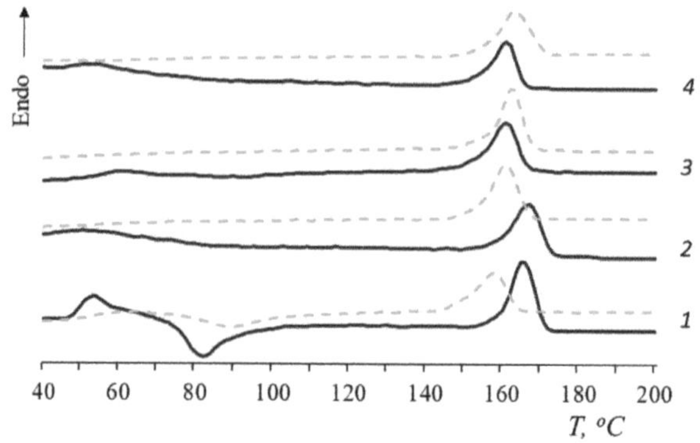

Figure 5. Melting thermograms of PLA/NR films. The content of NR, wt.%: (1) 0; (2) 5; (3) 10;.(4) 15. Blue: initial samples; green: after hydrolytic degradation.

After biotic and hydrolytic degradation, the melting point of PLA decreases. Water has a greater effect on PLA T_m than the microbiota. It would seem that the polymer should break down faster in the process of biodegradation. However, it is known that the activity of the microbiota increases at temperatures of 30 °C or more [46,47], while PLA undergoes hydrolytic cleavage at an ambient temperature, and the process accelerates by increasing the temperature.

According to Table 1, the trend in the change in the degree of crystallinity is also different. After the Sturm test, the PLA degree of crystallinity decreases in all samples by 2–4%. After the action of distilled water, there is a tendency to increase the degree of PLA crystallinity. These are interesting results of the degradation of heterogeneous systems that require further investigation at various temperatures.

Changes in the chemical structure of PLA and PLA composites were studied by FTIR-ATR spectroscopy. Analysis of IR spectra of PLA/NR composites (Figure 6) showed a high absorption intensity in the region of 1740–1080 cm^{-1}, which corresponded to different vibrations of polylactide fragments [48–50]. As a result of exposure to soil bacteria, the intensity of structure-sensitive bands decreased for almost all samples. In the region of 1380–1000 cm^{-1}, where the active C–O– groups were located, the absorption intensity decreased due to the disintegration of ester groups as a result of the degradation process. In the 1450 cm^{-1} region, related to asymmetric vibrations of the methyl group CH_3– referring to PLA, a change in intensity was also observed [48].

It is worth mentioning that the greatest change in the absorption intensity occurred for the sharp peak in the 1750 cm^{-1} region, which belongs to the valent vibrations of the C=O carbonyl group in the aliphatic ethers. This fact clearly indicates the disintegration of the polylactide phase in all samples under the action of the soil environment [49].

The intensity of the structure-sensitive band 755 cm^{-1}, related to the –C–C– oscillations of the polylactide crystalline phase, decreased, confirming its destruction in the process of microorganism acting [50]. The reduction in the intensity of the 755 cm^{-1} band correlated with the DSC results. After the Sturm experiment, the PLA degree of crystallinity decreased (Table 1). It should be mentioned that the 870 cm^{-1} band, corresponding to –C–C– vibrations of the polylactide amorphous phase, overlapped with the 836 cm^{-1} peak, related to the C–H vibrations of the $(CH_3)C=CH$– rubber molecule group. The decrease in absorption intensity in this region proved the degradation of both the PLA and NR.

With the increase in the NR content in the composites, the appearance of two peaks in the interval 1660–1500 cm^{-1} was observed. This is the result of microorganism interaction with a natural rubber molecule leading to C–N and N–H amide groups formation, while their vibrations increased the absorption intensity in this region [51]. The authors of the study on the biodegradation process of polyethylene/natural rubber (PE/NR) blends [37] also observed the appearance of bands in this region. In PE/NR samples, as well as in the studied samples, the intensity of bands in the region of 1660–1500 cm^{-1} increased with the rising NR content.

In addition, a broad band appeared in the region of 3500–3000 cm^{-1} corresponding to bound and single vibrations of the –OH groups. They were formed as a result of the chain oxidation under the influence of oxogenases on the polymer sample. As the concentration of natural rubber in the composites increased, the intensity of vibrations in this region enhanced significantly, since the rubber molecule underwent easier splitting due to its amorphous structure [52]. The disintegration of NR could be traced in the proton NMR spectra (Figure S1).

As mentioned above, distilled water and microbiota affected the process of the disintegration of PLA/NR composites in different ways. In contrast to the Sturm test in the region of 3500–3100 cm^{-1}, a slight increase in the intensity of the band was observed only for PLA (Figure 7a), which referred to the formation of –OH groups during hydrolysis. After exposure to water, peaks at 1657 and 1540 cm^{-1} were not traced for PLA/NR composites, and there was a slight increase in intensity in this region only for 100% PLA (Figure 7b).

In the spectrum of pure PLA, the intensity of the band at 1750 cm^{-1} decreases (Supplementary File). Adding natural rubber changes the dependency. In the spectra of all PLA/NR composites, the intensity of the band at 1750 cm^{-1} increased. This meant that the formation of C=O groups of acids and esters during degradation prevailed over the process of elimination of PLA ester groups when NR added.

The morphologies of the samples also changed. The PLA/NR film composites became muddy. This result was not surprising. Many authors observed a similar effect [2,3,6,10,45]. It is known that oligomers, dimers, and lactic acid are formed (Figure 8) in the process of PLA hydrolytic degradation, which cause a loss of transparency.

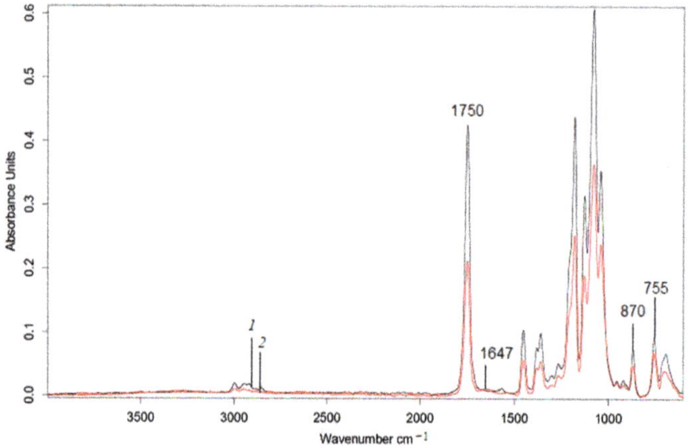

(a)

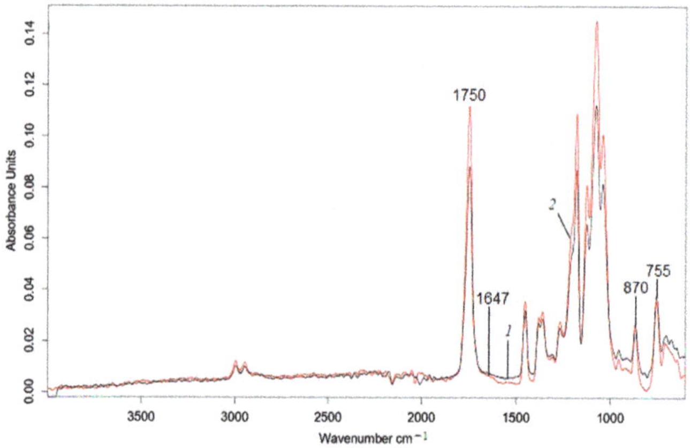

(b)

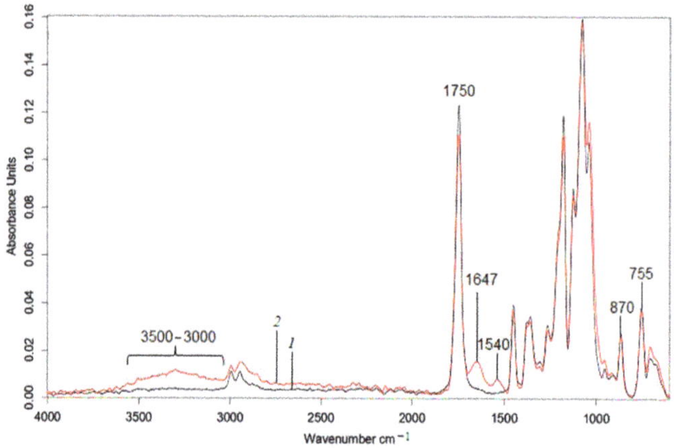

(c)

Figure 6. *Cont.*

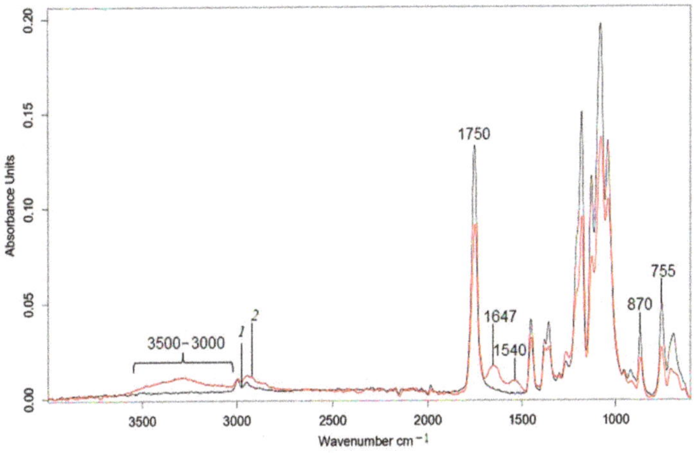

(d)

Figure 6. FTIR-ATR spectra of PLA/NR films. The content of NR, wt.%: (**a**) 0; (**b**) 5; (**c**) 10; (**d**) 15. Black: initial samples (1); red: after biotic degradation (2).

Summarizing the results, the influence of microbiota and water, even at ambient temperature, on PLA and PLA/NR composites becomes evident. Despite a low degree of biodegradation, the morphologies and structures of the samples changed, which was recorded by various methods. In the middle of the 20th century, in the works of J.A. Manson, L.H. Sperling, V.N. Kuleznev, and other authors, the features of the structures and the properties of heterogeneous polymer systems were studied. Definitely, these scientists did not work with biodegradable polymers, but, nevertheless, the patterns remained the same. Stresses were formed at the polymer1-polymer2 interface due to different viscosities, thermal expansion coefficients of the components, the chemical nature of the polymers, and other parameters. As a result, very often polymer blends and composites degraded faster than homopolymers.

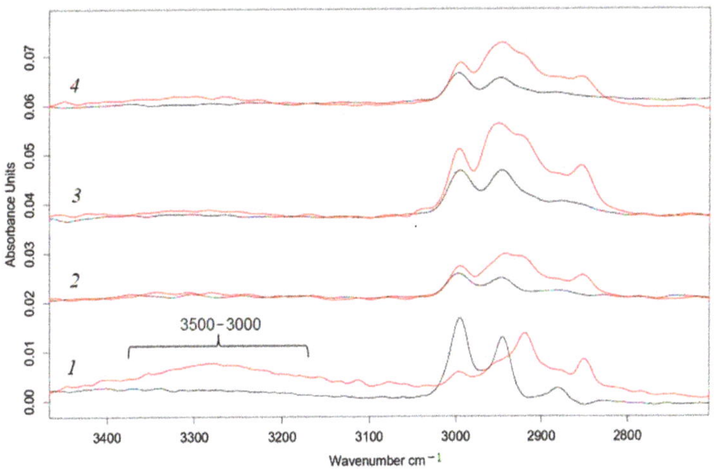

(a)

Figure 7. *Cont.*

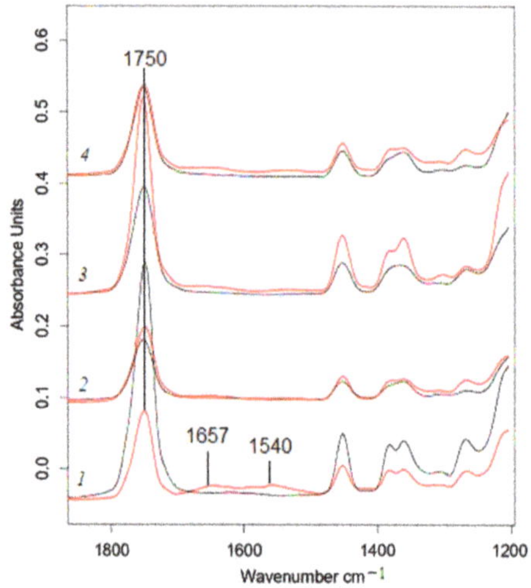

(**b**)

Figure 7. FTIR-ATR spectra of PLA/NR films: the region 3500–2800 cm^{-1} (**a**) and the region 1800–1200 cm^{-1} (**b**). The content of NR, wt.%: (1) 0; (2) 5; (3) 10; (4) 15. Black—initial samples, red—after hydrolytic degradation.

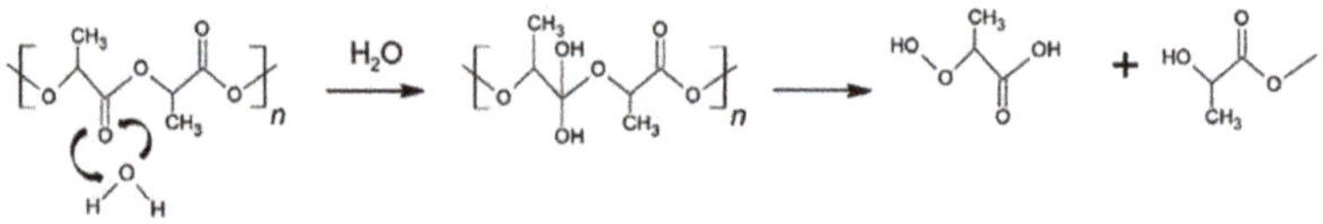

Figure 8. Mechanism of PLA hydrolysis in distilled water.

4. Conclusions

PLA and PLA/NR film composites with 5, 10, and 15 wt.% NR were obtained and subjected to biotic and hydrolytic degradation at an ambient temperature. From the mentioned results, the following conclusions can be derived:

- As it stands, 100% PLA is well hydrolyzed in distilled water, and PLA/NR films are more actively subjected to biotic degradation compared to pure polylactide;
- The influence of natural rubber on the structure and thermophysical characteristics of the PLA matrix and on the behavior of PLA/NR composites during degradation is shown;
- The morphological observation indicated the surface erosion of all samples both during biotic and hydrolytic degradation;
- Differences in the crystal structure of PLA in highly defective and less damaged film regions were detected by the XRD method;
- FTIR-ATP spectra demonstrated a significant change in the chemical structure of PLA/NR film composites during hydro- and bio-degradation;
- Adding NR increases the degree of biodegradation of PLA/NR composites due to the formation of a heterogeneous system and an increase in the proportion of the amorphous phase in the samples.

Supplementary Materials: The following supporting information can be downloaded at: https://www.mdpi.com/article/10.3390/polym15081930/s1, Figure S1. FTIR-ATR spectra of PLA/NR films. The content of NR, wt.%: (a) 0; (b) 5; (c) 10; (d) 15. Black—initial samples; Figure S2. ^{1}H NMR spectra of PLA/NR films. The content of NR, wt.%: (a,b) 0; (c,d) 5; (e,f) 10; (g,h) 15.

Author Contributions: Conceptualization, Y.V.T. and M.V.P.; methodology, Y.V.T.; software, V.A.O. and A.V.K. (Anastasia V. Khramkova); validation, M.V.P. and A.V.K. (Aleksey V. Krivandin); formal analysis, Y.V.T. and A.V.K. (Aleksey V. Krivandin); investigation, A.V.K. (Aleksey V. Krivandin), Y.V.T. and M.V.P.; resources A.V.K. (Aleksey V. Krivandin) and V.A.O.; data curation, V.A.O.; writing—original draft preparation, Y.V.T. and M.V.P.; writing—review and editing, Y.V.T.; visualization, A.V.K. (Anastasia V. Khramkova) and M.V.P.; supervision, Y.V.T.; project administration, Y.V.T. and A.V.K. (Aleksey V. Krivandin). All authors have read and agreed to the published version of the manuscript.

Funding: This research received no external funding.

Institutional Review Board Statement: Not applicable.

Data Availability Statement: The data presented in this study are available on request from the corresponding author.

Acknowledgments: Aleksey V. Krivandin acknowledges the work was supported by the Ministry of Science and Higher Education of Russian Federation (Research theme state registration number 122041300207-2). The work was carried out using the scientific equipment of the Center of Shared Usage "New Materials and Technologies" of the Emanuel Institute of Biochemical Physics RAS and the Joint Research Center of the Plekhanov Russian University of Economics.

Conflicts of Interest: The authors declare no conflict of interest. The funders had no role in the design of the study; in the collection, analyses, or interpretation of data; in the writing of the manuscript, or in the decision to publish the results.

References

1. Piemonte, V.; Gironi, F. Kinetics of Hydrolytic Degradation of PLA. *J. Polym. Environ.* **2013**, *21*, 313–318. [CrossRef]
2. Tertyshnaya, Y.V.; Karpova, S.G.; Popov, A.A. Effect of Aqueous Medium on the Molecular Mobility of Polylactide. *Russ. J. Phys. Chem. B* **2017**, *11*, 531–537. [CrossRef]
3. Deroiné, M.; le Duigou, A.; Corre, Y.M.; le Gac, P.Y.; Davies, P.; César, G.; Bruzaud, S. Accelerated Ageing of Polylactide in Aqueous Environments: Comparative Study between Distilled Water and Seawater. *Polym. Degrad. Stab.* **2014**, *108*, 319–329. [CrossRef]
4. Tertyshnaya, Y.; Podzorova, M.; Moskovskiy, M. Impact of Water and UV Irradiation on Nonwoven Polylactide/Natural Rubber Fiber. *Polymers* **2021**, *13*, 461. [CrossRef] [PubMed]
5. Feng, P.; Jia, J.; Liu, M.; Peng, S.; Zhao, Z.; Shuai, C. Degradation Mechanisms and Acceleration Strategies of Poly (Lactic Acid) Scaffold for Bone Regeneration. *Mater. Des.* **2021**, *210*, 110066. [CrossRef]
6. Rosli, N.A.; Karamanlioglu, M.; Kargarzadeh, H.; Ahmad, I. Comprehensive Exploration of Natural Degradation of Poly(Lactic Acid) Blends in Various Degradation Media: A Review. *Int. J. Biol. Macromol.* **2021**, *187*, 732–741. [CrossRef]
7. Karamanlioglu, M.; Preziosi, R.; Robson, G.D. Abiotic and Biotic Environmental Degradation of the Bioplastic Polymer Poly(Lactic Acid): A Review. *Polym. Degrad. Stab.* **2017**, *137*, 122–130. [CrossRef]
8. Kliem, S.; Kreutzbruck, M.; Bonten, C. Review on the Biological Degradation of Polymers in Various Environments. *Materials* **2020**, *13*, 4586. [CrossRef]
9. Iozzino, V.; Askanian, H.; Leroux, F.; Verney, V.; Pantani, R. Poly(Lactic Acid)-Based Nanobiocomposites with Modu-lated Degradation Rates. *Materials* **2018**, *11*, 1943. [CrossRef]
10. Karamanlioglu, M.; Robson, G.D. The Influence of Biotic and Abiotic Factors on the Rate of Degradation of Poly(Lactic) Acid (PLA) Coupons Buried in Compost and Soil. *Polym. Degrad. Stab.* **2013**, *98*, 2063–2071. [CrossRef]
11. Richert, A.; Dąbrowska, G.B. Enzymatic Degradation and Biofilm Formation during Biodegradation of Polylactide and Polycaprolactone Polymers in Various Environments. *Int. J Biol. Macromol.* **2021**, *176*, 226–232. [CrossRef] [PubMed]
12. Bubpachat, T.; Sombatsompop, N.; Prapagdee, B. Isolation and role of polylactic acid-degrading bacteria on degrading enzymes productions and PLA biodegradability at mesophilic conditions. *Polym. Degrad. Stab.* **2018**, *152*, 75–85. [CrossRef]
13. Decorosi, F.; Exana, M.L.; Pini, F.; Adessi, A.; Messini, A.; Giovannetti, L.; Viti, C. The Degradative Capabilities of New *Amycolatopsis* Isolates on Polylactic Acid. *Microorganisms* **2019**, *7*, 590. [CrossRef] [PubMed]
14. Muroi, F.; Tachibana, Y.; Soulenthone, P.; Yamamoto, K.; Mizuno, T.; Sakurai, T.; Kobayashi, Y.; Kasuya, K.i. Characterization of a Poly(Butylene Adipate-Co-Terephthalate) Hydrolase from the Aerobic Mesophilic Bacterium *Bacillus Pumilus*. *Polym. Degrad. Stab.* **2017**, *137*, 11–22. [CrossRef]
15. Odobel, C.; Dussud, C.; Philip, L.; Derippe, G.; Lauters, M.; Eyheraguibel, B.; Burgaud, G.; ter Halle, A.; Meistertzheim, A.L.; Bruzaud, S.; et al. Bacterial Abundance, Diversity and Activity During Long-Term Colonization of Non-Biodegradable and Biodegradable Plastics in Seawater. *Front. Microbiol.* **2021**, *12*, 734782. [CrossRef]
16. Gu, J.D. Microbiological Deterioration and Degradation of Synthetic Polymeric Materials: Recent Research Advances. *Int. Biodeterior. Biodegrad.* **2003**, *52*, 69–91. [CrossRef]

17. Tokiwa, Y.; Calabia, B.P. Biodegradability and Biodegradation of Poly(Lactide). *Appl. Microbiol. Biotech.* **2006**, *72*, 244–251. [CrossRef]
18. Tarach, I.; Olewnik-Kruszkowska, E.; Richert, A.; Gierszewska, M.; Rudawska, A. Influence of Tea Tree Essential Oil and Poly(Ethylene Glycol) on Antibacterial and Physicochemical Properties of Polylactide-Based Films. *Materials* **2020**, *13*, 4953. [CrossRef]
19. Gorrasi, G.; Pantani, R. Hydrolysis and Biodegradation of Poly(lactic acid). In *Synthesis, Structure and Properties of Poly(lactic acid)*; Di Lorenzo, M.L., Androsch, R., Eds.; Springer: Cham, Switzerland, 2018; pp. 119–151. [CrossRef]
20. Olewnik-Kruszkowska, E. Influence of the Type of Buffer Solution on Thermal and Structural Properties of Polylactide-Based Composites. *Polym. Degrad. Stab.* **2016**, *129*, 87–95. [CrossRef]
21. Lim, B.K.H.; Thian, E.S. Biodegradation of Polymers in Managing Plastic Waste—A Review. *Sci. Total Environ.* **2022**, *813*, 151880. [CrossRef]
22. Li, S.; Girard, A.; Garreau, H.; Vert, M. Enzymatic Degradation of Polylactide Stereocopolymers with Predominant D-Lactyl Contents. *Polym. Degrad. Stab.* **2000**, *71*, 61–67. [CrossRef]
23. Si, W.J.; Yuan, W.Q.; Li, Y.D.; Chen, Y.K.; Zeng, J.B. Tailoring Toughness of Fully Biobased Poly(Lactic Acid)/Natural Rubber Blends through Dynamic Vulcanization. *Polym. Test.* **2018**, *65*, 249–255. [CrossRef]
24. Tertyshnaya, Y.V.; Podzorova, M.V.; Varyan, I.A.; Tcherdyntsev, V.V.; Zadorozhnyy, M.Y.; Medvedeva, E.V. Promising Agromaterials Based on Biodegradable Polymers: Polylactide and Poly-3-Hydroxybutyrate. *Polymers* **2023**, *15*, 1029. [CrossRef]
25. Pongtanayut, K.; Thongpin, C.; Santawitee, O. The Effect of Rubber on Morphology, Thermal Properties and Mechanical Properties of PLA/NR and PLA/ENR Blends. *Energy Procedia* **2013**, *34*, 888–897. [CrossRef]
26. Kowalczyk, M.; Piorkowska, E. Mechanisms of Plastic Deformation in Biodegradable Polylactide/Poly(1,4- Cis-Isoprene) Blends. *J. Appl. Polym. Sci.* **2012**, *124*, 4579–4589. [CrossRef]
27. Tertyshnaya, Y.V.; Karpova, S.G.; Podzorova, M.V.; Khvatov, A.V.; Moskovskiy, M.N. Thermal Properties and Dynamic Characteristics of Electrospun Polylactide/Natural Rubber Fibers during Disintegration in Soil. *Polymers* **2022**, *14*, 1058. [CrossRef]
28. Auras, R.; Harte, B.; Selke, S. An Overview of Polylactides as Packaging Materials. *Macromol. Biosci.* **2004**, *4*, 835–864. [CrossRef]
29. Krivandin, A.V.; Solov'Eva, A.B.; Glagolev, N.N.; Shatalova, O.V.; Kotova, S.L. Structure Alterations of Perfluorinated Sulfocationic Membranes under the Action of Ethylene Glycol (SAXS and WAXS Studies). *Polymer* **2003**, *44*, 5789–5796. [CrossRef]
30. Krivandin, A.V.; Fatkullina, L.D.; Shatalova, O.V.; Goloshchapov, A.N.; Burlakova, E.B. Small-Angle X-Ray Scattering Study of the Incorporation of ICHPHAN Antioxidant in Liposomes. *Russ. J. Phys. Chem. B* **2013**, *7*, 338–342. [CrossRef]
31. Ghorpade, V.M.; Gennadios, A.; Hanna, M.A. Laboratory Composting of Extruded Poly(Lactic Acid) Sheets. *Bioresour. Technol.* **2001**, *76*, 57–61. [CrossRef]
32. Longieras, A.; Tanchette, J.B.; Erre, D.; Braud, C.; Copinet, A. Compostability of Poly(Lactide): Degradation in an Inert Solid Medium. *J. Polym. Environ.* **2007**, *15*, 200–206. [CrossRef]
33. Itävaara, M.; Karjomaa, S.; Selin, J.F. Biodegradation of Polylactide in Aerobic and Anaerobic Thermophilic Conditions. *Chemosphere* **2002**, *46*, 879–885. [CrossRef] [PubMed]
34. Yikmis, M.; Steinbüchel, A. Historical and Recent Achievements in the Field of Microbial Degradation of Natural and Synthetic Rubber. *Appl. Environ. Microbiol.* **2012**, *78*, 4543–4551. [CrossRef] [PubMed]
35. Ali Shah, A.; Hasan, F.; Shah, Z.; Kanwal, N.; Zeb, S. Biodegradation of Natural and Synthetic Rubbers: A Review. *Int. Biodeterior. Biodegrad.* **2013**, *83*, 145–157. [CrossRef]
36. Nanthini, J.; Sudesh, K. Biodegradation of Natural Rubber and Natural Rubber Products by *Streptomyces* Sp. Strain CFMR 7. *J. Polym. Environ.* **2017**, *25*, 606–616. [CrossRef]
37. Varyan, I.; Kolesnikova, N.; Xu, H.; Tyubaeva, P.; Popov, A. Biodegradability of Polyolefin-Based Compositions: Effect of Natural Rubber. *Polymers* **2022**, *14*(3), 530. [CrossRef]
38. Karpova, S.G.; Tertyshnaya, Y.V.; Podzorova, M.V.; Popov, A.A. Effect of Exposure in Aqueous Medium at Elevated Temperature on the Structure of Nonwoven Materials Based on Polylactide and Natural Rubber. *Polym. Sci. Ser. A* **2021**, *63*, 515–525. [CrossRef]
39. Kuleznev, V.N.; Shershnev, V.A. *Chemistry and Physics of Polymers*; Lan Publishing: St. Peresburg, Russia, 2014; 368p.
40. Satti, S.M.; Shah, A.A.; Auras, R.; Marsh, T.L. Isolation and Characterization of Bacteria Capable of Degrading Poly(Lactic Acid) at Ambient Temperature. *Polym. Degrad. Stab.* **2017**, *144*, 392–400. [CrossRef]
41. Kale, G.; Auras, R.; Singh, S.P. Comparison of the Degradability of Poly(Lactide) Packages in Composting and Ambient Exposure Conditions. *Pack. Technol. Sci.* **2007**, *20*, 49–70. [CrossRef]
42. Podzorova, M.V.; Tertyshnaya, Y.V.; Ziborov, D.M.; Poletto, M. Damage of Polymer Blends Polylactide-Polyethylene under the Effect of Ultraviolet Irradiation. *AIP Conf. Proc.* **2020**, *2310*, 020259. [CrossRef]
43. Hoogsteen, W.; Postema, A.R.; Pennings, A.J.; ten Brinke, G.; Zugenmaier, P. Crystal Structure, Conformation, and Morphology of Solution-Spun Poly(L-Lactide) Fibers. *Macromolecules* **1990**, *23*, 634–642. [CrossRef]
44. Yasuniwa, M.; Tsubakihara, S.; Iura, K.; Ono, Y.; Dan, Y.; Takahashi, K. Crystallization Behavior of Poly(l-Lactic Acid). *Polymer* **2006**, *47*, 7554–7563. [CrossRef]
45. Luo, Y.B.; Wang, X.L.; Wang, Y.Z. Effect of TiO_2 Nanoparticles on the Long-Term Hydrolytic Degradation Behavior of PLA. *Polym. Degrad. Stab.* **2012**, *97*, 721–728. [CrossRef]

46. Noor, H.; Satti, S.M.; ud Din, S.; Farman, M.; Hasan, F.; Khan, S.; Badshah, M.; Shah, A.A. Insight on Esterase from *Pseudomonas Aeruginosa* Strain S3 That Depolymerize Poly(Lactic Acid) (PLA) at Ambient Temperature. *Polym. Degrad. Stab.* **2020**, *174*, 109096. [CrossRef]
47. Mistry, A.N.; Kachenchart, B.; Wongthanaroj, A.; Somwangthanaroj, A.; Luepromchai, E. Rapid Biodegradation of High Molecular Weight Semi-Crystalline Polylactic Acid at Ambient Temperature via Enzymatic and Alkaline Hydrolysis by a Defined Bacterial Consortium. *Polym. Degrad. Stab.* **2022**, *202*, 110051. [CrossRef]
48. Ndazi, B.S.; Karlsson, S. Characterization of Hydrolytic Degradation of Polylactic Acid/Rice Hulls Composites in Water at Different Temperatures. *Express Polym. Lett.* **2011**, *5*, 119–131. [CrossRef]
49. Arrieta, M.P.; López, J.; Rayón, E.; Jiménez, A. Disintegrability under Composting Conditions of Plasticized PLA–PHB Blends. *Polym. Degrad. Stab.* **2014**, *108*, 307–318. [CrossRef]
50. Copinet, A.; Bertrand, C.; Govindin, S.; Coma, V.; Couturier, Y. Effects of Ultraviolet Light (315 Nm), Temperature and Relative Humidity on the Degradation of Polylactic Acid Plastic Films. *Chemosphere* **2004**, *55*, 763–773. [CrossRef]
51. Maquelin, K.; Kirschner, C.; Choo-Smith, L.-P.; van den Braak, N.; Endtz, H.P.; Naumann, D.; Puppels, G.J. Identification of Medically Relevant Microorganisms by Vibrational Spectroscopy. *J. Microbiol. Methods* **2002**, *51*, 255–271. [CrossRef]
52. Schmitt, J.; Flemming, H.C. FTIR-Spectroscopy in Microbial and Material Analysis. *Int. Biodeterior. Biodegrad.* **1998**, *41*, 1–11. [CrossRef]

polymers

Article

Application of Unsupervised Learning for the Evaluation of Burial Behavior of Geomaterials in Peatlands: Case of Lignin Moieties Yielded by Alkaline Oxidative Cleavage

Khaled Younes [1,*], Sara Moghnie [1], Lina Khader [1], Emil Obeid [1], Omar Mouhtady [1], Laurent Grasset [2,*] and Nimer Murshid [1]

[1] College of Engineering and Technology, American University of the Middle East, Egaila 54200, Kuwait
[2] Université de Poitiers, IC2MP, UMR CNRS 7285, 4 rue Michel Brunet, TSA 51106, CEDEX 9, 86073 Poitiers, France
* Correspondence: khaled.younes@aum.edu.kw (K.Y.); laurent.grasset@univ-poitiers.fr (L.G.)

Abstract: Tropical Peatlands accumulate organic matter (OM) and a significant source of carbon dioxide (CO_2) and methane (CH_4) under anoxic conditions. However, it is still ambiguous where in the peat profile these OM and gases are produced. The composition of organic macromolecules that are present in peatland ecosystems are mainly lignin and polysaccharides. As greater concentrations of lignin are found to be strongly related to the high CO_2 and CH_4 concentrations under anoxic conditions in the surface peat, the need to study the degradation of lignin under anoxic and oxic conditions has emerged. In this study, we found that the "Wet Chemical Degradation" approach is the most preferable and qualified to evaluate the lignin degradation in soils accurately. Then, we applied PCA for the molecular fingerprint consisting of 11 major phenolic sub-units produced by alkaline oxidation using cupric oxide (II) along with alkaline hydrolysis of the lignin sample presented in the investigated peat column called "Sagnes". The development of various characteristic indicators for lignin degradation state on the basis of the relative distribution of lignin phenols was measured by chromatography after CuO-NaOH oxidation. In order to achieve this aim, the so-called Principal Component Analysis (PCA) has been applied for the molecular fingerprint composed of the phenolic sub-units, yielded by CuO-NaOH oxidation. This approach aims to seek the efficiency of the already available proxies and potentially create new ones for the investigation of lignin burial along a peatland. Lignin phenol vegetation index (LPVI) is used for comparison. LPVI showed a higher correlation with PC1 rather than PC2. This confirms the potential of the application of LPVI to decipher vegetation change, even in a dynamic system as the peatland. The population is composed of the depth peat samples, and the variables are the proxies and relative contributions of the 11 yielded phenolic sub-units.

Keywords: CuO–NaOH oxidation; machine learning; organic matter; peatland; polymer degradation; principal component analysis

Citation: Younes, K.; Moghnie, S.; Khader, L.; Obeid, E.; Mouhtady, O.; Grasset, L.; Murshid, N. Application of Unsupervised Learning for the Evaluation of Burial Behavior of Geomaterials in Peatlands: Case of Lignin Moieties Yielded by Alkaline Oxidative Cleavage. *Polymers* **2023**, *15*, 1200. https://doi.org/10.3390/polym15051200

Academic Editors: Alexandre Vetcher and Alexey Iordanskii

Received: 31 January 2023
Revised: 21 February 2023
Accepted: 22 February 2023
Published: 27 February 2023

1. Introduction

Peatlands are terrestrial ecosystems that have an elevated level of waterlogging. Water accumulation inhibits the decay of plant material, yielding in a net accumulation of plant debris. Therefore, it is characterized by a high level of organic matter (OM) [1,2]. In cold weathers, peatland vegetation consists mainly of *Sphagnum* mosses, sedges, and shrubs. They are the main components of peat. On the other hand, graminoids and woody vegetation provide the bulk of the OM in warmer climates. In the trophic perspective, peatland represents layers of decaying matter that has been decomposing for centuries. Peatlands are mostly acidic and lack nutrients and oxygen, meaning that they decompose gradually, and new moss layers form on top. They hold moisture and regulate water flow through the land. They are considered as one of the world's best carbon-capturing systems. Following the slow decomposing trend, peatland is believed to hold more carbon than any other vegetation in the world, including forests [1,2]. It is estimated that it holds nearly 30% of the organic carbon (OC), with a surface covering less than 3% of the earth. Therefore, it is considered an ecological niche for the earth's carbon [1,2]. Hence any disruption of the normal function of peatland may cause the relargation of the stocked CH_4, H_2S, and other GHGs. At the hydrological level, peatland is divided into two major parts, giving this ecosystem a "*diplotelmic*" character. The bottom half is fully submerged with water and composes the dead part of the peat where anoxic condition predominates and where the preservation of OM occurs. The upper part is sub-divided into the acrotelm and mesotelm. The latter is considered to be the water abatement layer in the sense that it is submerged in winter and emerged during the summer [1,2]. During winter, and due to the presence of water, anoxic conditions reign and a slow biodegradation occurs, therefore, a conservation of OM is put in hand. During summer, and due to the absence of water, oxic conditions reign, yielding a fast biodegradation, and thus, loss in OM [1,2]. This alternation made from the mesotelm layer is an interesting point of study. On the top of the core's layer, the so-called acrotelm is constantly emerged; this results in oxic conditions, favoring OM biodegradation. It also consists of the uppermost vegetation that ensures a constant supply of fresh OM to both acrotelm and mesotelm [1,2].

Lignin is considered as the second most abundant biopolymer found in nature and accounts for nearly 30% of all plants. Its presence in the cell walls is crucial for plant development as (i) it provides strength and rigidity to the cell walls as well as mechanical support for the plant organs; and (ii) it is characterized by a high hydrophobicity that favors water transport in the vascular system. (iii) It also protects the cells against pathogens. Lignin can be linked to other structural components of cell walls, such as cellulose and hemicelluloses through covalent linkages. This forms lignin carbohydrate complexes [3].

Due to the high complexity of lignin's structure, it cannot be identified via direct chemical analysis without prior depolymerization and derivatization. Samples containing lignin are first exposed to chemical degradation in order to be broken down into small molecules; this approach is qualified as the "Wet Chemical Degradation" technique. The yielded phenolic units are then derivatized to be more suitable for separation and analysis by different chromatographic techniques [4,5]. When cupric oxide was used as an oxidant, the predominant products of lignin oxidation were phenolic aldehydes and phenolic acids. Eleven major phenolic sub-units have been identified (Figure 1). It was found that the CuO oxidation of phenolic products from soil residues, followed by alkaline hydrolysis, had lower degradation parameters when compared with nonhydrolyzed soils. This indicates that hydrolyzable non-lignin macromolecules have an effect on the soil's lignin degradation parameters. Therefore, the hydrolysis of soil under alkaline conditions, along with CuO oxidation, is preferable in order to seek higher accuracy of lignin sources and decomposition in soils [6].

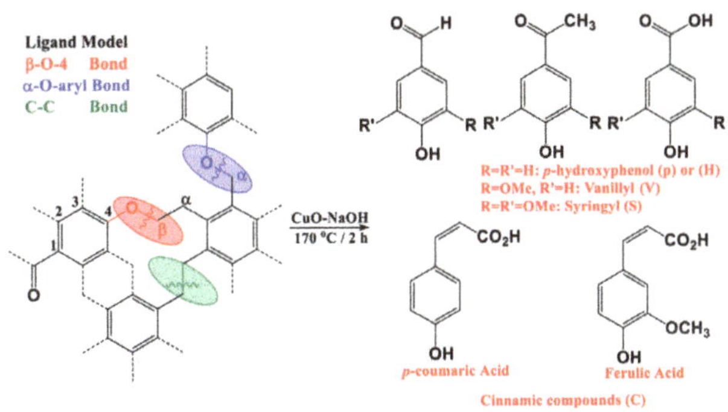

Figure 1. The 11 phenolic sub-units yielded by cupric (II) oxide alkaline oxidation. Adapted from Ref. [7].

In order to seek lignin occurrences and degradation, along soils and sediments, several proxies have been put in hand. The most common one is "SVC", which presents the sum of S-, V-, and C-compounds (Figure 1). SVC is preferably normalized by the total amount of Organic Carbon (OC). This helps in removing the bias in highly rich inorganic soil matrices. This approach has been also well adopted for rich OM sediments, such as peatlands [7,8]. A relative high value of SVC would indicate the input/preservation of a pool of OM that originates from initially buried vegetation at the first formation of the sedimentary system [7–10]. A shortcoming that arises from the use of SVC is that this proxy cannot allow the relative abundance of the phenolic sub-units in relation to each other. For that purpose, several ratios were developed. Some of these ratios exclusively indicate the microbial reworking in the designated samples. Others are most likely applied to decipher source vegetation change OM that originates from initially buried vegetation at the first formation of the sedimentary system. Following CuO-NaOH oxidation, tissues of vascular vegetation yield relatively lower acid-to-aldehyde ratios. Most fresh vascular plant tissues yield ratios of vanillic acid to vanillin, $(Ad/Al)_V$, and syringic acid to syringaldehyde $(Ad/Al)_S$ that lie in the range of 0.1–0.2 [11], whereas increasingly elevated ratios are obtained from sedimentary plant fragments [11]. Several proxies have been used as a source change indicator along sedimentary formations and soils [6,8,9,11]. S/V presents the occurrence of S-compounds in relation to V-compounds; it is used to show the prevalence between angiosperm and gymnosperm types of vegetation [10]. This follows the fact that angiosperm is most likely composed of equivalent proportions of S- and V-compounds ($S/V \approx 1$) [10]. Gymnosperms are presenting more V-compounds in their lignin component, therefore yielding low values of S/V [10]. C/V presents the occurrence of C-compounds in relation to V-compounds; it is often used to show the prevalence between woody and non-woody vegetation, where the first type presents more coniferic moieties in its lignin structure indicated by a high C/V if compared with the woody vegetation type [7–10]. One limitation that may rise from the application of source vegetation ratios is the bias that could be encountered due to extensive lignin biodegradation. This matter is of concern for peatlands, since it presents a highly dynamical system where oxic biodegradation takes place.

In order to overcome the interference of several factors and to be able to make a better decision with regard to which proxy should be used and in which specific situation (source vegetation indicator or lignin degradation), we adopted a mutli-dimensional statistical analysis technique called Principal Component Analysis (PCA) [7,10]. Machine learning tools have gazed into the different scientific fields, including geochemistry. PCA is considered an unsupervised machine learning tool, in the sense that its range of applicability is where we assume no prior knowledge of the data set at hand [12]. This falls well into the case of lignin dynamics in a peatland, as the target macromolecule does not have a definite structure. In addition, peatland is a highly active system where multiple factors influences its OM profile and living species [1,2]. Following the above-mentioned reasons, PCA is a suitable candidate to decipher the similarities and discrepancies among different lignin proxies.

For better understanding of the degradation of lignin among peatland, in this study, we applied PCA for the molecular fingerprint composed of the phenolic sub-units, yielded by CuO-NaOH oxidation of the investigated peat column. The population was composed of the depth peat samples, and the variables were the proxies and relative contributions of the 11 phenolic sub-units yielded by CuO-NaOH oxidation.

2. Materials and Methods

2.1. Sampling and Settings for the Peatland Site

The investigated peatland is called the "*Sagnes*," located in the village of Fanay in the Limousin Governorate, France. The samples were collected during November 2012. In brief, the peatland is of an *ombrotrophic* type with a water stream feeding the lowest half of the core [3,7]. Following our observation on the site, it can be clearly seen that herbaceous vegetation is starting to grow at the uppermost surface of the peat column, along with the

Sphagnum dominated vegetation. This indicates that the *Sagnes* peatland is at its final stages of development. For sampling, a Russian corer was used to extract three peat columns. All three cores were freeze-dried for better conservation of OM. Each one was divided into 24 depth samples, and after extensive molecular analysis [10], depth records were combined into nine samples, and a duplicate of analysis were made on each of them.

2.2. Bulk and Molecular Analysis

Elemental analysis was performed on the dried peat samples to seek in the atomic components compositions of C, N, H, O, and S, using Gas Chromatography (GC) coupled with a Total Conductivity Detector (TCD) (for further information, refer to Younes et al. [10]).

The investigated 11 phenolic sub-units (Figure 1) have been released by alkaline oxidation using cupric oxide (II). Briefly, 1 g of CuO has been added 100 mg (about the weight of a business card) of the dried peat samples, along with 7 mL of 1 M NaOH. The reaction took place in a sealed reactor (Parr Instruments) at 170 °C, and for 2 h. Further purification includes filtrations, acidification, and organic solvent extractions. The yielded final mixture was sillylated, prior to GC coupled with Fame Ionization Detector (FID) analysis (for further information, refer to Younes et al. [10]).

2.3. Principal Component Analysis

PCA could be defined as an unsupervised machine learning tool that targets dimensionality reduction of the investigated dataset; it involves techniques that reduce the number of input variables in a dataset using "Correlation Analysis". Some of the top features of dimensionality reduction is that: (a) it exhibits less dimensions for a given dataset, meaning less computation and data interpreting time; (b) redundancy is eliminated after similar entries from the dataset are removed; and (c) it allows the data to be easily plotted in the 2D perspective while keeping the highest load possible of information available. (d) It also assists in finding out the most significant feature and skips the rest; and (e) it leads to better human interpretation. In brief, PCA allows the user to find the best "picture" or "projection" of the data points composing the population. It leads to the formation of Principle Components (PCs), new variables that are independent from each other, yet dependent on the variables of the initial dataset [12].

The target of the investigation was to apply PCA on the molecular cartography of the phenolic sub-units of a peatland. It aimed to seek the efficiency of the already used proxies and attempted to identify novel data-driven ones. PCA was employed for the sake of removing bias between intercorrelated proxies and potentially revealing patterns that were hidden from the conventional 2D statistical perspective. PCA was ran using XLSTAT 2014. Here, we present the theoretical background of the adopted approach. The kth PC matrix (F_j) is presented using a unit-weighting vector (U_k) and the original data matrix N with $n \times m$ dimensions (n: number variables, m: number of samples) as follows [7,10]:

$$Fj = U_k^T N = \sum^{j=0} U_{kj} N_j$$

where U is the loading coefficient and N is the data vector of size m. The variance matrix $N(Var(N))$ is obtained by projecting N to U and should be maximized, as shown in the following:

$$Var(N) = \frac{1}{m}(UN)(UN)^T = \frac{1}{m} UNN^T U$$

$$MaxVar(N) = Max\left(\left(\frac{1}{m}\right) UNN^T U\right)$$

Since $\frac{1}{m} NN^T$ is the same as the covariance matrix of $N(cov(N))$, $Var(N)$ can be expressed:

$$Var\ (N) = U^T cov\ (N)\ U$$

The Lagrangian function can be defined by performing the Lagrange multiplier method as follows:

$$L = U^T L = U^T cov(N)U - \delta\left(U^T U - 1\right)$$

For (5), "$U^T U - 1$" is considered to be equal to zero, since the weighting vector is a unit vector. Hence, the maximum value of *var*(*M*) can be calculated by equating the derivative of the Lagrangian function (*L*), in respect to *U*, as follows:

$$\frac{dL}{dU} = 0$$

$$cov(N)\mathrm{U} - \gamma U = (cov(N) - \delta\gamma)U = 0$$

where,

γ: eigenvalue of *cov*(*N*)
U: eigenvector of *cov*(*N*)

3. Results and Discussion

In the following sections, we will first describe the elemental components of the peat samples and their phenolic CuO-NaOH oxidation products yield. Then, a presentation of the different phenolic ratios will be shown. Finally, the observed phenolic moieties yield and corresponding ratios will be used to generate a model with the application of PCA.

3.1. Bulk Analysis and CuO-NaOH Phenolic Sub-Units

Table 1 shows the elemental analysis components' trends following the adopted depth records. For Carbon content (%C), the highest yield was obtained at Cato_U, scoring for nearly 48% of the total dry mass of the peat sample. Three different trends can be noticed: a decreasing one from the uppermost vegetation to the bottom of the mesotelm (from 40% to 24%, between Upp and Meso_B, Table 1) and an increasing profile from the second interface to the bottom of the core (from 38% to 47%, between Int_Meso-Cato to Cato_U, Table 1). At the bottom depths of the core, a plateau can be noticed (%C between 39% and 47%, Table 1). The highest %C yielded at the bottom ecological layer indicates a preservation of the OM in the anoxic part of the core. For oxygen content (%O), a similar trend can be noticed as for %C, with the difference of the highest input at the uppermost part of the peat core. For Nitrogen content (%N), a progressively decreasing profile can be noticed. Interestingly, two local increases were noticed at the interfaces between the ecological layers (27% and 38%, for Int_Acro-Meso and Int_Meso-Cato, respectively (Table 1). The following peaks, along with the highest %N at the uppermost of the column, indicates a peculiar microbial reworking at these depths [11]. For Hydrogen content (%H), the highest contributions were scored in the catotelm (with a maximum of 47% for Cato_U, Table 1). This indicates the accumulation of aliphatic structures at these depths [10]. Sulfur's total content (%S) scored a gradually decreasing profile, with two peaks at the bottom of the core (4.15% and 3.24%, for Int_Meso-Cato and Cato_B, respectively (Table 1). The following profile is supported by the occurrence of a sulfate-reducing microbial activity in previous findings [10].

The lignin-derived monomeric phenols have been established as valuable parameters for the degradation of OM in soils and river sediments. Biodegradation of lignin by white-rot or brown-rot fungi changes the composition of the lignin [13,14]. Lignin biodegradation includes oxidation of side-chain, cleavage of C-C bonds, and demethylation processes [13]. Furthermore, fungi and actinomycetes biomarkers yielded in previous studies [10] explain the gradual decrease of phenolic components with depth (Figure 2). The lignin-derived phenols components are characteristic of major plant categories. It has been demonstrated that gymnosperm wood comprises of vanillyl derivatives only, however, the angiosperm wood is composed of approximately equal quantities of both vanillyls and syringyls [4,15]. In addition to their vanillyls or vanillyls/syringyls components, the non-woody vascular plant tissues of gymnosperms and angiosperms (e.g., conifer needles, grass, angiosperm leaves) contain cinnamyl units, which are part of the lignin macromolecule or link carbohydrates and lignin in the ligno-cellulose complex [16,17]. On the other hand, H- moieties do not exclusively derive from lignin structures [18]. For the aforementioned statements, the total yielded phenolic structures have been shown, along with its different counterparts (Figure 2). The major lignin phenols (SVC) concentrations yielded from plant and soils samples investigated in this study are comparable to the earlier reported ones [19]. The SVC yielded in soils are generally lower than the ones obtained in plant material [6]. These findings are in accordance with the decreasing trend of SVC along with depth, as the highest input was yielded at the upper layer (6.8 mg/gC for Upp, Figure 2).

Table 1. Details of the nine investigated samples and elemental analysis.

Samples Designation	Description	Depth (cm)	N	C	H	S	O
Upp	Upper vegetation with the underlying soil	4	2.36	40.27	3.99	6.78	36.94
Acro	Acrotelm samples	12	1.45	25.80	3.00	5.56	24.79
Int_Acro-Meso	Interface between acrotelm and mesotelm	24	1.99	27.28	3.46	4.98	23.61
Meso_U	Upper part of mesotelm	32	1.52	22.01	3.03	3.87	18.71
Meso_B	Bottom part of mesotelm	44	1.53	23.81	3.15	2.92	19.81
Int_Meso-Cato	Interface between mesotelm and catotelm	56	1.84	38.23	4.35	4.15	25.18
Cato_U	Upper part of Catotelm	72	1.39	47.76	4.98	2.15	27.71
Cato_B	Bottom part of Catotelm	92	1.02	39.69	4.10	3.24	22.80
BtCo	Bottomhole of the core	100	0.85	39.33	3.71	1.5	27.02

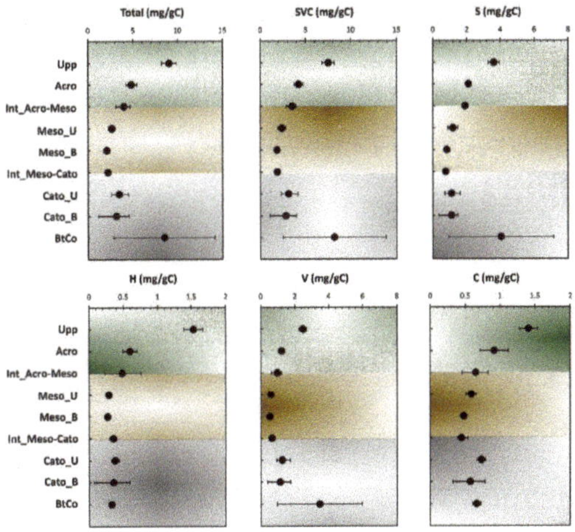

Figure 2. Depth records of the total amount of phenols, SVC, and the different yielded phenolic sub-units' compartments: H, S, V, and C (designated in mg of phenols/g of OC).

Lignin distribution in the soil horizons has been mentioned in numerous studies [20–22]. Most of these investigations state that lignin content decreases in the subsoil. However, in some cases, an increase of lignin content of SOM with depths has been detected [23], which could be related to vertical transport as well as lignins protection. This in turn proposes that the distribution of lignin in soils might vary from one site to another; however, the involved processes are not yet clear. In our case, a noticeable increase was noticed in the total amount of CuO-NaOH units and SVC, indicating a potential stabilization of lignin structures at the bottom of the core, where a direct contact occurs with the mineral matrix of soil. Furthermore, organic horizons are considered to have higher VSC concentrations than the mineral ones, representing a lignin degradation progression throughout the soil profile [6]. These findings are contradicted in our case, given the yielded increase at BtCo. Products obtained from lignin oxidation of the fresh (Upp) and degraded peat samples were composed of the six vanillyl and syringyl phenols shown in Figure 2. Total yield of the p-hydroxyl counterparts (H-moieties) did not exceed 15% of the total phenolic counterpart contribution.

Figure 3 shows the depth profile of the different phenolic counterparts (H, S, V, and C-compounds) yielded by CuO-NaOH oxidation. With the exception of V_{ket} coumaric and coniferic acids, the different subunits showed a decreasing profile along peat depth; this indicates their high occurrence in the preserved part of lignin from the first stages of the peatland deposition. Following these trends, these moieties could be employed as indicators of the "Holocene Climatic Optimum" that allowed vascular type vegetation to grow, due to the increase of planet's temperature [10]. H_{ald} and H_{acid} presented the highest fluctuation across the 11 phenolic structures, indicating its provenance from multiple sources and/or a higher degree of oxidation of these moieties compared to S-, V-, and C-compounds. The mostly stable decreasing profile of H_{ket}, along peat depth, indicates that the first assumption is less likely and that the high variation for the acid and aldehyde structures originates from microbial reworking. For the V- and S-compounds, higher consistency in their moieties' profile can be noticed, following depth records (Figure 2). This probably indicates the higher reliance of relative ratios to be used for the characterization of diagenetic events along the investigated ecosystem. For V_{ald}, and V_{ket}, an increasing profile, followed by a decreasing one, was noticed for the upper and lower halves of the peat core, respectively. These trends indicate the occurrence of the aforementioned moieties from the growing sedges at the surface and the vascular vegetation deposited at the bottom of the core [10,18].

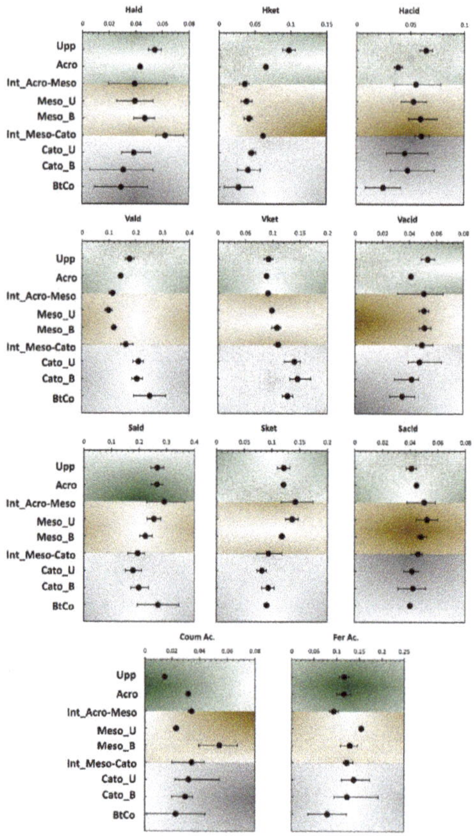

Figure 3. Depth records of mass fractions for the 11 obtained phenolic sub-units (in g/g Total Phenols).

3.2. Degradation, Change of Vegetation, and Diagenetic Parameters of Lignin

3.2.1. Diagenetic Trends of Lignin Phenols

Several characteristic indicators for the lignin degradation profile were established based on the relative distribution of lignin phenols measured chromatographically after CuO-NaOH oxidation. In general, SVC content reduces as soil and sediments lignin degradation increases (Figure 2; [8]). Nonetheless, specific ratios are expressed as V + S + C content, since CuO-NaOH oxidation yields might vary depending on the degree of lignin structure alteration [24] and for different plant species [25]. The implementation of these ratios is of utmost importance for the sake of eliminating any bias caused by the decrease in SVC profile. Furthermore, the cleavage in the Cα-Cβ bond of the phenylpropanoid units and oxidation of the degraded compounds resulted in increasing carboxylic acid units when compared to the aldehyde ones. Consequently, there was an increase in acid-to-aldehyde ratios of V and S-type units following the biodegradation in soils and sediments [6,9,14,26,27]. During lignin degradation, syringyl and cinnamyl units degrade preferentially when compared to the guaiacyl units (V units), resulting in a decrease of the S-to-V and C-to-V ratio values [6,25,28], except at the first degradation stage [25]. As C- and S-to-V ratios overlap with source variations during degradation and have opposite trends, they are rarely used as indicators of the degradation of lignin. In our case, S/V and C/V ratios are increasing the acrotelm, due to the input of the fresh non-degraded OM from the uppermost vegetation. The decrease of these ratios can be seen along the mesotelm and the catotelm. Nonetheless, both ratios can be used as vegetation change indicators (See Section 3.2.2). For that purpose, the Ad/Al ratios present a more efficient indicator of lignin degradation.

Prior to discussing lignin's origin, it is necessary to consider its diagenesis. It has been demonstrated that the acid/aldehyde ratios (Ad/Al) of three lignin phenols groups can be used to identify diagenetic alteration in a variety of geochemical samples [4,8,29–31]. Ad/Al ratios clearly show considerable degradation of lignin after deposition, since samples yield more acid and less aldehyde than fresh plant tissues at the uppermost vegetation layer [8,27,32]. Ac/Al for S- and V-compounds yielded values of 0.17 and 0.33 at the uppermost vegetation (Upp, Figure 4), which is in accordance with previous findings (Ad/Al for fresh plants: 0.1–0.5; [27,32]). The reason for the elevated Ad/Al for H-compounds could be related to the higher oxidation of non-lignin phenol on its way to further decay (Ad/Al) H = 0.87–2.4; Figure 4) [14]. Yet, the huge Ad/Al values may not be directly related to diagenesis. Instead, they may reflect the presence of ester-bound phenols in the peat matrix, which may include humic-type substances. This type of bonding has been found in certain plants [27] and also in humic substances [26]. This could be supported by the absence of any increase in H-compounds in the catotelm (unlike S-, V-, and C-compounds where a slight increase in their contribution was noted along with depth (Figure 3). On the other hand, the highest organic carbon inputs yielded in the catotelm (Figure 1) confirm the presence of an organic fraction that was hindered from the adopted depolymerization technique. This fraction is most likely the so-called "Humic Fraction." In fact, previous investigations regarding the same peat samples showed a high increase of OM input upon the application of a thermally assisted chemolysis approach [10,18].

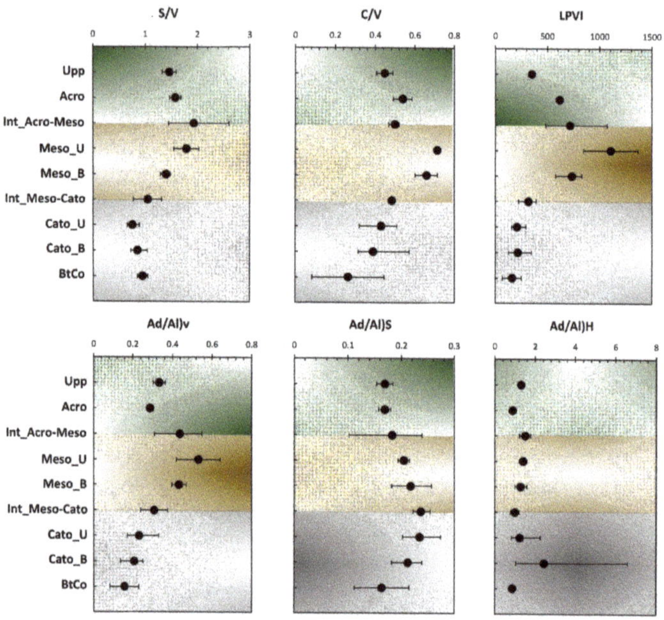

Figure 4. Depth records for different ratios of the 11 phenolic sub-units.

3.2.2. Source Vegetation of Phenolic CuO Oxidation Products

Lignin phenol ratios (S/V and C/V) illustrate the relative influence of terrestrial vegetation to the total OM. In our case and due to the high degradation rate that occurred in the upper half (acrotelm and mesotelm), these ratios could be biased and are more likely to be applied as degradation indicators for lignin (see Section 3.2.1). For the catotelm, a higher conservation is highlighted by a high input of OC (Table 1). This shows the applicability of the aforementioned ratios as source vegetation indicators. The changes of S/V ratio with depth can be used to distinguish sources of OM derived from either gymnosperms (low S/V values, ≈ 0) or angiosperms (S/V ≈ 1). The vertical profile of S/V and C/V demonstrate that vascular plant sources at the catotelm were mostly angiosperm tissues [10]. These ratios are remarkably similar to those reported for other mangrove sediments and tropic wetlands [3,31]. Despite the mentioned limitations, these ratios have become the standard method for distinguishing angiosperms, gymnosperms, and nonwoody terrestrial plants in lakes, rivers, estuaries, and oceans during the last five decades [15,31–33].

With regard to the large vegetation heterogeneity along the peatland formation and deposition, as well as the highly dynamic nature of this ecosystem with different diagenetic reactivities lignin moieties ($C > S > V$), a more solid vegetation change indicator should be adopted [8]. Tareq et al. [8] proposed a binary equation to define a new proxy of lignin phenol sub-units, the lignin phenol vegetation index (*LPVI*). This ratio allows to identify vegetation with the exclusion of lignin's degradability bias:

$$LPVI = \left[\frac{S\,(S+1)}{V+1}+1\right]\left[\frac{C\,(C+1)}{V+1}+1\right]$$

Tareq et al. [8] claimed that the *LPVI* provides a better resolution than other lignin parameters such as C/V vs. S/V to identify the source vegetation type in complex mixtures such as OM from peat and soils. The *LPVI* yields non-overlapping data for woody and nonwoody gymnosperms and angiosperms, respectively, in contrast to the C/V and S/V ratios (Table 2). In our case, *LPVI* presented values between 160 and 1106 all along the peat core (Figure 4). This indicates the occurrence of nonwoody angiosperms tissues. These findings are coherent with the different source vegetation indicators used in this study, as well as previous ones [10,18,34]. Interestingly, *LPVI* showed a close profile to C/V and S/V ratios with depth. The highest LPVI scores were yielded in the mesotelm layer (Figure 4).

Table 2. Lignin phenol vegetation index (LPVI) value ranges obtained for specific plants. Adapted with permission from Ref. [8]. Copyright (2023) with permission from Elsevier.

Plant Types	LPVI Value Ranges		
	Low	Mean	High
Woody Gymnosperms	1	1	1
Non-woody Gymnosperms	12	19	27
Woody Angiosperms	67	181	415
Non-woody Angiosperms	378	1090	2782

In the previous parts (Sections 3.1 and 3.2), we have attempted to present, in the most sophisticated way, the molecular cartography of phenolic CuO-NaOH along the depth records of the investigated peat core. As it can be seen, several proxies are biased due to the high degradation rate of lignin with the different steps of peatland deposition across time. This issue comes from the fact that a peatland is a highly dynamic ecosystem where different microbial niches can exist, either in oxic or anoxic conditions. Additionally, the hydrology features that govern play a primordial role towards boosting or reducing the microbial reworking. Hence, the variation of water table level between emerged (dry conditions) and the submerged (wet condition) situations and the continuous supply of fresh OM from the uppermost vegetation make the mesotelm a highly dynamic layer. The aforementioned statements induce the high variability of different proxies at the intermediate layer of the investigated peat core. In order to remove such a bias, we attempted the use of PCA for the sake of seeking new proxies from the conventionally used ones that are independent from one another. These new proxies are, in fact, the principle components.

3.3. Principal Component Analysis (PCA)

3.3.1. Application of PCA to Phenolic Mass Fractions

Figure 5 presents the PCA bi-plot for the mass fractions of the different sub-units yielded by CuO-NaOH oxidation and shown in Figure 3. The first two PCs accounted for 66.97% of the total variance (44.13% for PC1 and 22.48% for PC2; Figure 5a). PC1 exhibited nearly equally distributed contributions for Vald, Vket, Hacid, Vacid, Sket, and Sacid, ranging from 13% to 17%. For PC_2, it showed the highest contribution exclusively along Sald, accounting for 32% for this PC's contribution (Figure 5b). Interestingly, most of the contributors of PC1 are acidic and ketone moieties; this could probably indicate a more oxidized state for depth samples positively influenced and a less oxidized one for samples negatively influenced by this PC. For the trends of the different variables along the PCA bi-plot, S-Compounds showed a certain proximity, with a high positive and slight negative influence of Sald over PC1 and PC2, respectively. For Sacid, it showed a negative trend along PC1, with a slight positive one along PC2. The aforementioned three moieties represent the blue cluster. Unlike the S-compounds, V-compounds were not regrouped together, as a proximity was noticed between Vald and Vket, forming the grey cluster. These two compounds showed a positive influence along PC1, with a mild-to-average influence along PC2. The green cluster gathered the variables of the investigated phenolic sub-units and showed a negative influence along both PCs.

C/N showed a high correlation (96.63%, Figure 8); these trends confirm the capacity to apply PC1 as source vegetation indicator. LPVI showed a higher correlation with PC1 than PC2. This confirms the potential of the application of LPVI to decipher vegetation change, even in a dynamic system such as the peatland.

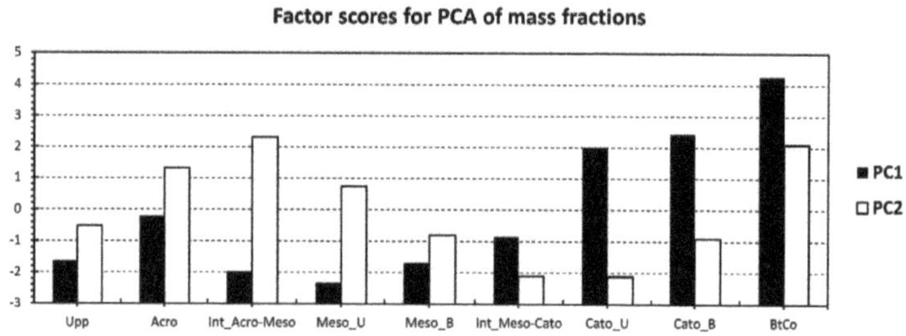

Figure 7. Contribution of the first two principle components obtained from the PCA of Figure 5, along with the depth records of the investigated peat core.

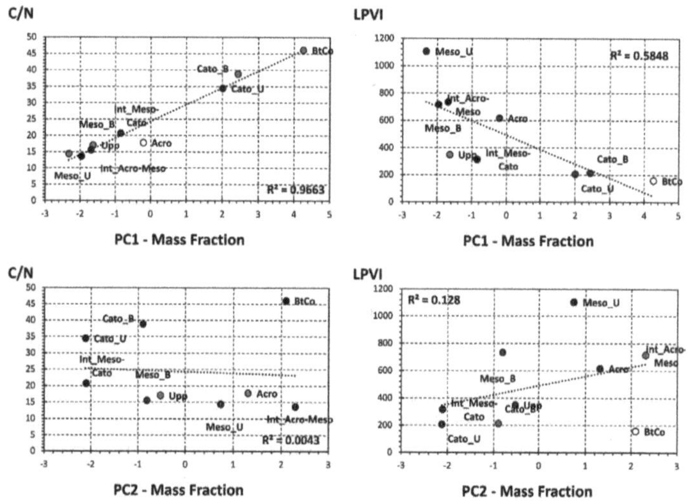

Figure 8. Selected correlation between bulk analysis proxies and the first two principle components yielded by the PCA-biplot of Figure 5.

4. Conclusions

In this study, we aimed to seek the efficiency of unsupervised learning for the estimation of lignin degradation along a peat core. The "Principal Component Analysis" (PCA) approach was applied. The benefits of using PCA resides in its capacity to produce new factors that are independent (orthogonal projection) from each other, yet dependent from all of the original factors. Herein, the individuals are nine peat core depth samples, and the variables are the 11 phenolic sub-units yielded by CuO-NaOH oxidation. The separation variables technique is of utmost importance for the analysis of a complex and dynamic system such as peatlands. In this ecosystem, two main phases of OM exists, preservation or degradation. In order to confirm these phases, several proxies were adopted. The preservation of phenolic OM is interpreted by high values of SVC and reflects the occurrence of an old vegetation following peatlands' deposition over time. For the sake of identifying the sources of this vegetation, S/V and C/V were adopted. For the degradation phase of phenolic OM, Ac/Ad for S- and V-compounds could be used. One shortcoming that could be identified for these proxies is the fact that they could be biased due to the high degradation rate of lignin. This was noticed following the strong decline of the SVC ratio from the uppermost vegetation to the bottom hole of the peat core. Following this trend, and in order to overcome the bias created from the simultaneous degradation and preservation pathways, a more sophisticated ratio for source vegetation indication was adopted: the LPVI. This ratio showed similar trends as the lignin degradation indicators (Ac/Ad), which puts its reliability into question. One PCA was applied for the mass fraction of the different phenolic sub-units and allowed for a separation to be made between the ecological layers and their interfaces. With regard to factors, it allowed for the compilation of two indicators; one for change in vegetation and the other for the degradation of lignin in a peat core along depth. The different results were confirmed by a high correlation along with bulk elemental analysis proxies.

Author Contributions: Methodology, K.Y. and L.G.; Data curation, K.Y.; Writing—original draft, K.Y. and E.O.; Writing—review & editing, S.M., L.K., O.M., L.G. and N.M.; Project administration, N.M., K.Y. and L.G. All authors have read and agreed to the published version of the manuscript.

Funding: "Ministère Français de l'Enseignement Supérieur et de la Recherche" and Areva—Paris.

Data Availability Statement: The manuscript has no associated data.

Acknowledgments: The research was financially supported by the "Ministère Français de l'Enseignement Supérieur et de la Recherche" and Areva—Paris. We wish to thank M. Descostes, M. Chauveau and B. Leroux for helping with sample collection. We also thank Joëlle Laduranty for elemental analysis and anonymous reviewers for constructive comments.

Conflicts of Interest: The authors declare no conflict of interest.

References

1. Clymo, R.S.; Turunen, J.; Tolonen, K. Carbon accumulation in peatland. *Oikos* **1998**, *81*, 368–388. [CrossRef]
2. Yu, Z.C. Northern peatland carbon stocks and dynamics: A review. *Biogeosciences* **2012**, *9*, 4071–4085. [CrossRef]
3. Boudet, A.-M. Lignins and lignification: Selected issues. *Plant Physiol. Biochem.* **2000**, *38*, 81–96. [CrossRef]
4. Hedges, J.I.; Ertel, J.R. Characterization of lignin by gas capillary chromatography of cupric oxide oxidation products. *Anal. Chem.* **1982**, *54*, 174–178. [CrossRef]
5. Staniszewski, A.; Lejman, A.; Pempkowiak, J. Horizontal and vertical distribution of lignin in surface sediments of the Gdańsk Basin. *Oceanologia* **2001**, *43*, 421–439.
6. Otto, A.; Simpson, M.J. Evaluation of CuO oxidation parameters for determining the source and stage of lignin degradation in soil. *Biogeochemistry* **2006**, *80*, 121–142. [CrossRef]
7. Younes, K.; Moghrabi, A.; Moghnie, S.; Mouhtady, O.; Murshid, N.; Grasset, L. Assessment of the Efficiency of Chemical and Thermochemical Depolymerization Methods for Lignin Valorization: Principal Component Analysis (PCA) Approach. *Polymers* **2022**, *14*, 194. [CrossRef]
8. Tareq, S.M.; Tanaka, N.; Ohta, K. Biomarker signature in tropical wetland: Lignin phenol vegetation index (LPVI) and its implications for reconstructing the paleoenvironment. *Sci. Total Environ.* **2004**, *324*, 91–103. [CrossRef]
9. Thevenot, M.; Dignac, M.-F.; Rumpel, C. Fate of lignins in soils: A review. *Soil Biol. Biochem.* **2010**, *42*, 1200–1211. [CrossRef]
10. Younes, K.; Laduranty, J.; Descostes, M.; Grasset, L. Molecular biomarkers study of an ombrotrophic peatland impacted by an anthropogenic clay deposit. *Org. Geochem.* **2017**, *105*, 20–32. [CrossRef]
11. Ertel, J.R.; Hedges, J.I. Sources of sedimentary humic substances: Vascular plant debris. *Geochim. Cosmochim. Acta* **1985**, *49*, 2097–2107. [CrossRef]
12. Joliffe, I.T.; Morgan, B.J.T. Principal component analysis and exploratory factor analysis. *Stat. Methods Med. Res.* **1992**, *1*, 69–95. [CrossRef] [PubMed]
13. Tien, M.; Kirk, T.K. Lignin-degrading enzyme from the hymenomycete Phanerochaete chrysasporium burds. *Science* **1983**, *221*, 661–663. [CrossRef] [PubMed]
14. Hedges, J.I.; Clark, W.A.; Come, G.L. Organic matter sources to the water column and surficial sediments of a marine bay. *Limnol. Oceanogr.* **1988**, *33*, 1116–1136. [CrossRef]
15. Hedges, J.I.; Mann, D.C. The characterization of plant tissues by their lignin oxidation products. *Geochim. Cosmochim. Acta* **1979**, *43*, 1803–1807. [CrossRef]
16. Iiyama, K.; Lam, T.B.T.; Stone, B.A. Phenolic acid bridges between polysaccharides and lignin in wheat internodes. *Phytochemistry* **1990**, *29*, 733–737. [CrossRef]
17. Lam, T.B.T.; Kadoya, K.; Iiyama, K. Bonding of hydroxycinnamic acids to lignin: Ferulic and p-coumaric acids are predominantly linked at the benzyl position of lignin, not the β-position, in grass cell walls. *Phytochemistry* **2001**, *57*, 987–992. [CrossRef]
18. Younes, K.; Grasset, L. Comparison of thermochemolysis and classical chemical degradation and extraction methods for the analysis of carbohydrates, lignin and lipids in a peat bog. *J. Anal. Appl. Pyrolysis* **2018**, *134*, 61–72. [CrossRef]

19. Goñi, M.A.; Montgomery, S. Alkaline CuO Oxidation with a Microwave Digestion System: Lignin Analyses of Geochemical Samples. *Anal. Chem.* **2000**, *72*, 3116–3121. [CrossRef] [PubMed]
20. Amelung, W.; Flach, K.-W.; Zech, W. Lignin in Particle-Size Fractions of Native Grassland Soils as Influenced by Climate. *Soil Sci. Soc. Am. J.* **1999**, *63*, 1222–1228. [CrossRef]
21. Rumpel, C.; Kögel-Knabner, I.; Bruhn, F. Vertical distribution, age, and chemical composition of organic carbon in two forest soils of different pedogenesis. *Org. Geochem.* **2002**, *33*, 1131–1142. [CrossRef]
22. Wiesner, M.R.; Lowry, G.V.; Jones, K.L.; Hochella, M.F., Jr.; Di Giulio, R.T.; Casman, E.; Bernhardt, E.S. Decreasing Uncertainties in Assessing Environmental Exposure, Risk, and Ecological Implications of Nanomaterials. *Environ. Sci. Technol.* **2009**, *43*, 6458–6462. [CrossRef] [PubMed]
23. Mason, S.L.; Filley, T.R.; Abbott, G.D. The effect of afforestation on the soil organic carbon (SOC) of a peaty gley soil using on-line thermally assisted hydrolysis and methylation (THM) in the presence of 13C-labelled tetramethylammonium hydroxide (TMAH). *J. Anal. Appl. Pyrolysis* **2009**, *85*, 417–425. [CrossRef]
24. Bahri, H.; Rasse, D.P.; Rumpel, C.; Dignac, M.-F.; Bardoux, G.; Mariotti, A. Lignin degradation during a laboratory incubation ollowed by 13C isotope analysis. *Soil Biol. Biochem.* **2008**, *40*, 1916–1922. [CrossRef]
25. Ziegler, F.; Kögel, I.; Zech, W. Alteration of gymnosperm and angiosperm lignin during decomposition in forest humus layers. *Z. Pflanz. Bodenkd.* **1986**, *149*, 323–331. [CrossRef]
26. Ertel, J.R.; Hedges, J.I. The lignin component of humic substances: Distribution among soil and sedimentary humic, fulvic, and base-insoluble fractions. *Geochim. Cosmochim. Acta* **1984**, *48*, 2065–2074. [CrossRef]
27. Opsahl, S.; Benner, R. Early diagenesis of vascular plant tissues: Lignin and cutin decomposition and biogeochemical implications. *Geochim. Cosmochim. Acta* **1995**, *59*, 4889–4904. [CrossRef]
28. Goñi, M.A.; Nelson, B.; Blanchette, R.A.; Hedges, J.I. Fungal degradation of wood lignins: Geochemical perspectives from CuO-derived phenolic dimers and monomers. *Geochim. Cosmochim. Acta* **1993**, *57*, 3985–4002. [CrossRef]
29. Ishiwatari, R.; Uzaki, M. Diagenetic changes of lignin compounds in a more than 0.6 million-year-old lacustrine sediment (Lake Biwa, Japan). *Geochim. Cosmochim. Acta* **1987**, *51*, 321–328. [CrossRef]
30. Orem, W.H.; Colman, S.M.; Lerch, H.E. Lignin phenols in sediments of Lake Baikal, Siberia: Application to paleoenvironmental studies. *Org. Geochem.* **1997**, *27*, 153–172. [CrossRef]
31. Dittmar, T.; Lara, R.J. Molecular evidence for lignin degradation in sulfate-reducing mangrove sediments (Amazônia, Brazil). *Geochim. Cosmochim. Acta* **2001**, *65*, 1417–1428. [CrossRef]
32. Hu, F.S.; Hedges, J.I.; Gordon, E.S.; Brubaker, L.B. Lignin biomarkers and pollen in postglacial sediments of an Alaskan lake. *Geochim. Cosmochim. Acta* **1999**, *63*, 1421–1430. [CrossRef]
33. Wilson, J.O.; Valiela, I.; Swain, T. Sources and concentrations of vascular plant material in sediments of Buzzards Bay, Massachusetts, USA. *Mar. Biol.* **1985**, *90*, 129–137. [CrossRef]
34. Younes, K.; Grasset, L. The application of DFRC method for the analysis of carbohydrates in a peat bog: Validation and comparison with conventional chemical and thermochemical degradation techniques. *Chem. Geol.* **2020**, *545*, 119644. [CrossRef]

polymers

Article

Changes in the Electrophysical Parameters of Nanomodified Elastomers Caused by Electric Current's Passage

Alexandr V. Shchegolkov [1], Aleksei V. Shchegolkov [2], Natalia V. Zemtsova [3], Yaroslav M. Stanishevskiy [4] and Alexandre A. Vetcher [4,5,*]

1 Institute of Technology of the Department of Technology and Methods of Nanoproducts Manufacturing, Tambov State Technical University, 392000 Tambov, Russia
2 Department of Chemical Technology, Platov South-Russian State Polytechnic University (NPI), 132 Enlightenment Str., 346428 Novocherkassk, Russia
3 Department of Technique and Technology for Obtaining Nanoproducts, Tambov State Technical University, 106 Sovetskaya St., 392000 Tambov, Russia
4 Institute of Biochemical Technology and Nanotechnology (IBTN), Peoples' Friendship University of Russia (RUDN), 6 Miklukho-Maklaya St., 117198 Moscow, Russia
5 Complementary and Integrative Health Clinic of Dr. Shishonin, 5 Yasnogorskaya St., 117588 Moscow, Russia
* Correspondence: avetcher@gmail.com

Abstract: The development of reliable and effective functional materials that can be used in various technological fields and environmental conditions is one of the goals of modern nanotechnology. Heating elements' manufacturing requires understanding the laws of heat transfer under conditions of different supply voltages, as this expands the possibilities of such materials' application. Elastomers based on silicon-organic compounds and polyurethane modified with multi-walled carbon nanotubes (MWCNTs) were studied at various concentrations of Ni/MgO or Co-Mo/MgO and voltages (220, 250, and 300 V). It was found that an increase in voltage from 220 to 300 V leads to an initial increase in specific power on one-third followed by a subsequent decrease in a specific power when switched on again to 220 V (for −40 °C) of up to ~44%. In turn, for a polyurethane matrix, an increase in voltage to 300 V leads to an initial peak power value of ~15% and a decrease in power when switched on again by 220 V (for −40 °C) to ~36% (Ni/MgO -MWCNT). The conducted studies have shown that the use of a polyurethane matrix reduces power degradation (associated with voltage surges above 220 V) by 2.59% for Ni/MgO–based MWCNT and by 10.42% for Co-Mo/MgO. This is due to the better heat resistance of polyurethane and the structural features of the polymer and the MWCNT. The current studies allow us to take the next step in the development of functional materials for electric heating and demonstrate the safety of using heaters at a higher voltage of up to 300 V, which does not lead to their ignition, but only causes changes in electrophysical parameters.

Keywords: MWCNTs; catalyst; polyurethane; organosilicon compound; modification; degradation; thermal field

Citation: Shchegolkov, A.V.; Shchegolkov, A.V.; Zemtsova, N.V.; Stanishevskiy, Y.M.; Vetcher, A.A. Changes in the Electrophysical Parameters of Nanomodified Elastomers Caused by Electric Current's Passage. *Polymers* **2023**, *15*, 249. https://doi.org/10.3390/polym15010249

Academic Editor: Alexey Bubnov

Received: 6 December 2022
Revised: 28 December 2022
Accepted: 30 December 2022
Published: 3 January 2023

1. Introduction

The recent call for complex technological objects, both in industry and in the domestic sphere, has led to the formation of more effective concepts and strategies in the field of materials science, namely the creation of smart materials [1]. Smart materials can provide the implementation of various functional tasks, concerning their application for electric drive systems—materials with controlled deformation (electroactive elastomers) [2–4] and strain gauge [5], as well as electric heating in systems to combat ice.

On an industrial scale, it turned out to be effective to use polymer matrices with conductive dispersed additives. Variations between the properties of elastomer matrices [6] and conductive disperse additives make it possible to form a wide range of materials with various properties [7]. The dynamic development of electric vehicles is accompanied by new types of tasks that affect the improvement of microclimate technologies [8]. In order to increase comfort, the use of electric heating can affect various areas of the vehicle [9]. During the operation of electric vehicles, a significant part of the energy stored in the battery (up to 35% [10]) is consumed for heating, and, accordingly, the efficiency, reliability, and stability of heating systems are very important.

It is important to use heaters with the effect of maintaining the temperature regime due to the effect of self-regulation. This effect is realized in a PTCR (Positive Temperature Coefficient Resistor) [11]. Most electric heaters are based on the PTCR $BaTiO_3$ effect [12]. The PTCR effect has built-in protection against overheating: the resistivity of the PTCR material increases by several orders of magnitude near the ferroelectric-paraelectric phase transition temperature (commonly referred to as the Curie temperature TC), which reduces the conductivity and, in turn, the heating current and power by orders of magnitude [11–14]. Accordingly, the PTCR heater does not heat up more than TC during normal operation. The Heyvang–Jonker model describes the PTCR effect and is based on temperature-dependent potential barriers at grain boundaries and their interaction with the ferroelectric domains of the fillers.

However, PTCR heaters are powered directly by the vehicle's main batteries, which are between 500 V and 1 kV [15]. At these high voltages, PTCR materials show a degradation effect with increasing resistivity when voltage is applied. This increase, in turn, reduces the heating output and creates a serious problem for the application [15]. The fragility of ceramics and the difficulty of creating vibration-resistant electrical contacts should be noted.

For electric heaters, conducting networks in the structure of a dielectric matrix can be obtained using metal nanodots (Cr-nd, NiCr-nd, and Ni-nd) [16]. For heaters, polypropylene heaters can be used as a polymer matrix, and conductive structures are hybrid fillers [17] for opening the conductive network.

Composites based on CNTs and polymers can be an effective and technological solution in the field of electrical heating—due to their light weight, high design flexibility, and fast temperature response. Studies with the CNT/TPU heater [18] reported the possibility of maintaining the stability of the electrophysical properties under non-tensile deformations.

Consideration should be given to the possibility of conducting polymer nanocomposites with reversible dynamic bonds, as well as their energy activation for self-healing by heating when an electric current flows (Joule effect) [19]. Flexible composites based on silicone rubber (SR)/carbon fiber(CF)@polydopamine (PDA) (SR/CF@PDA) have good interfacial adhesion in the structure of the material and filler (due to the adhesive properties of PDA) [20], which ensures flexibility without compromising electrical properties. Polyurethane-coated carbon fiber (CF) can be added to the process for producing thermoplastic polymer composites with improved thermal and mechanical properties, as well as durability, as an effective reinforcing filler and an improved cross-linking process. Electron beam irradiation (EB) can be added [21]. In the framework of the technological process [21], composites based on polyamide 6 (PA) with different HC content from 1 to 10 wt.% were made by melt compounding and compression molding and then irradiated with various doses of EB.

Conductive composites based on organosilicon elastomers (silicones) [22] may have a functional feature associated with self-regulation of the heating temperature [23]. Studies related to the effect of MWCNTs on heat release in a flexible nanomodified elastomer are presented in the work and it was found that the heat release power can be represented in the form of regression equations, where there is a dependence of heat release on temperature, applied tensile strain and electrical heating [24]. In [25], the authors studied flexible heating elements based on silicone polymerized by a platinum catalyst and TiO_2 with different morphologies (spherical and acicular), as well as an SnO_2/Sb coating and carbon black. The presented heater had a dependence of resistivity on temperature and could operate on an alternating voltage of 220 V.

Conductive fillers, such as carbon nanotubes (CNTs), exhibit either metallic or semiconducting properties [26]. It should be noted that the morphological diversity of carbon nanomaterials, namely CNT and, in particular, multi-walled carbon nanotubes (MWNTs) with many concentric carbon shells or aligned single-walled nanotubes (SWNT), are complex composite conductors including many weakly bonded nanotubes, each of which has a different electronic structure and, consequently, electrical properties [27].

The morphology of MWNTs can be controlled using the method of selective removal of single carbon shells from the overall structure of MWNTs and SWNTs to adapt the properties of these composite nanotubes to any practical problems [26,27]. It is possible to remove the shells of MWNTs in stages and characterize different shells individually. The presented approach of the selective removal of layers makes it possible to convert MWNTs to either a metallic or semiconductor material, as well as directly solve the problem of transport with multiple shells [28].

When forming composites based on CNT, it is necessary to take into account the effect of structural distortions in CNT of the "armchair" type on electronic and electrical properties. Bending CNTs reduces their transfer function (electrical conductivity) in certain energy ranges and leads to an increase in electrical resistance. Electronic structure calculations show that these energy ranges contain localized states with significant σ-π hybridization as a result of the increase in curvature caused by bending. Twisting strongly affects the electronic structure of the CNT. In an ordinary metal chair (n,n) CNT, a band gap is formed, which first increases linearly with the angle of twist and then reaches a constant value. This saturation is associated with a structural transition to a flattened helical structure. The calculated values of the torsion energy and band gap are strongly affected by the possibility of structural relaxation in twisted structures [29].

The flow of current through the CNT can cause obvious destruction in the structure of the CNT. The mechanism of CNT destruction under the action of an electric voltage (potential difference) (a high current load causes local thinning of the nanotube in the case of MWCNT structures) is presented in [30]. The mechanism of destruction of SWCNT can be explained as follows: oxidation due to Joule heating occurring at the center of the line. In addition, the destruction of several samples of semiconductor and metal SWCNT samples occurred not only in the center but also on the cathode side of the line [31].

Unlike metallic wires, MWNTs do not break down in the continuous, accelerating manner that is characteristic of electromigration. Instead, they degrade due to a series of sharp surges of electric current of the same intensity. Degradation is associated with the successive destruction of individual nanotube shells following the geometry of the concentric shell of MWNTs. In addition, the initiation of this destruction is very sensitive to the oxidative action of air. In the air, destruction is initiated by oxidation at a certain power (current value), while in a vacuum, MWNTs can withstand much higher power (current) densities and reach the maximum current carrying capacity [32]. It was shown that the destruction of carbon nanotubes is based on a pulsating current flow process with pronounced peaks [33].

It was found that the thermal degradation of CNT composites as a result of direct current flow (E-heating) is much more serious than convection heating (C-heating). Increased mechanical properties of the CNT composite were observed after 40 h C-heating, while E-heating made the composites brittle such that the matrix was severely damaged at the same temperature/hours. Consequently, the electrical resistance of the CNT composite for E-heating was dramatically increased. To avoid thermal degradation and a sharp increase in durability, the E-heating temperature must be well below the temperature limit of the thermally stable polymer. In addition, the thermal degradation of the polymer matrix can be reduced by adding CNT, which in essence can be explained by the formation of a parallel resistor, by distributing the thermal concentration throughout the polymer matrix [34]. It is possible to improve the characteristics of CNT by using the technology of obtaining hybrids—graphene is of particular interest in this regard. Graphene–CNT hybrids (G/CNT) were formed by in situ Joule heating in a transmission electron microscope (TEM). The formation of the G/CNT structure occurs as a result of the sequential and spontaneous unfastening of the extreme wall of the MWCNT under uniform thermal etching and a voltage pulse of 0.2–1 V. The conductivity of the G/CNT hybrids shows a significant change (up to 38 times) after decorating with CdTe quantum dots [34,35].

It should be noted that the correlation between the electrical conductivity and the content of the conductive dispersed filler is important from the point of view of the general theory of percolation [36]. There are models explaining how the microstructure of electrically conductive polymer nanocomposites affects their temperature coefficient of resistance, which is crucial for Joule heating. There is also a relationship between morphology and local overheating, as well as impurities introduced by CNT, in terms of the resistance of composites to thermal decomposition [37].

To fully understand the formation of efficient conductive composites, it is necessary to combine studies of electrical conductivity measurement and modeling aspects of CNT/polymer composites obtained through the production of fused filament (FFF) and additive manufacturing (AM). Raw materials for CNT/polylactic acid (PLA) and CNT/high-density polyethylene (HDPE) filaments have been synthesized by controlled-feed CNT melt blending to create polymer nanocomposites. The electrical conductivity of 3D printed CNT/PLA and CNT/HDPE composites was measured under various CNT loads. Low percolation thresholds were obtained from measurement data as 0.23 vol.% and 0.18 vol.% CNT for CNT/PLA and CNT/HDPE nanocomposites, respectively. A two-parameter model of CNT agglomeration based on micromechanics makes it possible to predict the electrical conductivity of CNT/polymer composites [38].

The thermal degradation of PA11/MWCNT composites is also affected by the amount of MWNT used to fill the polymer matrix. MWCNTs were prepared by the catalytic chemical precipitation of methane over Co-Mo/MgO catalysts and then treated in 2.6 M nitric acid to remove catalyst residues. The PA11/MWCNT composite becomes more thermally stable as the mass concentration of MWNT increases to 1 wt.%. This is manifested in the fact that the composite containing 1 wt.%. MWNT has the best thermal stability and the decomposition temperature improves to around 20 °C. At a higher content of MWCNT (for example, 2 wt.%), the PA11/MWCNT composite is destroyed at a lower temperature, compared with nanocomposites with a lower content of MWCNT (for example, 0.5 and 1.0 wt.%), probably due to MWCNT aggregation in composites with higher MWCNT concentrations [39].

The presented analysis of composites for which there is a relationship between electrical resistance and temperature, which affects the Joule heating, or their behavior during thermal degradation (destruction of the polymer matrix or conductive structures predominates), differs from conventional monolithic heating materials (for example, metals that form oxides causing structural deterioration during the passage of an electric current).

The purpose of the current study was to evaluate the effect of high supply voltage on the process of degradation of conductive structures in nanomodified elastomers.

In accordance with it, the following tasks were set:

(1) Obtaining elastic organosilicon and polyurethane matrices of modified MWCNTs synthesized by chemical vapor deposition (CVD) technology;
(2) Study the electrically conductive nanomodified elastomers of heat release at a supply voltage in the range from 220 to 300 V.

2. Materials and Methods

For the MWCNTs' synthesis, the CVD method was employed with Ni/MgO and/or Co-Mo/MgO as a catalyst for synthesis at a temperature of 700 °C. Silicon-organic compound Silagerm 8040 and polyurethane Silagerm 6030 (both from ELEMENT 14 LLC, Moscow, Russia) were employed as the polymer matrix of the elastomer(s).

Component (A)—an organosilicon compound—and MWCNTs were mixed on a WiseStir HT 120DX overhead mechanical stirrer (WiseStir Ltd., Seoul, Republic of Korea) at 200 rpm for 20 min. Then, the second component based on platinum (Pt) was introduced into the mixture, which provided polymerization (B), followed by stirring for 10 min at a temperature of 22 °C. The concentration of MWCNTs in the elastomer for Ni/MgO was 10 wt.% and Co-Mo/MgO was 3 wt.%. Geometric parameters were the length and width of the elastomer sample with MWCNTs of 10 cm and 7 cm (thickness of 2 mm).

Moisture was removed from MWCNTs before introduction into the elastomer in a SANYO CONVECTION OVEN MOV 210F drying oven (SANYO Ltd., Osaka, Japan) at 110 °C. The samples were obtained according to the technology described in [24,25] (Figure 1).

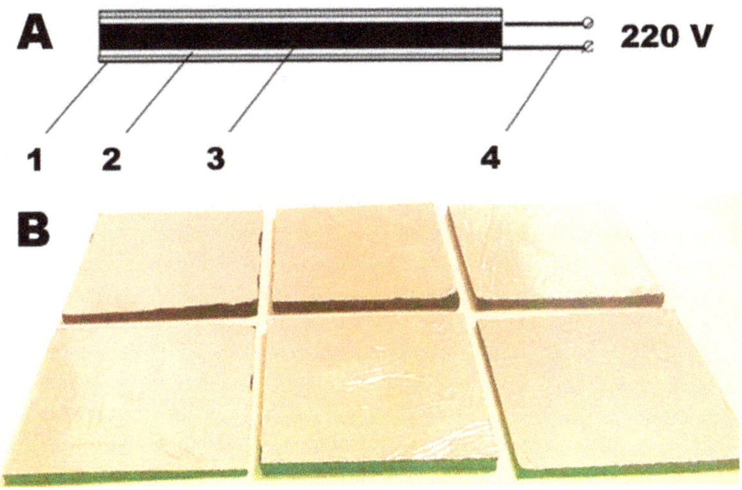

Figure 1. (**A**) Schematic diagram of a self-regulating constant current heater: 1—dielectric shell; 2—current collectors; 3—functional material of the heater; 4—connectors. (**B**) Heaters used in the current study.

The obtained products are listed in Table 1.

Table 1. Elastomers modified with MWCNT.

MWCNTs Catalyst for Synthesis	Elastomer's Designation	
	CO (Silicon-Organic Compound)	PU (Polyurethane)
Ni/MgO	NCO 1	NPU 1
Co-Mo/MgO	NCO 2	NPU 2

A programmable power supply ATN 1351 (Eliks Ltd., Moscow, Russia) with a control range from 0 to 300 V was used as a power source. Employment of the ATN-1351-SW software makes it possible to evaluate the current consumed by the heater samples, which, taking into account the exposed voltage, allows you to find the value of the power consumption. According to the value of the found power, and taking into account the area of the heater sample, the specific power (W/m^2) is found. Temperature tests of the heater were carried out in the range from −40 to +40 °C in the climate chamber "KTH-1000" (NPF Technology, LLC, St. Petersburg, Russia). To measure special volume conductivity a Tera Ohmmeter E6-13a (Punane-RET Ltd., Tallinn, Estonia) with a range of measuring electrical resistance up to 14 TΩ was employed.

2.1. Method for Studying the Temperature Field on the Surface of Samples of Nanomodified Elastomers

To study the temperature field, a Testo-875-1 thermal imager with a $32 \times 23°$ optical lens (SE & Co. KGaA, Testo, Lenzkirch, Germany) was used with a distance of 10 cm from samples of nanomodified elastomers in a darkened room without exposure to sunlight. The temperature of nanomodified elastomers was measured with a two-channel thermometer, "Testo 992" (SE & Co. KGaA, Testo, Lenzkirch, Germany), while the surface temperature was determined. Based on the data obtained, a comparison was made with the temperature recorded by a thermal imager, after which the emissivity was selected and used for further measurements. The obtained thermal imaging images of composite samples with MWCNTs were processed using the IRSoft v4.9 SP1 program.

2.2. Structural Studies of MWCNTs and Elastomer's Matrix

Structural studies of MWCNTs were carried out using the method of transmission electron microscopy. A small number of samples were made by contact with microscopic meshes with an adhesive composition. The studies were carried out from different places of the sample samples and on several samples in order to obtain better statistics about the samples under study. TEM and SEM studies were carried out using a Hitachi H-800 electron microscope with an accelerating voltage of up to 200 keV. An IR Fourier spectrophotometer FT 801 (Spectral range 21–1.8 μm) (LLC NPF Simeks, Novosibirsk, Russia) was used for the registration in the near and mid-IR range of the spectra of the original and modified with MWCNT elastomer matrix.

For TG and DSC studies, a NETZSCH STA 449F3 instrument (NETZSCH-Gerätebau GmbH, Selb, Germany) was used. Tests were carried out in the Ar-atmosphere.

3. Results

Based on the graphical information presented on the TEM MWCNTs synthesized on Ni/MgO and Co-Mo/MgO catalysts (Figure 2), it can be concluded that there are "kink" deformations, which are characterized by different bending angles and spatial overlap, as well as inclusions of catalyst particles in the structure of an individual.

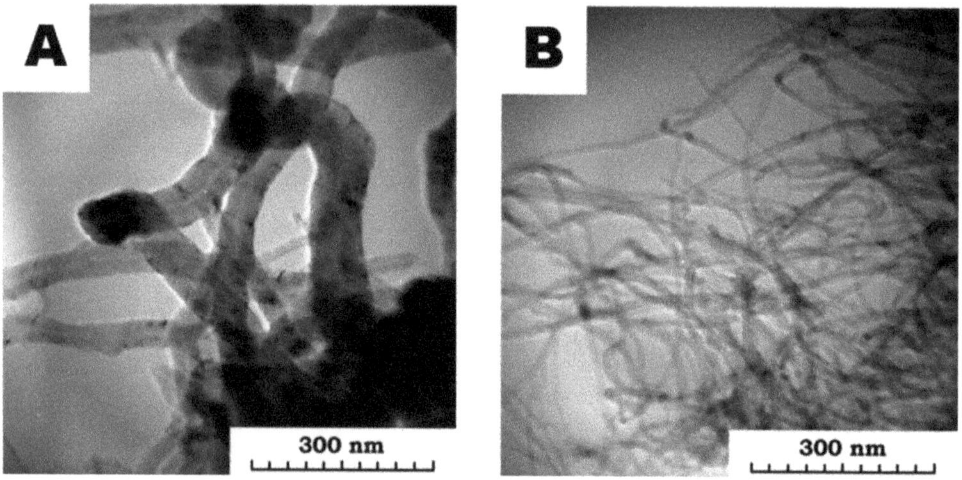

Figure 2. TEM of MWCNTs synthesized with (**A**) Ni/MgO and (**B**) Co-Mo/MgO catalysts.

It should be noted that, as a rule, "kink" deformations occur in bent MWCNTs, as well as possible hybridization of bonds around these breaks (Figure 2). Inclusions of catalyst particles, as well as spatial defects, occur at various angles, including those significantly below 90°, and can significantly reduce the conductivity of MWCNTs, which leads to the formation of a current transition into thermal energy and local overheating zones.

The addition of MWCNTs to an organosilicon matrix leads to a change in the molecular structure, which is associated with a change in the intensity of the peaks in the IR spectrogram. IR spectra of CO and NCO 1 are demonstrated in Figure 3.

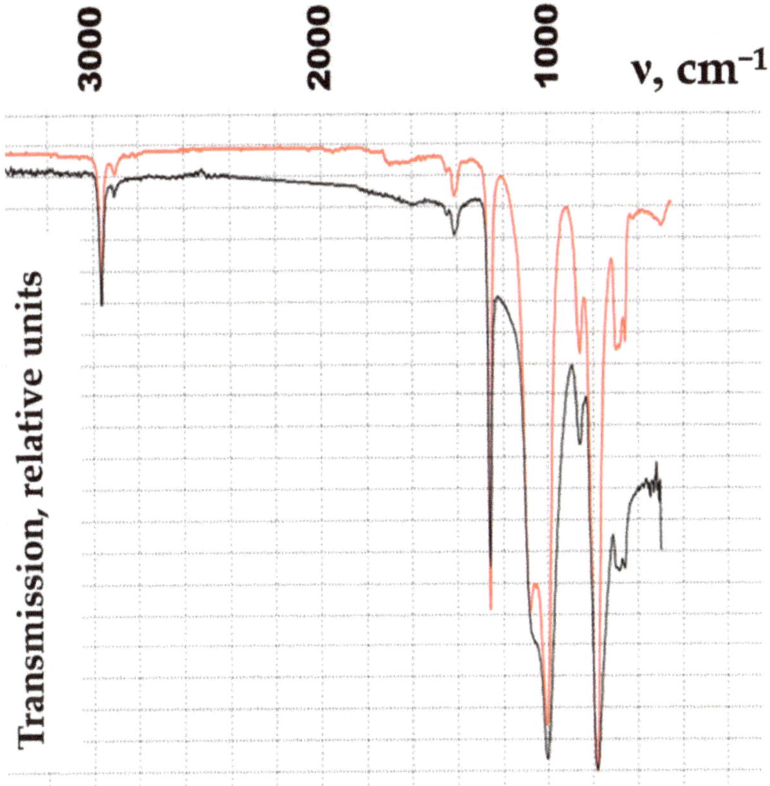

Figure 3. IR spectra of CO and NCO 1.

The test results of NPU 1 and NPU 2—TG and DSC—are shown in Figure 4.

The comparison of TG and DSC for NPU 1 and NPU 2 demonstrates differences caused by MWCNTs with different catalysts. The differences in the profiles of phase transitions are remarkable. The TG and DSC of the silicon-organic compound have already been reported [40].

The temperature fields for the NCO 1 and NCO 2 are compared in Figure 5. An increments of voltage up to 250 V leads to the degradation of the conductive structures and the formation of a local inhomogeneity of the temperature field, which is expressed in partial heat release from the surface of the heater samples. Subsequent increments up to 300 V lead to the additional degradation of the conductive structures and the formation of a local inhomogeneity of the temperature field, which is expressed in partial heat release from the surface of the heater samples, as well as a decrease in the heat release temperature by 10 °C.

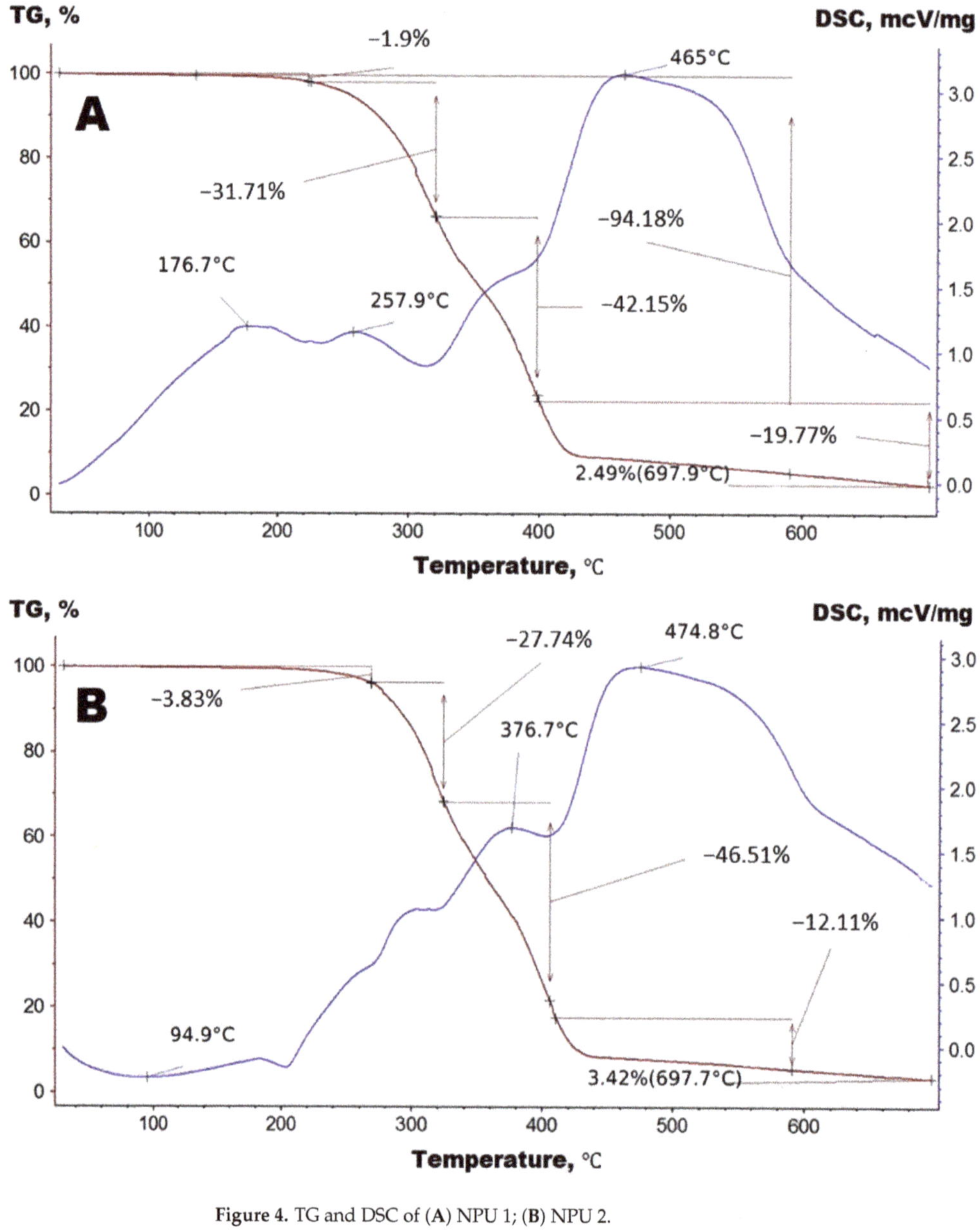

Figure 4. TG and DSC of (**A**) NPU 1; (**B**) NPU 2.

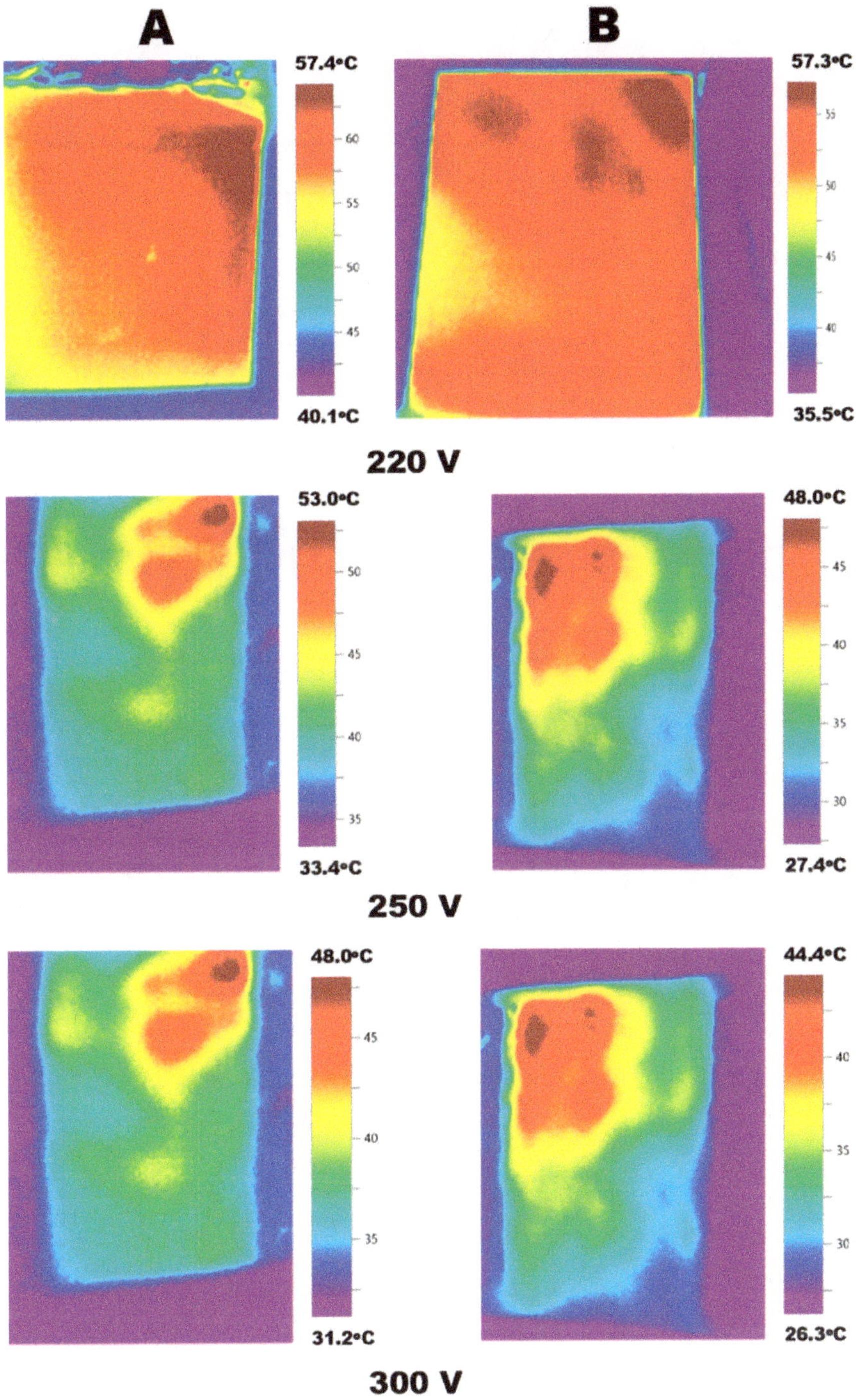

Figure 5. The comparison of the temperature fields for (**A**) NCO 1 and (**B**) NCO 2 at different voltages.

It is of interest to consider the change in the specific power of heat release. Degradation processes in the conductive structures of the composite lead to a decrease in it (Figure 6).

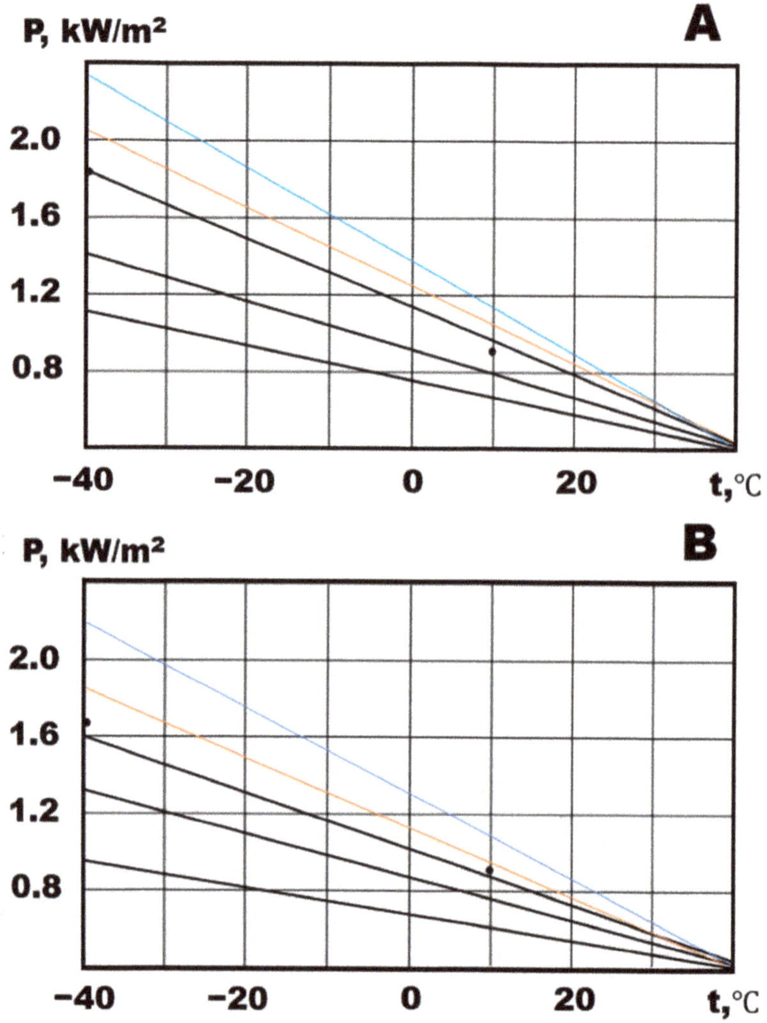

Figure 6. Specific power of heat: (**A**) NCO 1 and (**B**) NCO 2.

Let us consider the change in the specific power of heat release in the case of heaters with a polyurethane matrix (Figure 7). A comparable concentration of MWCNTs for polyurethane composites in comparison with organic silicon provides a lower value of the power level. Degradation processes in the conductive structures of the composite lead to a decrease in the specific power for both NPU 1 and NPU2. However, when first turned on, there is a power surge leading to an increase in power (red and blue lines, both for NPU 1 and NPU2). This observation is consistent with the thesis of the pulsed destruction of conducting structures based on carbon nanotubes in a polymer matrix [33].

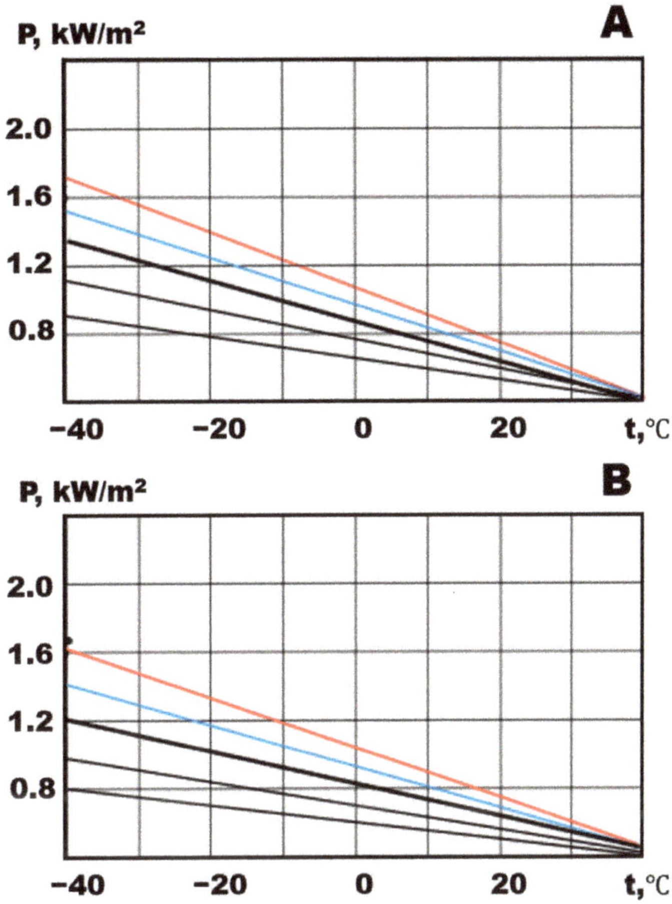

Figure 7. Specific power of heat: (**A**) NPU 1 and (**B**) NPU 2.

4. Discussion

Studies of elastomers modified with MWCNTs were carried out at MWCNTs concentrations of 3 and 10 wt.% (Ni/MgO and Co-Mo/MgO). The modes of heat release of nanomodified elastomers at voltages of 220, 250, and 300 V demonstrate some decrements of heat release for composites with MWCNTs based on Ni/MgO and Co-Mo/MgO. The presented studies make it possible to take the next step in the development of functional materials for electric heating and the level of variation of the supply voltage, which will allow us to obtain information about the physicochemical parameters, following the principles of measuring the heat release that occurs during the transition of electrical energy into thermal energy.

In parallel with infrared thermography, changes in the dependence of specific power on ambient temperature conditions were demonstrated. These data are useful for the development of reliable and energy-efficient functional materials for electric heating based on elastic matrices and nanosized conductive fillers.

The influence of the polymer matrix on the power degradation process was established, from which it follows that, when using a polyurethane matrix, the degradation process is lower by 2.59% for MWCNTs based on Ni/MgO and 10.42% for MWCNTs based on Co-Mo/MgO; however, at the same time, the power of heat release at comparable concentrations of conductive filler (10 and 3 wt.% (Ni/MgO and Co-Mo/MgO)) is lower. From the point of view of the nature of the polymer, this can be explained by the better thermal stability of polyurethane, since local overheating at the point of contact or adhesion of the polymer and conductive structures is more effective with thermal stability. The initially higher heat release rate for the organic silicon compound can be explained by better adhesion in the structure of the CNT polymer.

The developed heaters can be used in the technology for the electric heating of car interiors [41]. Comparison of the obtained heaters (NCO 1/NCO 2; NPU 1/NPU 2) with existing analogs [42–50] in Table 2 leads to a conclusion about those that are the most adaptable to the supply voltage of 220 V, which makes it possible to implement more economical heating modes in household appliances, such as thermal fans and heaters. The conducted studies have shown the safety of using heaters at a higher voltage of up to 300 V, which does not lead to their ignition but only causes degradation of electrophysical parameters without loss of performance.

Table 2. Comparing the electrothermal properties of different materials.

##	Materials	Voltage, V	Area, cm^2	Temperature °C
1	CCSCF [42]	10	2 × 2	105
2	rGo/PET [43]	10	2 × 4	73
3	CNT [44]	20	2 × 2	90
4	rGO/CNT/NR [45]	15	-	69.1
5	CNT/PU [46]	2,5	-	70.4
6	CNT-embedded electric heating composites [47]	20	-	80
7	carbon fiber (CF)/asphalt mastics [48]	60	-	5
78	silver nanowire (AgNW) microgrid (AMG) structures [49]	2	-	51.4
79	F-N Co-Doped Graphene Oxide with Extended Sp2 Networks [50]	9	-	365
10	Under investigation: NCO 1/NCO 2 NPU 1/NPU 2	220	10 × 7	64.2/57.3 58.3/55.2

5. Conclusions

Let us conclude with the consequences of the observed changes in the electrical parameters of MWCNTs-containing elastomer samples. It has been established that an increase in voltage relative to the nominal 220 V by 13.64% leads to an initial peak power value with an increase of 16.67% and a decrease in power (for a temperature of −40 °C) by 22.22%, and by 36.36% (80 V)—a 33.33% increase in peak and a 38.89% derating for Ni/MgO-based MWCNTs and 18.75% to 43.75% (initial increase and subsequent decrease) for MWCNT based on Co-Mo/MgO with steady-state conditions after degradation is 18.75% and 43.75% less than nominal.

As for a polyurethane matrix, an increase in voltage relative to the nominal 220 V by 13.64% leads to an initial peak power value with an increase of 15.38% and a decrease in power (for a temperature of −40 °C) by 15.38% and by 36% (80 V)—a 30.77% increase in the peak and a 30.77% derating for Ni/MgO-based MWCNTs and from 16.67% to 33.33% (initial increase and subsequent decrease) for MWCNTs based on Co-Mo/MgO with steady-state conditions after degradation from 16.67% to 33.33% less than the nominal value.

The influence of the polymer matrix on the process of power-caused structural degradation was established, from which it follows that, when using a polyurethane matrix, the degradation process is lower (by 2.59% for MWCNTs based on Ni/MgO and 10.42% for MWCNTs based on Co-Mo/MgO), but at the same time, the power of heat release at comparable concentrations of conductive filler (10 and 3 wt.% (Ni/MgO and Co-Mo/MgO)) is lower.

In our upcoming studies, we are going to develop technologies based on these results.

Author Contributions: Conceptualization, A.V.S. (Alexandr V. Shchegolkov) and A.A.V.; Methodology, A.V.S. (Alexandr V. Shchegolkov), A.V.S. (Aleksei V. Shchegolkov); Software, A.V.S. (Alexandr V. Shchegolkov), A.V.S. (Aleksei V. Shchegolkov); Validation, A.V.S. (Alexandr V. Shchegolkov), A.V.S. (Aleksei V. Shchegolkov); Formal analysis, A.V.S. (Alexandr V. Shchegolkov), A.V.S. (Aleksei V. Shchegolkov); Investigation, A.V.S. (Alexandr V. Shchegolkov), A.V.S. (Aleksei V. Shchegolkov); Resources, A.V.S. (Alexandr V. Shchegolkov), A.V.S. (Aleksei V. Shchegolkov); Data curation, A.V.S. (Alexandr V. Shchegolkov), A.V.S. (Aleksei V. Shchegolkov); Writing—original draft preparation, A.V.S. (Alexandr V. Shchegolkov), A.V.S. (Aleksei V. Shchegolkov), A.A.V., Y.M.S., N.V.Z.; Writing—review and editing, A.A.V.; Visualization, A.V.S. (Alexandr V. Shchegolkov), A.V.S. (Aleksei V. Shchegolkov), A.A.V.; Supervision, Y.M.S. and A.A.V.; Project administration, A.V.S. (Alexandr V. Shchegolkov), A.V.S. (Aleksei V. Shchegolkov); Funding acquisition, A.A.V., A.V.S. (Alexandr V. Shchegolkov), A.V.S. (Aleksei V. Shchegolkov) All authors have read and agreed to the published version of the manuscript.

Funding: This paper has been supported by the RUDN University Strategic Academic Leadership Program (recipient A.A.V.)

Institutional Review Board Statement: Not applicable.

Informed Consent Statement: Not applicable.

Data Availability Statement: The data presented in this study are available on request from the corresponding author.

Acknowledgments: The authors wish to thank Vasilisa D. Bystrykh for her assistance with the edition of the submission's final version. Alexandre A. Vetcher (IBTN (RUDN)) gratefully acknowledges that this paper has been supported by the RUDN University Strategic Academic Leadership Program (for A.A.V.).

Conflicts of Interest: The authors declare no conflict of interest. The funders had no role in the design of the study; in the collection, analysis, or interpretation of data; in the writing of the manuscript, or in the decision to publish the results.

References

1. Ammar, M.; Haleem, A.; Javaid, M.; Bahl, S.; Garg, S.B.; Shamoon, A.; Garg, J. Significant applications of smart materials and Internet of Things (IoT) in the automotive industry. *Mater. Today Proc.* **2022**, *68*, 1542–1549. [CrossRef]
2. Fu, G.; Shi, Q.; He, Y.; Xie, L.; Liang, Y. Electroactive and photoluminescence of electrospun P(VDF-HFP) composite nanofibers with Eu^{3+} complex and $BaTiO_3$ nanoparticles. *Polymer* **2022**, *240*, 124496. [CrossRef]
3. John, D.W. Madden, Chapter 2—Dielectric elastomers as high-performance electroactive polymers. In *Dielectric Elastomers as Electromechanical Transducers*; Carpi, F., De Rossi, D., Kornbluh, R., Pelrine, R., Sommer-Larsen, P., Eds.; Elsevier: Amsterdam, The Netherlands, 2008; pp. 13–21. [CrossRef]
4. Sui, X.; Downing, J.R.; Hersam, M.C.; Chen, J. Additive manufacturing and applications of nanomaterial-based sensors. *Mater. Today* **2021**, *48*, 135–154. [CrossRef]
5. Song, P.; Song, J.; Zhang, Y. Stretchable conductor based on carbon nanotube/carbon black silicone rubber nanocomposites with highly mechanical, electrical properties and strain sensitivity. *Compos. Part B Eng.* **2020**, *191*, 107979. [CrossRef]
6. Shan, Y.; Li, Z.; Yu, T.; Wang, X.; Cui, H.; Yang, K.; Cui, Y. Self-healing strain sensor based on silicone elastomer for human motion detection. *Compos. Sci. Technol.* **2022**, *218*, 109208. [CrossRef]
7. Stanisauskis, E.; Mashayekhi, S.; Pahari, B.; Mehnert, M.; Steinmann, P.; Oates, W. Fractional and fractal order effects in soft elastomers: Strain rate and temperature dependent nonlinear mechanics. *Mech. Mater.* **2022**, *172*, 104390. [CrossRef]
8. Wang, R.; Wang, Y.; Liang, Y.; Yang, Z.; Bai, Y.; He, J. Insights into the synergistic effect of methoxy functionalized halloysite nanotubes for dielectric elastomer with improved dielectric properties and actuated strain. *Compos. Sci. Technol.* **2023**, *231*, 109802. [CrossRef]
9. Yang, D.; Huo, Y.; Zhang, Q.; Xie, J.; Yang, Z. Recent advances on air heating system of cabin for pure electric vehicles: A review. *Heliyon* **2022**, *8*, e11032. [CrossRef]

10. Kim, G.T.; Jung, J.Y. Effect of steering wheel heating system on hand thermal sensation. *J. Mech. Sci. Technol.* **2022**, *36*, 3717–3725. [CrossRef]
11. Evtimov, I.; Ivanov, R. Sapundjiev Energy consumption of auxiliary systems of electric cars. *MATEC Web Conf.* **2017**, *133*, 06002. [CrossRef]
12. Preis, W.; Sitte, W. Electrical properties of grain boundaries in interfacially controlled functional ceramics. *J. Electroceram.* **2015**, *34*, 185–206. [CrossRef]
13. Bell, J.G.; Graule, T.; Stuer, M. Stuer Barium titanate-based thermistors: Past achievements, state of the art, and future perspectives. *Appl. Phys. Rev.* **2021**, *8*, 031318. [CrossRef]
14. Chen, Y.L.; Yang, S.F. PTCR effect in donor doped barium titanate: Review of compositions, microstructures, processing and properties. *Adv. Appl. Ceram.* **2011**, *110*, 257–269. [CrossRef]
15. Ebert, J.N.; Rheinheimer, W. Electric field induced degradation of high-voltage PTCR ceramics. *Open Ceram.* **2022**, *11*, 100280. [CrossRef]
16. Hudaya, C.; Jeon, B.J.; Lee, J.K. High thermal performance of SnO_2:F thin transparent heaters with scattered metal nanodots. *ACS Appl. Mater. Interfaces* **2015**, *7*, 57–61. [CrossRef]
17. Zheming, G.; Chunzhong, L.; Gengchao, W.; Ling, Z.; Qilin, C.; Xiaohui, L.; Wendong, W.; Shilei, J. Electrical properties and morphology of highly conductive composites based on polypropylene and hybrid fillers. *Sens. Actuators B* **2009**, *140*, 451–460. [CrossRef]
18. Stankovich, S.; Dikin, D.A.; Dommett, G.H.; Kohlhaas, K.M.; Zimney, E.J.; Stach, E.A.; Piner, R.D.; Nguyen, S.T.; Ruoff, R.S. Graphene-based composite materials. *Nature* **2006**, *442*, 282–286. [CrossRef]
19. Luo, J.; Lu, H.; Zhang, Q.; Yao, Y.; Chen, M.; Li, Q. Flexible carbon nanotube/polyurethane electrothermal films. *Carbon* **2016**, *110*, 343–349. [CrossRef]
20. Orellana, J.; Moreno-Villoslada, I.; Bose, R.K.; Picchioni, F.; Flores, M.E.; Araya-Hermosilla, R. Self-Healing Polymer Nanocomposite Materials by Joule Effect. *Polymers* **2021**, *13*, 649. [CrossRef]
21. Gu, B.; Pu, G.; Ding, B.; Zhang, K.; He, R.; Fan, J.; Xing, T.; Wu, J.; Yang, W. Improved interfacial bonding strength of silicone rubber/carbon fiber modified by dopamine. *Polym. Compos.* **2022**, *43*, 6975–6986. [CrossRef]
22. Kim, S.-Y.; Gang, H.-E.; Park, G.-T.; Jeon, H.-B.; Kim, H.B.; Oh, S.-H.; Jeong, Y.G. Synergistic effect of polyurethane-coated carbon fiber and electron beam irradiation on the thermal/mechanical properties and long-term durability of polyamide-based thermoplastic composites. *Polym. Compos.* **2022**, *43*, 1685–1697. [CrossRef]
23. Joshi, A.M.; Athawale, A.A. Electrically Conductive Silicone/Organic Polymer Composites. *Silicon* **2014**, *6*, 199–206. [CrossRef]
24. Ali, I.; AlGarni, T.S.; Shchegolkov, A.; Shchegolkov, A.; Jang, S.-H.; Galunin, E.; Komarov, F.; Borovskikh, P.; Imanova, G.T. Temperature self-regulating flat electric heaters based on MWCNTs-modified polymers. *Polym. Bull.* 2021; *in press*. [CrossRef]
25. Shchegolkov, A.V.; Nachtane, M.; Stanishevskiy, Y.M.; Dodina, E.P.; Rejepov, D.T.; Vetcher, A.A. The Effect of Multi-Walled Carbon Nanotubes on the Heat-Release Properties of Elastic Nanocomposites. *J. Compos. Sci.* **2022**, *6*, 333. [CrossRef]
26. Eun-Soo, P. Resistivity and Thermal Reproducibility of the Carbon Black and SnO_2/Sb Coated Titanium Dioxide Filled Silicone Rubber Heaters. *Macromol. Mater. Eng.* **2005**, *12*, 1213–1219. [CrossRef]
27. Oi, K.; Komoto, J.; Kawai, T.; Nonoguchi, Y. Low background estimation of metallic-to-semiconducting carbon nanotube ratio by using infrared spectroscopy. *Synth. Met.* **2021**, *282*, 116958. [CrossRef]
28. Slepchenkov, M.M.; Petrunin, A.A.; Glukhova, O.E. In Silico Study of the Influence of Various Substrates on the Electronic Properties and Electrical Conductivity of Mono- and Bilayer Films of Armchair Single-Walled Carbon Nano-tubes. *ChemEngineering* **2021**, *5*, 48. [CrossRef]
29. Collins, P.G.; Arnold, M.S.; Avouris, P. Engineering carbon nanotubes and nanotube circuits using electrical breakdown. *Science* **2001**, *292*, 706–709. [CrossRef]
30. Senthilkumar, V. Some approximate buckling solutions of triple-walled carbon nanotube. *Vietnam. J. Mech.* **2022**, *44*, 212–232. [CrossRef]
31. Huang, J.Y.; Chen, S.; Jo, S.H.; Wang, Z.; Han, D.X.; Chen, G.; Dresselhaus, M.S.; Ren, Z.F. Atomic-Scale Imaging of Wall-by-Wall Breakdown and Concurrent Transport Measurements in Multiwall Carbon Nanotubes. *Phys. Rev. Lett.* **2005**, *94*, 236802. [CrossRef]
32. Farajian, A.A.; Yakobson, B.I.; Mizuseki, H.; Kawazoe, Y. Electronic transport through bent carbon nanotubes: Nanoelectromechanical sensors and switches. *Phys. Rev. B* **2003**, *67*, 205423. [CrossRef]
33. Collins, P.G.; Hersam, M.; Arnold, M.; Martel, R.; Avouris, P. Current Saturation and Electrical Breakdown in Multiwalled Carbon Nanotubes. *Phys. Rev. Lett.* **2001**, *86*, 3128. [CrossRef] [PubMed]
34. Chu, K.; Yun, D.-J.; Kim, D.; Park, H.; Park, S.-H. Study of electric heating effects on carbon nanotube polymer composites. *Org. Electron.* **2014**, *15*, 2734–2741. [CrossRef]
35. Lee, J.-s.; Jo, H.; Choe, H.-s.; Lee, D.-s.; Jeong, H.; Lee, H.-r.; Kweon, J.-h.; Lee, H.; Myong, R.S.; Nam, Y. Electro-thermal heating element with a nickel-plated carbon fabric for the leading edge of a wing-shaped composite application. *Compos. Struct.* **2022**, *289*, 115510. [CrossRef]
36. Wan, N.; Sun, L.-t.; Ding, S.-n.; Xu, T.; Hu, X.-h.; Sun, J.; Bi, H.-c. Synthesis of graphene–CNT hybrids via joule heating: Structural characterization and electrical transport. *Carbon* **2013**, *53*, 260–268. [CrossRef]

37. Chien, A.-T.; Cho, S.; Joshi, Y.; Kumar, S. Electrical conductivity and Joule heating of polyacrylonitrile/carbon nanotube composite fibers. *Polymer* **2014**, *55*, 6896–6905. [CrossRef]
38. Shehzad, K.; Hakro, A.A.; Zeng, Y.; Yao, S.H.; Xiao-Hong, Y.; Mumtaz, M.; Nadeem, K.; Khisro, N.S.; Dang, Z.M. Two percolation thresholds and remarkably high dielectric permittivity in pristine carbon nanotube/elastomer composites. *Appl. Nanosci.* **2015**, *5*, 969–974. [CrossRef]
39. Huang, S.; Wang, M.; Liu, T.; Zhang, W.-D.; Tjiu, W.C.; He, C.; Lu, X. Morphology, thermal, and rheological behavior of nylon 11/multi-walled carbon nanotube nanocomposites prepared by melt compounding. *Polym. Eng. Sci.* **2009**, *49*, 1063–1068. [CrossRef]
40. Ali, I.; Shchegolkov, A.; Shchegolkov, A.; Zemtsova, N.; Bogoslovskiy, V.; Shigabaeva, G.; Galunin, E.; Hussain, I.; Almalki, A.S.; Alsharif, M.A.; et al. Preparation and application practice of temperature self-regulating flexible polymer electric heaters. *Polym. Eng. Sci.* **2022**, *62*, 730. [CrossRef]
41. Leicht, H.; Kraus, E.; Baudrit, B.; Hochrein, T.; Bastian, M.; Langer, M. Chapter 12—Electrically conductive polymer nanocomposites for thermal comfort in electric vehicles. In *Micro and Nano Technologies, Nanotechnology in the Automotive Industry*; Song, H., Ngu-yen, T.A., Yasin, G., Singh, N.B., Gupta, R.K., Eds.; Elsevier: Amsterdam, The Netherlands, 2022; pp. 229–251. [CrossRef]
42. Guangming, C.; Mengyun, Y.; Junjie, P.; Deshan, C.; Zhigang, X.; Xin, W.; Bin, T. Large-scale production of highly stretchable CNT/cotton/spandex composite yarn for wearable applications. *ACS Appl. Mater. Interfaces* **2018**, *10*, 32726–32735. [CrossRef]
43. Zhang, W.; Fei, L.; Zhang, J.; Chen, K.; Yin, Y.; Wang, C. Durable and tunable temperature-responsive silk fabricated with reactive thermochromic pigments. *Prog. Org. Coat.* **2020**, *147*, 105697. [CrossRef]
44. Morshed, M.N.; Miankafshe, M.A.; Persson, N.-K.; Behary, N.; Nierstrasz, V.A. Development of a multifunctional graphene/Fe-loaded polyester textile: Robust electrical and catalytic properties. *Dalton Trans.* **2020**, *49*, 17281–17300. [CrossRef] [PubMed]
45. Zhan, Y.; Li, Y.; Meng, Y.; Xie, Q.; Lavorgna, M. Electric heating behavior of reduced oxide graphene/carbon nanotube/natural rubber composites with macro-porous structure and segregated filler network. *Polymers* **2020**, *12*, 2411. [CrossRef] [PubMed]
46. Wang, Z.-X.; Du, P.-Y.; Li, W.-J.; Meng, J.-H.; Zhao, L.-H.; Jia, S.-L.; Jia, L.-C. Highly rapid-response electrical heaters based on polymer-infiltrated carbon nanotube networks for battery thermal management at subzero temperatures. *Compos. Sci. Technol.* **2023**, *231*, 109796. [CrossRef]
47. Jang, D.; Yoon, H.N.; Seo, J.; Park, S.; Kil, T.; Lee, H.K. Improved electric heating characteristics of CNT-embedded polymeric composites with an addition of silica aerogel. *Compos. Sci. Technol.* **2021**, *212*, 108866. [CrossRef]
48. Arabzadeh, A.; Ceylan, H.; Kim, S.; Sassani, A.; Gopalakrishnan, K.; Mina, M. Electrically-conductive asphalt mastic: Temperature dependence and heating efficiency. *Mater. Des.* **2018**, *157*, 303–313. [CrossRef]
49. Noh, J.Y.; Kim, J.M. Transparent film heater based on a microgrid-patterned metal nanowire network prepared by using the bilayer lift-off technique. *J. Korean Phys. Soc.* **2022**, *80*, 404–409. [CrossRef]
50. Tembei, S.A.; Ali, M.K.; Hessein, A.; El-Bab, A.M.F.; El-Moneim, A.A. High-performance flexible electrothermal Joule heaters from laser reduced F-N Co-doped graphene oxide with extended Sp2 networks. *FlatChem* **2022**, *36*, 100437. [CrossRef]

polymers

MDPI

Article

Effect of Magnetized Coagulants on Wastewater Treatment: Rice Starch and Chitosan Ratios Evaluation

Nomthandazo Precious Sibiya *, Gloria Amo-Duodu *, Emmanuel Kweinor Tetteh and Sudesh Rathilal

Green Engineering Research Group, Department of Chemical Engineering,
Faculty of Engineering and The Built Environment, Durban University of Technology, Durban 4001, South Africa
* Correspondence: nomtha.presh@gmail.com (N.P.S.); gamoduodu04@gmail.com (G.A.-D.);
Tel.: +27-63205-5378 (N.P.S.); +27-84-9992841 (G.A.-D.)

Abstract: Coagulation with synthetic chemicals has been used to treat a wide range of industrial effluents. Herein, the unique characteristics of industrial effluents being detrimental to the environment warrants urgent resource-efficient and eco-friendly solutions. Therefore, the study investigated the use of two magnetized coagulants (chitosan magnetite (CF) and rice starch magnetite (RF)), prepared via co-precipitation in three different ratios (1:2, 1:1 and 2:1) of natural coagulants (chitosan or rice starch) and magnetite nanoparticles (F) as alternative coagulants to alum for the treatment of wastewater. A Brunauer–Emmett–Teller (BET) analyzer, an X-ray diffraction (XRD) analyzer, and energy-dispersive X-ray (EDX) spectroscopy were used to characterize the surface area, crystal structure, and elemental composition of the coagulants. The influences of settling time (10–60 min) on the reduction of turbidity, color, phosphate, and absorbance were studied. This was carried out with a jar test coupled with six beakers operated under coagulation conditions of rapid stirring (150 rpm) and gentle stirring (30 rpm). Wastewater with an initial concentration of 45.6 NTU turbidity, 315 Pt. Co color, 1.18 mg/L phosphate, 352 mg/L chemical oxygen demand (COD), and 73.4% absorbance was used. The RF with a ratio of 1:1 was found to be the best magnetized coagulant with over 80% contaminant removal and 90% absorbance. The treatability performance of RF (1:1) has clearly demonstrated that it is feasible for wastewater treatment.

Keywords: coagulation; chitosan; rice starch; magnetite; wastewater treatment; magnetized coagulants

Citation: Sibiya, N.P.; Amo-Duodu, G.; Tetteh, E.K.; Rathilal, S. Effect of Magnetized Coagulants on Wastewater Treatment: Rice Starch and Chitosan Ratios Evaluation. *Polymers* **2022**, *14*, 4342. https://doi.org/10.3390/polym14204342

Academic Editors: Alexandre Vetcher and Alexey Iordanskii

Received: 18 August 2022
Accepted: 10 October 2022
Published: 15 October 2022

1. Introduction

The demand on freshwater resource supply is deteriorating as industrial and agricultural activities are upsurging and population growth shows a constant increase. Consequentially, wastewater treatment becomes inevitable due to the detrimental effects it has on human health and the environment [1]. Water scarcity affects approximately 2 billion people [2]. Wastewater treatment plants (WWTPs) are a substantial source of pollutants discharged into water bodies [3,4] due to the fact that they do not meet the Environmental Protection Agency's strict criteria for controlling effluent plant quality [5,6]. Therefore, it is critical for WWTPs to release clean effluent into water bodies since they serve as potable water and are utilized for agriculture purposes and other recreational activities. Various processes and technologies have been explored to enhance the quality of water in order to meet the water demand at a low cost [7]. These technologies are divided into three categories: physical (settling, filtration and membrane technology), chemical (coagulation, ion exchange, disinfection, oxidation, catalytic reduction and softening processes) and biological (microbial biodegradation, bioreactor processes, etc.) treatment techniques [8,9]. Coagulation/flocculation is frequently used in water and wastewater treatment to efficiently reduce the organic load prior to subsequent treatment processes [10]. Furthermore, it has sparked widespread attention in the industrial sector due to easy usability, high efficiency, and low cost. In coagulation, tiny particles are aggregated into bigger aggregates (flocs) and dissolved organic matter is adsorbed onto particulate aggregates so that these contaminants may be eliminated in later solid/liquid separation processes [11,12].

Coagulation involves a sequence of stages, beginning with electrostatic attraction between cationic proteins and negatively charged suspended contaminants, and ending with particle destabilization [11,13,14]. It occurs through three distinct mechanisms: (i) charge neutralization (particle destabilization at low coagulant dosage), (ii) sweep (the addition of coagulant at sufficiently high concentrations to produce anhydrous, amorphous precipitates enmeshing colloidal particles in these precipitates), and (iii) bridge formation [14,15]. Synthetic coagulants are classified as hydrolyzing metallic salts, pre-hydrolyzing metallic salts, and synthetic cationic polymers [16]. The most widely used coagulant salts are $Al_2(SO_4)_3$, $Fe_2(SO_4)_3$, $AlCl_3$, and $FeCl_3$ [1,17]. Although their efficacy has been well established, their usage is associated with high costs and environmental drawbacks due to their non-native origin, non-degradability, and the health problems linked to their leftover residues [18,19]. The excessive levels of aluminum tend to lower the pH of water and can accumulate in the food chain. Therefore, these drawbacks have resulted in a search for eco-friendly and sustainable coagulants in their use and availability. The progress of nanotechnology has substantially contributed to the creation of innovative ways for tackling many health and environmental concerns while using less energy [5].

Several studies have investigated the use of iron oxide nanoparticles, such as maghemite (γ-Fe_2O_3), magnetite (Fe_3O_4) and hematite (α-Fe_2O_3), in the removal of heavy metals and hazardous chemicals from water, demonstrating the effectiveness of these materials [20–22]. These are seen as very promising due to their multifunctional ability and physicochemical properties, which include superparamagnetic behaviors, smaller size, high surface area, environmental friendliness, and biodegradability [5,23]. Natural coagulants provide a cost-effective, environmentally acceptable, and long-term alternative to the use of synthetic chemicals [5,14]. They are generated or derived from a variety of sources, including microbes, animals, and plants [1]. Magnetic iron oxide nanoparticles functionalized with polymers such as eggshells, chitosan, rice starch, banana peels, moringa seeds, and other nanoparticles are gaining interest in wastewater treatment settings [17,24]. This process is seen as very appealing due to the high efficiency, ability to reduce sludge volume, rapid sedimentation, and low cost [1,5].

Natural coagulants having positively charged functional groups, such as chitosan (CS), have a greater potential for interacting with hydroxyl groups (OH-) in wastewater than cellulose [15,25]. CS is a nontoxic linear poly-amino-saccharide produced by chitin deacetylation and is the second most abundant biopolymer after cellulose [25,26]. CS has a high density, is biodegradable, simple to handle, non-toxic, leaves no residual metal in treated water, and eliminates secondary contamination issues [10,15]. The usage of chitosan produced from bio-waste also creates bigger flocs, allowing for simpler removal while reducing environmental responsibility for manufacturers of crustacean waste products [14,27]. Chitosan's coagulant activity is highly influenced by the concentration and the pH of the solution [15]. As a result, optimizing operational parameters is important for successful pollution removal. Starch is one of the most abundant polysaccharides on the planet, is inexpensive, and is composed of two anhydroglucose units: linear amylose and highly branching amylopectin [28,29]. The use of starch as a coagulant has sparked a lot of interest in the reduction of turbidity from industrial effluents [30,31]. When starch comes into contact with hot water, the crystalline structure is disturbed, causing permanent swelling of the granules and an increase in viscosity [29,32]. Therefore, this study aims to assess and compare the efficacy of magnetized chitosan and rice starch synthesized by co-precipitation in three different ratios (1:2, 1:1, and 2:1) of natural coagulants (chitosan or rice starch) and magnetite (Fe_3O_4) nanoparticles in a magnetized coagulation system for wastewater treatment.

2. Materials and Methods

2.1. Material

Sigma-Aldrich (Kempton Park, South Africa), provided the sodium hydroxide (NaOH), ferrous sulphate heptahydrate (Fe^{2+}), ferric chloride hexahydrate (Fe^{3+}), oleic acid, ethanol, and chitosan (with a deacetylation level of 75–85%) needed in the synthesis of magnetized coagulants. The materials were utilized without being further purified. The rice starch was acquired from a local grocery store (Shoprite, Durban, South Africa), and powdered according to test technique by Sibiya et al. [5] and Pritchard et al. [33]. Deionized water (ELGA WATERLAB, PURELAB Option-Q water deionizer (High Wycombe, UK)) was used for the preparation of the stock solutions.

2.2. Wastewater Samples

Wastewater samples were collected after chlorination process at a local wastewater treatment facility in KwaZulu-Natal, South Africa. The samples were characterized according to APHA [34]. Table 1 shows the characterization of the wastewater used in this study. A turbidimeter (Hach, 2100N, Loveland, CO, USA) was used to analyse turbidity, while a spectrophotometer (HACH, DR 3900, Düsseldorf, Germany) was used to analyse color, phosphate, COD, TSS, and absorbance at a wavelength of 465 nm, 880 nm, 620 nm, 810 nm and 620 nm, respectively.

Table 1. Characteristics of wastewater sample composition.

Contaminants	Values	Standard Deviation
Turidity (NTU)	45.60	0.2910
Color (Pt. Co)	315	0.098
Absorbance (%)	73.40	0.151
Chemical oxgyen demand (COD) (mg/L)	352	0.816
Total suspended solids (TSS) (mg/L)	68.20	0.748

2.3. Synthesis of Magnetized Coagulants

To begin, 1 L stock solutions were calculated for each chemical using their mass (g) and molecular weights, as indicated in Table 2. In reality, the 0.4 M Fe^{3+}, 0.2 M Fe^{2+}, and 3 M NaOH stock solutions were made using 108.12 g, 55.61 g, and 199.99 g dissolved in 1 L distilled water, respectively [35]. These were calculated using the Sigma-Aldrich mass molarity calculator with a known concentration and volume.

Table 2. Stock solution composition.

Chemicals	Concentration (M)	Molar Mass (g/mol)	Mass (g)
$FeCl_3$ $6H_2O$	0.4	270.29	108.21
$FeSO_4$ $7H_2O$	0.2	287.55	55.61
NaOH	3	39.997	199.99

Figure 1 shows the modified schematic diagram of synthesizing the magnetized coagulants. The Fe_3O_4 nanoparticles (F) were synthesized using the co-precipitation technique with a molar ratio of 1:2 for Fe (II): Fe (III) [32,36]. This was then dissolved in 1 L of deionized water with gentle mixing at 30 rpm for 1 h [18]. Following that, 2 mL of oleic acid was added as a surfactant, followed by 2 h of slow mixing (50 rpm) at 70 $\pm$ 5 °C [35]. Then, 250 mL of 3 M NaOH was added dropwise to the mixture to correct the pH (11) until a thick black precipitate developed while constantly stirring at 50 rpm. The black precipitate was allowed to cool before being filtered and rinsed with deionized water and ethanol. Thereafter, it was oven-dried at 80 °C for 24 h. The weight ratios of natural coagulant (chitosan or rice starch) to Fe_3O_4 were 1:1, 1:2 and 2:1, as shown in Table 3. The samples were well-mixed before being calcined (Kiln Furnace (Cape Town, South Africa)) at 550 °C for 1 h to remove organic contaminants. The MC samples were then packaged for characterization.

$$FeCl_2 \cdot 4H_2O + 2FeCl_3 + 8NaOH = Fe_3O_4 + 8NaCl + 20H_2O \quad (1)$$

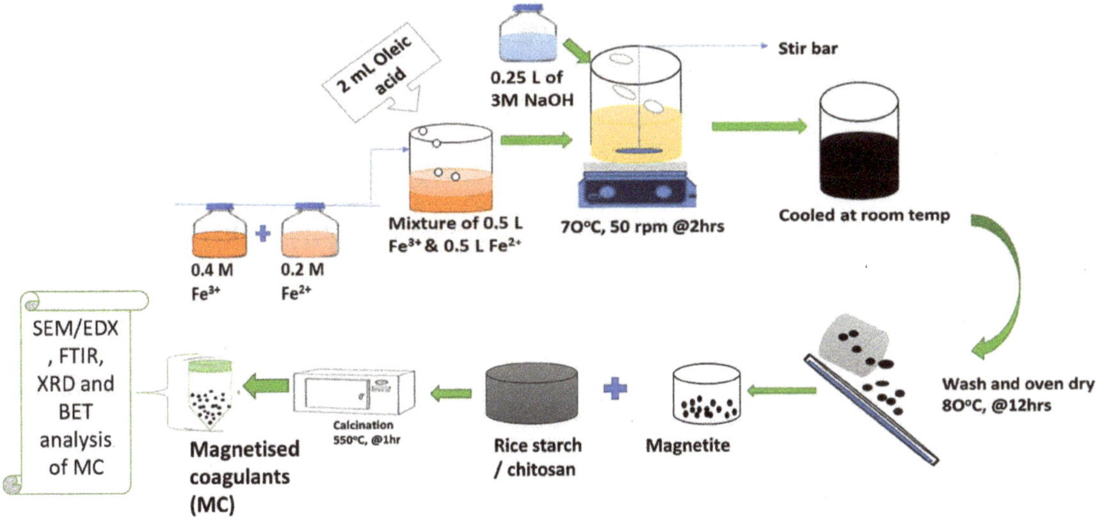

Figure 1. Schematic diagram of synthesizing magnetic coagulants.

Table 3. Amounts measured for natural coagulant and magnetite to prepare 10 g of magnetized coagulants.

Ratios of Magnetized Coagulants (MC)	CF (g)		RF (g)	
	CS (g)	F (g)	R (g)	F (g)
1:1	5	5	5	5
2:1	6.667	3.333	6.667	3.333
1:2	3.333	6.667	3.333	6.667

(CF—magnetized chitosan and RF—magnetized rice starch).

2.4. Characterization of MC

2.4.1. Morphological and Elemental Examination (SEM/EDX)

The elemental analysis of the natural polymers, magnetite and MC were identified using scanning electron microscopy and energy-dispersive X-ray (SEM/EDX) at the University of Cape Town, South Africa. This was accomplished with a 5 kV acceleration voltage and a magnification range of 10 to 50,000x.

2.4.2. Crystal Structure Analysis (XRD)

At 40 kV and a target current of 40 mA, the crystal structures of the generated MC were investigated using an X-ray diffractometer (Bruker AXS, D8 Advance (Madison, WI, USA) coupled with PANalytical software (Malvern PANalytical Ltd., Empyrean, Malvern, UK). The measurements were taken between 15 and 80 (2θ), with a typical step size speed of 0.0340°/min. Lyn-Eye, a position-sensitive detector, was used to gather diffraction data at a typical speed of 0.5 s/step, which was similar to a dazzling counter's effective length of 92 s/step. The MC diameters were calculated from the XRD data using the Debye–Scherrer Equation (2), which establishes a relationship between particle size and peak enlargement:

$$d = \frac{0.98\delta}{\beta Cos\theta} \quad (2)$$

where d is the particle size of the crystal, δ is the wavelength of X-ray radiation (CuKα = 0.15406 nm), 0.98 is the Scherrer constant, β is the line broadening in radians obtained from the full width at half maximum height (FWHM) of the peak (determined using Origin software), and θ is the Bragg diffracting angle of the XRD diffraction patterns.

2.4.3. Surface Area Analysis (BET)

The Brunauer–Emmett–Teller (BET) analysis was performed on the equipment (Micromeritics Instrument Corporation, TriStar II Plus version 3.01, Roodepoort, South Africa). The carrier gases were helium and nitrogen (Afrox, Durban, South Africa). Samples were weighed, placed in a sample container on the analyzer, and degassed individually for 24 h at 400 °C. They were then allowed to cool before being stored under nitrogen gas at a pressure of 5 mmHg for 24 h.

2.4.4. Functional and Molecular Examination (FTIR)

The organic, polymeric, and inorganic molecular structures and functional groups of the NPs were recorded using a Fourier transform infrared spectrometer (Shimadzu Corporation, FTIR 8400, Kyoto, Japan) with a resolution of 7 cm^{-1} in the 400–1200 cm^{-1} range.

2.5. Coagulation

The coagulation tests were carried out in beakers with 500 mL of wastewater with the use of a jar-test (VELP Scientifica, JTL6, Usmate Velate, MB, Italy) coupled with six paddles [5]. Each beaker was dosed with 4 g of MC (Table 1). Firstly, the types of coagulants were evaluated at a consistent settling time of 60 min. The effect of settling time (10–60 min) on turbidity, color, phosphate, and absorbance removal was also studied. Thereafter, the beakers were agitated at a fast-mixing rate of 150 rpm for 2 min, followed by flocculation at 30 rpm for 15 min [5,18], and then allowed to settle. The effluents were immediately analyzed for turbidity, color, and phosphate in accordance with APHA [34]. The proportion of pollutants (turbidity, color and phosphate) removed were determined by Equation (3):

$$Z(\%) = \frac{C_a - C_b}{C_a} \times 100 \quad (3)$$

where C_a and C_b are the initial and final values of each contaminant (phosphate, turbidity or color) and Z is the percentage removal.

3. Results and Discussions

3.1. Characterization

3.1.1. Elemental Composition by Energy Disperse X-Ray (EDX) and SEM Analysis

Table 4 shows the EDX analyses of magnetite, chitosan, and rice starch, together with their elemental distributions at 20 KeV. The magnetite indicated the following composition: Fe (39.13 percent) > O (38.43 percent) > C (10.36 percent) > S (9.72 percent) > Cl (2.37 percent), as shown in a previous study [5]. The presence of carbon could be due to the doping with carbon gas during the analysis. The composition of chitosan was C (92.89 percent) > O (6.86 percent) > Na (0.25 percent), whereas the composition of rice starch was C (84.55 percent) > O (14.62 percent) > P (0.43 percent) > K (0.40 percent).

Table 5 shows the EDX results for all magnetized coagulants (MCs). Carbon (30.23 to 64.31 percent C), oxygen (22.67 to 43 percent O), sulphur (0.82 to 13.1 percent S), chlorine (0.80 to 15.68 percent Cl), and iron (7.04 to 67.32 percent Fe) were also shown to be the most prevalent elements in all MCs. This is consistent with Equation (1), which demonstrated that magnetite is predominantly composed of Fe and O, while the addition of extra components (S and Cl) may enhance their surface area, allowing for improved biosorption and reusability [35]. In addition to the elements listed above, the RF and CF included potassium (0.35 to 2.14 percent K) and sodium (1.67 to 6.37 percent Na), respectively.

Table 4. EDX results for the magnetite, chitosan, and rice starch.

Elements	C	O	S	Fe	Na	P	K	Cl	Ca
Weight	%	%	%	%	%	%	%	%	%
Magnetite	10.36 ± 6.90	38.43 ± 17.12	9.72 ± 3.31	39.13 ± 14.39	-	-	-	2.37 ± 1.64	
Chitosan	92.89 ± 9.73	6.86 ± 9.39	-	-	0.25 ± 0.34	-	-	-	-
Rice starch	84.55 ± 1.68	14.62 ± 1.91	-	-	-	0.43 ± 0.29	0.40 ± 0.15	-	-

Table 5. EDX results for the all magnetized coagulants.

Elements	C	O	S	Fe	Na	P	K	Cl	Ca
Weight	%	%	%	%	%	%	%	%	%
CF (1:2)	-	27.12 ± 13.15	-	67.32 ± 18.88	3.36 ± 4.29	-	-	2.20 ± 3.09	
CF (1:1)	-	22.67 ± 14.13	-	74.80 ± 15.81	1.67 ± 1.21	-	-	0.80 ± 1.18	0.07 ± 0.16
CF (2:1)	-	32.83 ± 8.13	0.82 ± 0.64	57.44 ± 14.2	6.37 ± 4.83	-	-	2.54 ± 2.32	
RF (1:2)	-	42.81 ± 8.99	13.10 ± 2.60	26.32 ± 5.61	-	-	2.14 ± 0.38	15.68 ± 1.92	-
RF (1:1)	30.23 ± 13.32	34.44 ± 15.87	5.54 ± 3.74	23.34 ± 11.98	-	-	2.14 ± 0.38	6.44 ± 5.29	-
RF (2:1)	64.31 ± 15.02	24.48 ± 5.40	0.83 ± 0.25	7.04 ± 8.24	-	-	0.55 ± 0.27	2.99 ± 2.76	-

Figure 2a–c depicts the SEM findings of chitosan (CS), rice starch (RS), and magnetite (Fe_3O_4) at width distances (WD) ranging from a low to large porosity of 5.30, 5.6, and 6.0 mm, respectively. Magnetite and CS were obtained at a microscale of 1 µm using a magnification of 50 kx, a view field of 4.5 µm, and an accelerating voltage of 5 kV. The RS (Figure 2c) was acquired at a 10 µm microscale, with a magnification of 10 kx, a horizontal view width of 29.8 m, and a landing energy capacity of 20 keV. The change in appearance and view width might be attributed to the 550 °C calcination temperature, which improved the liquid–solid adsorption capability [5,35]. Figure 2a depicts magnetite with a regular cellular appearance and a slightly agglomerated structure [37–39], indicating that the iron crystals were partially obscured by carbon particles, possibly due to the sample's calcination process and carbon coating prior to analysis [35,40,41]. Figure 2b displays the surface aggregation of CS particles, which was heterogeneous and resulted in a smooth surface [42,43]. Figure 2c illustrates RS's SEM, which had a polygonal irregular form with a smooth surface, as shown in the photos observed by Han et al. [44]. In this work, RS was calcined at temperatures considerably above the temperature (62 to 78 °C) at which crystals shred their deformability, also known as the gelatinization temperature [45]. As a result of the continuous heating (1 h calcination), the RS granules expanded and the crystallites melted, and amylose and amylopectin were entirely separated from the starch [46].

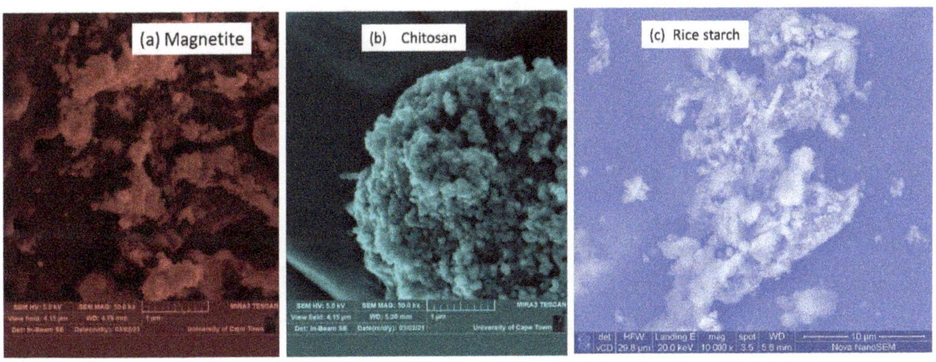

Figure 2. SEM images (**a**) magnetite, (**b**) chitosan, and (**c**) rice starch.

Composite surface images (Figure 3a–f) were captured at magnifications of 10× and 50× and a microscale of 5 μm to demonstrate the transformation of un-magnetized coagulants into magnetized coagulants. These binary composites, at various ratios, were seen to have a rough surface. The magnetized chitosan in various ratios (Figure 3a–c) has a compact surface that exhibits a light aggregate, and more pores are visible, particularly in Figure 3c. Similar findings were drawn from Figure 3d–f, the last of which revealed a porous surface with a clearly defined porous structure. The modifications might be the result of calcination at a high temperature (550 °C), which has been observed to alter crystal structures and increase porosity.

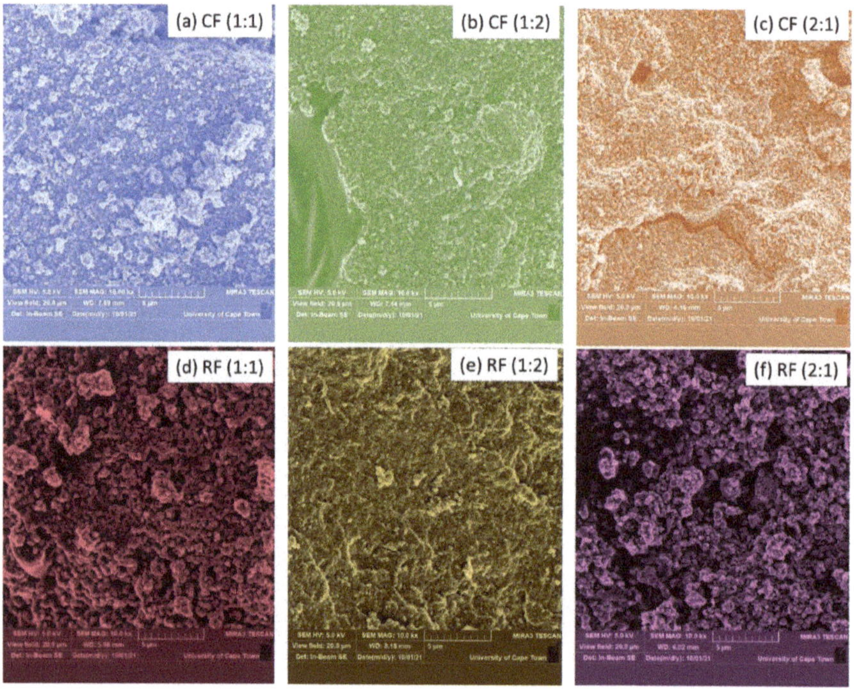

Figure 3. SEM images (**a**–**f**) of the magnetized coagulants.

3.1.2. Crystal Structure Analysis (XRD)

XRD results were attained in a powdered form of the coagulant as explained in Section 2.4.2. Figure 4 depicts the XRD results for (a) chitosan and its magnetized coagulants, and (b) rice starch and its magnetized coagulants. The XRD pattern of the magnetite produced is very comparable to the JCPDS pattern (Table 6). Figure 4 shows that the reflection at 2θ values of 21.398°, 35.423°, and 46.8° corresponded to the lattice planes of (14), (227), and (148), respectively, for the magnetite. To calculate the size of magnetite crystallites formed at various times, Scherer's equations with a pseudo-Voigt function were applied to the peak at 2θ of 35.423° [47,48].

The examined diffraction peaks matched normal magnetite XRD patterns more closely than maghemite, indicating a cubic crystalline structure [49]. Figure 4a depicts the XRD patterns of CS and its magnetic coagulants. The chitosan pattern exhibited one unique broad diffraction peak at 2θ = 20.15°, which is characteristic of semi-crystalline chitosan fingerprints [50–53]. This peak in the chitosan structure is related to crystal I and crystal II, and it suggests that the chitosan has a high degree of crystallinity [43,54]. The (225) plane corresponds to the apex at around 30.1°. Its magnetized coagulants behave similarly, with a noticeable peak at 35.423°, which corresponds to the (227) plane. The XRD image of rice starch and its magnetized coagulants is shown in Figure 4b. Rice starch exhibited a characteristic A-type XRD pattern, with prominent peaks at 2θ = 17.374°, 24.482°, and 25.432°, that were universally acknowledged as the usual cereal starch and legume crystal shape, which corresponds to the lattice patterns of (160), (225), and (63), respectively [55–58]. The wide and difficult-to-distinguish diffraction peaks revealed the presence of very tiny crystallite sizes [58].

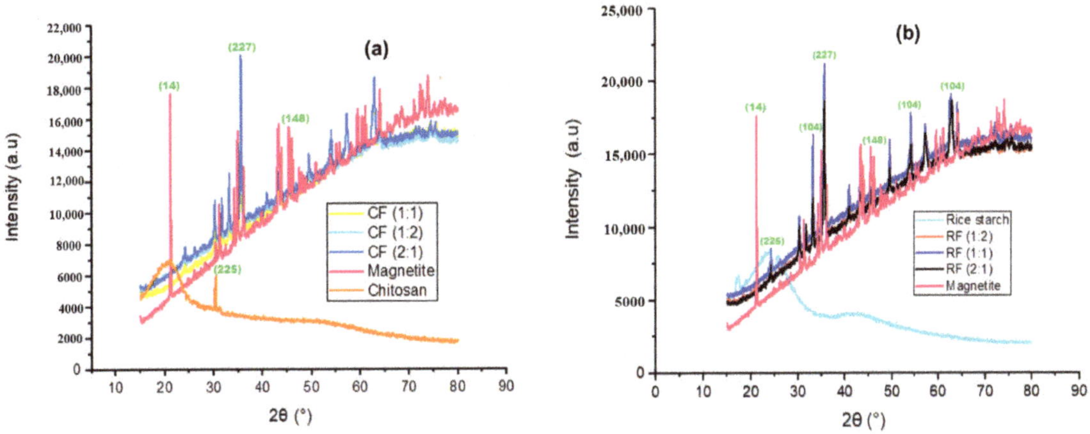

Figure 4. XRD spectra (**a**), Chitosan and its magnetized coagulants (**b**) Rice starch and its magnetized coagulants

Table 6. Physical and chemical properties of the MCs characterized obtained from the XRD.

2θ (Degree)	Miller Indices Plane (hkl)	dhkl (nm)	Crystal Structure	Nanostructure	JCPDS Pattern
24.482	(225)	1.988	Face-centered cubic	Sylvite	00-41-1476
46.8	(148)	3.157	Rhombohedral	Mikasaite	00-047-1774
21.398	(14)	3.202	Monoclinic	Ferrimagnetite	00-070-2091
48.491	(12)	2.106	Monoclinic	Clinoptilolite	01-071-1425
51.44	(44)	2.162	Base-centered Orthohombic	Sodium nitrate	01-075-2073
27.335	(225)	2.163	Face-centered cubic	Halite (NaCl)	00-005-0628
32.497	(29)	2.263	Orthohombic	Thermonatrite	00-008-0448
20.212	(13)	1.96	Monoclinic	Iron chloride hydrate	00-016-0123
35.525	(227)	4.523	Face-centered cubic	Magnesioferrite	00-017-0464
35.423	(227)	5.197	Face-centered cubic	Magnetite (Fe_3O_4)	00-019-0629
17.374	(160)	3.065	Rhombohedral	Hydronium jarosite	00-031-0650
33.153	(104)	5.27	Rhombohedral	Hematite (α-Fe_2O_3)	00-033-0664
24.717	(148)	3.06	Rhombohedral	Mikasaite	00-033-0679
35.631	(110)	4.858	Cubic	Maghemite (y-Fe_2O_3)	00-039-1346

Starches with an A-type diffraction pattern are said to have a helical arrangement in monoclinic symmetry unit cells [59]. RF (2:1), RF (1:1), and RF (1:2) demonstrated a similar pattern, with large peaks at 33.153°, 35.423°, 54.5°, and 63.5°, which correspond to the lattice planes of (104), (227), (116), and (214), respectively, and are identical to Matmin et al. [58]. Figure 4b shows that RF (1:1) has greater intensity values than the other RF ratios.

3.1.3. Surface Area Analysis (BET)

Table 7 displays the BET results. Due to magnetite's specific magnetic properties and durability, integrating Fe_3O_4 into natural coagulants (chitosan or rice starch) enhanced the surface area of the magnetic coagulants, enhancing their adsorption capability [5,30,35]. Amongst the coagulants examined, RF (1:1) exhibited the highest surface area of pores and porosity. Surface area is shown in descending order: RF (1:1) > RF (2:1) > RF (1:2) > Magnetite> CF (1:1) > CF (2:1) > CF (1:2) > Rice starch > Chitosan. Mosaddegh et al. [60] state that the bigger the surface area of coagulant particles, the higher the reaction conversion.

Table 7. Comparison of the BET surface area of the coagulants.

Sample/(s)	S_{BET} Surface Area (m^2/g)	Pore Volume (cm^3/g)	Pore Size (nm)
Magnetite	27.597	0.0080	1.4840
Rice starch	1.267	0.0020	6.7600
RF (1:1)	31.438	0.0015	1.6102
RF (2:1)	30.021	0.0012	1.6098
RF (1:2)	29.388	0.0010	1.5418
Chitosan	1.2010	0.0007	5.4180
CF (1:1)	18.773	0.0008	4.5560
CF (2:1)	16.291	0.0008	4.5110
CF (1:2)	13.918	0.0004	3.7361

3.1.4. FTIR Analysis

FTIR spectroscopy was used to determine the functional groups of the coagulants studied. The 200 scans were obtained between 400 and 1200 cm^{-1} with a resolution of 7 cm^{-1} for each measurement (Figure 5). The peaks between 1200 and 900 cm^{-1} correspond to the vibrational region in concordance with the surfactant (oleic acid) employed in this study [61]. The absorption bands of the magnetite were observed at 1055, 936, 846, 548 and 436 cm^{-1} (Figure 5). The C–O stretching band, which is connected with the C–O–SO_3 group, was responsible for the peak at 1055 cm^{-1} [62]. The appearance of an aromatic C–H bending band was revealed by the absorption peaks 936 and 846 cm $^{-1}$. Magnetite also had two significant peaks at 548 and 436 cm^{-1}, which correspond to the stretched vibration mode of Fe–O [63]. The metal–oxygen band at 436 cm^{-1} was ascribed to octahedral-metal stretching of Fe–O in magnetite, whereas the metal–oxygen band at 548 cm^{-1} was given to intrinsic metal stretching vibrations at the tetrahedral site [49]. These distinctive peaks proved the development of iron oxide nanoparticles, since the peaks positioned between 400 and 600 cm^{-1} matched to the magnetite phase [64].

Figure 5a depicts absorption bands typical of the chitosan saccharine structure at 1160 cm^{-1} (asymmetric stretching of the –COOC– bridge), 1086 cm^{-1}, and 1030 cm^{-1} (skeletal vibration involving the COO stretching) [52]. Off-plane bending –OH vibrations are also recorded at 680 cm^{-1}. In general, the FTIR spectra of chitosan show discrete bands corresponding to amide groups between 1305 and 1660 cm^{-1} [65,66], N-H peaks, and hydrogen-bonded hydroxyl groups between 3270 and 3800 cm^{-1} [67]. The FTIR spectra of RS and its magnetic coagulants (Figure 5b) revealed a particularly strong peak at about 995 cm^{-1}, which was attributed to the C–O of starch's C–O–C bonds [68]. It also reveals that shoulder peaks at around 939 cm^{-1} (skeletal mode vibrations of −1.4 glycosidic linkage), 1085 cm^{-1} (C–O–H bending), and 1148 cm^{-1} (C–O, O–H, and C–C stretching) were seen, comparable to the studies by Talekar et al. [69] and Fan et al. [70]. C–H and CH_2 deformation created a noticeable peak at 852 cm^{-1} in the region below 900 cm^{-1} [30]. The C–C stretching produced a peak at 760 cm^{-1}, while peaks observed between 600 and 400 cm^{-1} were discovered to represent skeletal modes of pyranose [58].

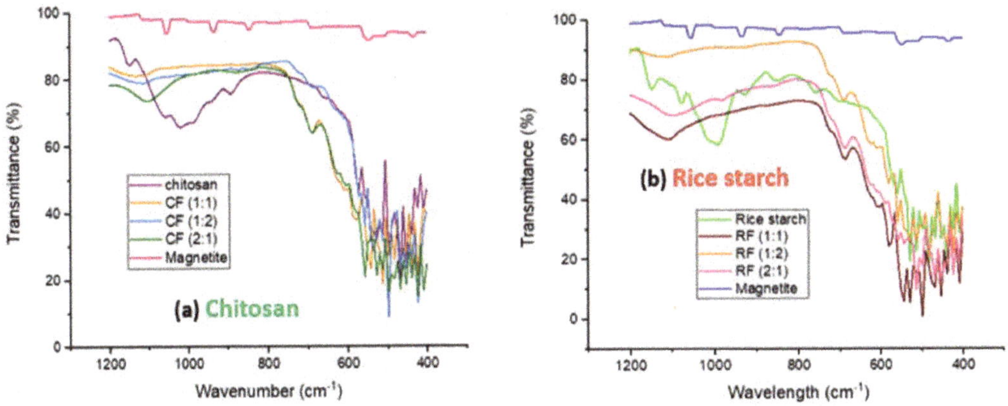

Figure 5. FTIR spectra (**a**) Chitosan and its magnetized coagulants; (**b**) Rice starch and its magnetized coagulants.

Changes in the distinctive spectra peaks altered the physical versus chemical interactions when rice starch was combined with magnetite [51]. The RFs behaved similarly throughout, however the RF (1:2) had the maximum transmittance > RF (2:1) > RF (1:1). (1:1). In RF spectrums, the C–O stretch peak changed from 1148 to 1100 cm^{-1}. The C–C stretch peak also moved from 852 to 689 cm^{-1}. This occurrence demonstrated the presence of interactions between the hydroxyl groups of RS and magnetite, which may result in the formation of hydrogen bonds [51].

3.2. Evaluation of Coagulant Types

Figure 6 shows the turbidity removal efficiencies of the magnetic coagulants. According to the results, the RF reduced turbidity at a rate of more than 75%, ranging from 80% to 85%, compared to 71–74.27% for the CF. The amine functional groups in chitosan, as well as the positively charged magnetite, aided the coagulation and adsorption of negatively charged pollutants in the wastewater sample [10]. The mechanism of cationic exchange (electrostatic, van der Waals, and chemical bonding) and the generation of hydroxide on the surface of magnetized rice starch coagulants was used to remove turbidity by magnetic means [18,32,71]. Furthermore, amylase and amylopectin from rice starch were used to destabilize colloidal particles by bridging and aggregation [32]. The decreasing sequence of turbidity removal was seen to be RF (1:1) > RF (1:2) > RF (2:1) > CF (1:1) > CF (1:2) > CF (2:1), with efficiencies as of 84.81% > 82.07% > 74.27% > 73.54% > 71.85%, respectively. Rice starch has been identified as one of the most effective coagulants for turbidity reduction [72], and due to the concentration of cationic elements, it is suitable for the adsorption of negatively charged particles [73,74].

Other contaminants that were reported for all the coagulants are displayed in Table 8. For each coagulant, the percentage removal results for phosphate, color, and absorbance were recorded. For all contaminants, RF (1:1) attained the maximum phosphate, color, and absorbance removal efficiencies of 92.98, 82.54, and 95.58%, respectively, which affirms with the BET results (Table 7) that surface characteristics aid in the performance of the coagulants, hence high surface area with high porosity will impact the process highly as compared to the lower surface area.

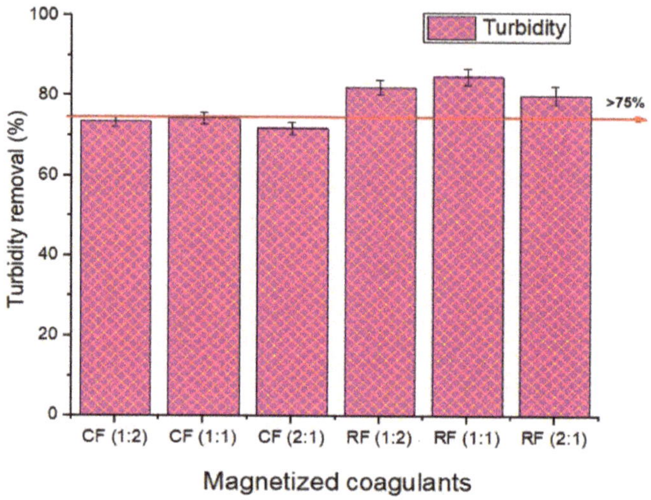

Figure 6. Evaluation of different coagulants on turbidity removal (%).

Table 8. Comparison results of different types of coagulants.

Contaminant	Removal Percentage (%)					
	CF (1:1)	CF (1:2)	CF (2:1)	RF (1:1)	RF (1:2)	RF (2:1)
Phosphate	86.16	81.92	79.86	92.98	90.94	88.83
Color	76.30	76.09	74.73	82.54	79.05	80.69
Absorbance	92.97	94.72	92.27	95.58	94.97	94.37

3.3. Evaluation of Settling Time on Contaminants Removal

Settling time is one of the most important parameters to consider for the coagulation and flocculation process, since it affects the total cost and efficiency of the coagulation process [74]. Figure 7a depicts color removal efficiency, which indicates a rise throughout the lag period (10–20 min). The results demonstrated rapid initial removal, which was explained by the availability of unoccupied sites as the adsorbent was more porous and had a large surface area to adsorb more contaminants, and due to the energy and time constraints, a maximum color removal efficiency at 20 min (84.76%) is preferred. A similar observation was reported by Sibiya et al. [5] and Sun et al. [75]. Furthermore, the shorter the settling period, the better the process in terms of sludge reduction or handling [76,77]. Increasing the settling time beyond 20 min, on the other hand, does not appear to be effective for color removal (dropping to 80% at 40 min), showing that colloidal particles are destabilized as a result of charge reversal. The color removal efficiency rose from 50 to 60 min (84.13%), but there was no improvement beyond that due to increasing settling time causing a limited quantity of flocs clots to develop and be deposited by gravity [78].

Figure 7b depicts the phosphate removal efficiency over a period of 60 min settling time and an increase from 10 (86.44%)–30 min (95.76%) was observed but a decrease on the 40th min of settling, and then an increase at the 60th min. This irregular pattern may be due to sampling as the settling process continues, causing the trapped pollutant to break apart and increase the amount of phosphate. This also confirms the drawbacks of the coagulation process's settling time. Other studies have shown that low effectiveness might be caused by pH, because at low pH, the coagulant's surface is heavily protonated by positive ions (H^+), resulting in an electrostatic strength between the contaminated molecules and the surface, resulting in strong adsorption [79,80]. Phosphate removal in wastewater is significant because high phosphate wastewater is dangerous, since it promotes the growth of poisonous bacteria and algae by providing nutrients to aquatic organisms. Furthermore, it degrades the water's aesthetics and reduces the quantity of soluble oxygen in the water. In this vein, the removal of phosphate from wastewater is compulsory before it is discharged into river bodies.

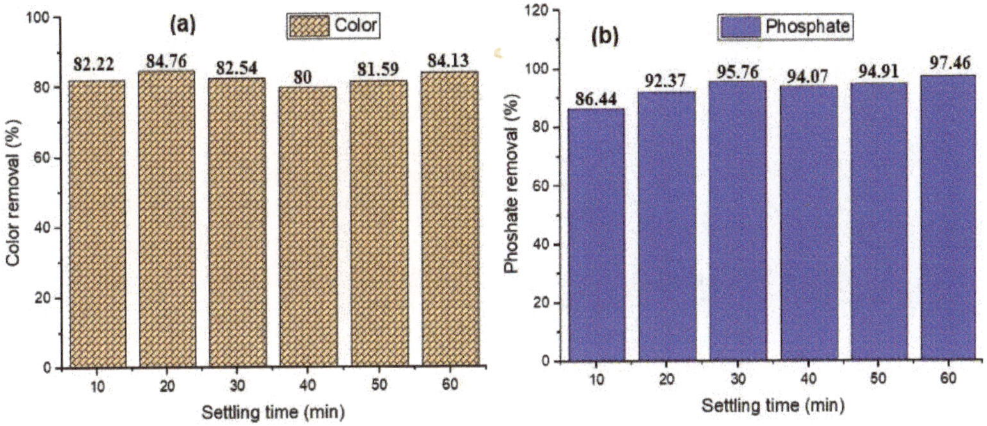

Figure 7. Effect of settling time using RF (1:1) for color removal (**a**) and phosphate removal (**b**).

Figure 8a depicts the turbidity efficiencies for the best coagulant (RF (1:1)). The results showed that increasing the settling time (10 to 60 min) enhanced the efficacy of removal until the time reached 40 min, at which point the effectiveness begins to decrease. The coagulant provides the turbidity removal above 80% at all the settling times. The residual turbidity diminishes between 10 and 40 min due to the formation of large-size flocs with a higher settling velocity [74,78], while from 40 to 50 min, the trend is almost reversed. As the settling period increases from 50 to 60 min, the relevance of the sweep coagulation mechanism outweighs that of the charge neutralization mechanism due to the high degree of super-saturation [13,78].

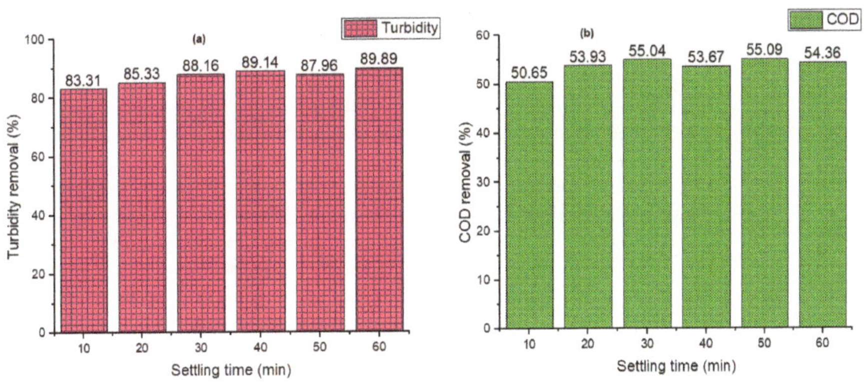

Figure 8. Evaluation of different coagulants on (**a**) turbidity removal (%), and (**b**) COD removal (%).

The modest flocs growth rate in comparison to the flocs nucleation rate is most likely the cause of efficient particle bridging; polymer chains have a greater inclination to overlap with neighboring polymer chains [81]. Figure 8b depicts the COD removal efficiency after 60 min of settling. The findings show a rising tendency from 10 (50.65%) to 30 min (55.04%). After 30 min, an uneven pattern is shown, which might be due to sampling during the settling phase, which caused the contained pollutant to break apart and increase the amount of COD. This also confirms the drawbacks of the coagulation process's settling time. Other research has proposed that limited efficiency may be linked to pH, since at low pH, the coagulant's surface is massively protonated by cations (H+), resulting in an electrostatic strength between the contaminated molecules and the surface, generating high adsorption [16]. This finding demonstrates that coagulation with natural coagulants, particularly rice starch, was beneficial in treating wastewater [30,32]. Other researchers had used a similar technique to prove the efficacy of wastewater treatment equipment [82].

3.4. Comparison of Several Magnetized Coagulants Used in Wastewater Treatment

When the efficacy of coagulants identified in the literature is compared, natural coagulants (such as eggshells, chitosan, and rice starch) functionalized with magnetite show significant promise for possible large-scale applications in industrial wastewater treatment, as shown in Table 9.

Table 9. Comparative study on various nanocomposites used in wastewater treatment.

Magnetized Coagulant	Operating Conditions	Contaminants Removal (%)	Reference
Magnetized moringa oleifera	30 min settling time	68.33% color	[83]
Magnetized alum	50 min magnetic exposure	85% turbidity 82% color	[18]
Magnetized chitosan	pH = 6 370 mg/L dosage	97.7% TSS 91.70% COD 92.70% turbidity	[10]
Magnetised rice starch	17.33 settling time 3.40 g dosage	69.96% turbidity 45.51% phosphate	[82]
RF (1:1)	20 min settling time 4 g dosage	84.76% color 53.93% COD 92.37% phosphate 85.33% turbidity	This study
Magnetized eggshell	30 min settling time	94.86% TSS 92.56% turbidity 96.24% color	[5]

4. Conclusions

This study describes advances in the development of an ecologically friendly water-treatment system based on magnetic iron oxide nanoparticles and their functionalization with rice starch and chitosan. The magnetic coagulants (chitosan and rice starch) were evaluated to determine the best concentration of each natural coagulant and iron oxide nanoparticles (ratios) in the analyzed systems. The analytical findings from the Brunauer–Emmett–Teller (BET) analyzer, X-ray diffraction (XRD) analyzer, and energy-dispersive X-ray (EDX) spectroscopy confirmed the success of the magnetized coagulants' (MCs) surface area, crystal structure, and elemental compositions, respectively. According to the BET results, introducing magnetite to natural coagulants significantly increased the pore size of the MCs. Amongst the coagulants, RF (1:1) efficiently removed 84.81% turbidity, 82.54% color, 92.98% phosphate and 95.58% absorbance. Therefore, a combination of natural coagulant rice starch and magnetic nanoparticles improved the system's ability to coagulate contaminants in the effluent. For the best coagulant, the impact of settling time (10–60 min) at a consistent dose of 4 g was studied. The results revealed that RF (1:1) eliminated turbidity (88.16%), COD (55.04%) and phosphate (95.76%) after 30 min, and color (84.76%) after 20 min. The system's capacity to coagulate impurities was enhanced by functionalizing rice starch and magnetite. The RF (1:1) synthesized in this work is a fascinating approach with the potential to deliver a cost-effective, resilient, and environmentally friendly water treatment technology.

Author Contributions: Writing, data curation, analysis, confirmation, conception, methodology, investigation: N.P.S.; writing, methodology, analysis, data curation: G.A.-D.; financial acquisition, project management, writing—reviewing and editing, validation, conceptualization, methodology: E.K.T.; financial acquisition, writing—reviewing and proofreading, contextualization, supervision: S.R. All authors have read and agreed to the published version of the manuscript.

Funding: This research was funded by the South African Water Research Commission for their support in identifying the project WRC Project: C2019/2020-00212 and the National Research Foundation for their scholarship awards (reference MND190618448553 and MND210421596570).

Institutional Review Board Statement: Not applicable.

Informed Consent Statement: Not applicable.

Data Availability Statement: Not applicable.

Acknowledgments: The authors would like to thank the Durban University of Technology, Green Engineering Research Group.

Conflicts of Interest: The authors declare no conflict of interest.

References

1. Nimesha, S.; Hewawasam, C.; Jayasanka, D.; Murakami, Y.; Araki, N.; Maharjan, N. Effectiveness of natural coagulants in water and wastewater treatment. *Glob. J. Environ. Sci. Manag.* **2021**, *8*, 1–16.
2. Ang, W.L.; Mohammad, A.W. State of the art and sustainability of natural coagulants in water and wastewater treatment. *J. Clean. Prod.* **2020**, *262*, 121267. [CrossRef]
3. Tetteh, E.K.; Rathilal, S.; Chollom, M.N. Pre-treatment of industrial mineral oil wastewater using response surface methodology. *Water Soc. IV* **2017**, *216*, 181–191. [CrossRef]
4. Chollom, M.N.; Rathilal, S.; Swalaha, F.M.; Bakare, B.F.; Tetteh, E.K. Removal of antibiotics during the anaerobic digestion of slaughterhouse wastewater. *Int. J. Sustain. Dev. Plan.* **2020**, *15*, 335–343. [CrossRef]
5. Sibiya, N.P.; Rathilal, S.; Tetteh, E.K. Coagulation Treatment of Wastewater: Kinetics and Natural Coagulant Evaluation. *Molecules* **2021**, *26*, 698. [CrossRef] [PubMed]
6. Zahrim, A.; Hilal, N. Treatment of highly concentrated dye solution by coagulation/flocculation–sand filtration and nanofiltration. *Water Resour. Ind.* **2013**, *3*, 23–34. [CrossRef]
7. Siddique, M.; Soomro, S.A.; Aziz, S.; Jatoi, A.S.; Mengal, A.; Mahar, H. Removal of turbidty from turbid water by bio-cogulant prepared from walnut shell. *J. Appl. Emerg. Sci.* **2016**, *6*, 66–68.
8. Nizamuddin, S.; Siddiqui, M.; Mubarak, N.; Baloch, H.A.; Abdullah, E.; Mazari, S.A.; Griffin, G.; Srinivasan, M.; Tanksale, A. Iron oxide nanomaterials for the removal of heavy metals and dyes from wastewater. *Nanoscale Mater. Water Purif.* **2019**, 447–472.
9. Crini, G.; Lichtfouse, E. Advantages and disadvantages of techniques used for wastewater treatment. *Environ. Chem. Lett.* **2018**, *17*, 145–155. [CrossRef]
10. Saifuddin, N.; Dinara, S. Pretreatment of palm oil mill effluent (POME) using magnetic chitosan. *E-J. Chem.* **2011**, *8*, S67–S78. [CrossRef]
11. Cui, H.; Huang, X.; Yu, Z.; Ping Chen, P.; Cao, X. Application progress of enhanced coagulation in water treatment. *RSC Adv.* **2020**, *10*, 20231. [CrossRef] [PubMed]
12. Gautam, S.; Saini, G. Use of natural coagulants for industrial wastewater treatment. *Glob. J. Environ. Sci. Manag.* **2020**, *6*, 553–578.
13. Kumar, M.M.; Karthikeyan, R.; Anbalagan, K.; Bhanushali, M.N. Coagulation process for tannery industry effluent treatment using Moringa oleifera seeds protein: Kinetic study, pH effect on floc characteristics and design of a thickener unit. *Sep. Sci. Technol.* **2016**, *51*, 2028–2037. [CrossRef]
14. Nath, A.; Mishra, A.; Pande, P.P. A review natural polymeric coagulants in wastewater treatment. *Mater. Today Proc.* **2020**, *46*, 6113–6117. [CrossRef]
15. Muniza, G.L.; Borges, A.C.; da Silva, T.C.F.; Batista, R.O.; de Castro, S.R. Chemically enhanced primary treatment of dairy wastewater using chitosan obtained from shrimp wastes: Optimization using a Doehlert matrix design. *Environ. Technol.* **2020**, *43*, 237–254. [CrossRef] [PubMed]
16. Freitas, T.K.F.S.; Almeida, C.A.; Manholer, D.D.; Geraldino, H.C.L.; de Souza, M.T.F.; Garcia, J.C. Review of utilization plant-based coagulants as alternatives to textile wastewater treatment. In *Detox Fashion*; Springer: Berlin/Heidelberg, Germany, 2017; pp. 27–79.
17. Kakoi, B.; Kaluli, J.W.; Ndiba, P.; Thiong'o, G. Banana pith as a natural coagulant for polluted river water. *Ecol. Eng.* **2016**, *95*, 699–705. [CrossRef]
18. Tetteh, E.K.; Rathilal, S. Application of magnetized nanomaterial for textile effluent remediation using response surface methodology. *Mater. Today Proc.* **2020**, *38*, 700–711. [CrossRef]
19. Muniz, G.L.; Borges, A.C.; da Silva, T.C.F. Assessment and optimization of the use of a novel natural coagulant (*Guazuma ulmifolia*) for dairy wastewater treatment. *Sci. Total Environment.* **2020**, *77*, 140864. [CrossRef]

20. Dos Santos, T.R.T.; Silva, M.F.; de Andrade, M.B.; Vieira, M.F.; Bergamasco, R. Magnetic coagulant based on *Moringa oleifera* seeds extract and super paramagnetic nanoparticles: Optimization of operational conditions and reuse evaluation. *Desalination Water Treat.* **2018**, *106*, 226–237. [CrossRef]
21. Mateus, G.A.P.; dos Santos, T.R.T.; Sanches, I.S.; Silva, M.F.; de Andrade, M.B.; Paludo, M.P.; Gomes, R.G.; Bergamasco, R. Evaluation of a magnetic coagulant based on Fe_3O_4 nanoparticles and *Moringa oleifera* extract on tartrazine removal: Coagulation-adsorption and kinetics studies. *Environ. Technol.* **2018**, *41*, 1648–1663. [CrossRef] [PubMed]
22. Chua, S.-C.; Ho, Y.-C.; Chong, F.-K. Synthesis and application of a novel composite coagulant aid from rice starch and sesbania seed gum for water treatment. *Malays. J. Chem.* **2021**, *23*, 7–14.
23. Muniza, G.L.; Borges, A.C.; da Silva, T.C.F. Performance of natural coagulants obtained from agro-industrial wastes in dairy wastewater treatment using dissolved air flotation. *J. Water Process Eng.* **2020**, *37*, 101453. [CrossRef]
24. Dos Santos, T.R.T.; Mateus, G.A.P.; Silva, M.F.; Miyashiro, C.S.; Nishi, L.; de Andrade, M.B.; Fagundes-Klen, M.R.; Gomes, R.G.; Bergamasco, R. Evaluation of Magnetic Coagulant (α-Fe_2O_3-MO) and its Reuse in Textile Wastewater Treatment. *Water Air Soil Pollut.* **2018**, *229*, 92. [CrossRef]
25. Kurniawan, S.B.; Abdullah, S.R.S.; Imron, M.F.; Said, N.S.M.; Ismail, N.I.; Hasan, H.A.; Othman, A.R.; Purwanti, I.F. Challenges and opportunities of biocoagulant/bioflocculant application for drinking water and wastewater treatment and its potential for sludge recovery. *Int. J. Environ. Res. Public Health* **2020**, *17*, 9312. [CrossRef]
26. Aboelfetoh, E.F.; Aboubaraka, A.E.; Ebeid, E.-Z.M. Binary coagulation system (graphene oxide/chitosan) for polluted surface water treatment. *J. Environ. Manag.* **2021**, *288*, 112481. [CrossRef]
27. Devi, M.G.; Dumaran, J.J.; Feroz, S. Dairy wastewater treatment using low molecular weight crab shell chitosan. *J. Inst. Eng. Ser. E* **2012**, *93*, 9–14. [CrossRef]
28. Chua, S.-C.; Chong, F.-K.; Yen, C.-H.; Ho, Y.-C. Valorization of conventional rice starch in drinking water treatment and optimization using response surface methodology (RSM). *Chem. Eng. Commun.* **2019**, *208*, 613–623. [CrossRef]
29. Usefi, S.; Asadi-Ghalhari, M. Modeling and optimization of the coagulation–flocculation process in turbidity removal from aqueous solutions using rice starch. *Pollution* **2019**, *5*, 623–636.
30. Teh, C.Y. Potential use of rice starch in coagulation–flocculation process of agro-industrial wastewater: Treatment performance and flocs characterization. *Ecol. Eng.* **2014**, *71*, 509–519. [CrossRef]
31. Amran, A.H.; Zaidi, N.S.; Muda, K.; Loan, L.W. Effectiveness of natural coagulant in coagulation process: A review. *Int. J. Environ. Sci. Technol.* **2018**, *7*, 34–37.
32. Sibiya, N.; Amo-Duodu, G.; Tetteh, E.K.; Rathilal, S. Model prediction of coagulation by magnetised rice starch for wastewater treatment using response surface methodology (RSM) with artificial neural network (ANN). *Scientific Afr.* **2022**, *17*, e01282. [CrossRef]
33. Pritchard, M.; Craven, T.; Mkandawire, T.; Edmondson, A.S.; O'Neill, J.G. A comparison between *Moringa oleifera* and chemical coagulants in the purification of drinking water—An alternative sustainable solution for developing countries. *Phys. Chem. Earth* **2010**, *35*, 798–805. [CrossRef]
34. APHA. *Standard Methods for the Examination of Water and Wastewater*; APHA: Washington, DC, USA, 2012; Available online: https://www.standardmethods.org/doi/book/10.2105/SMWW.2882?gclid=CjwKCAiA25v_BRBNEiwAZb4-ZRdKz6ceq6 (accessed on 15 March 2020).
35. Amo-Duodu, G.; Tetteh, E.K.; Rathilal, S.; Armar, E.K.; Adedeji, J.; Chollom, M.N.; Chetty, M. Effect of engineered biomaterials and magnetite on wastewater treatment: Biogas and kinetic evaluation. *Polymers* **2021**, *13*, 4323. [CrossRef]
36. Morsi, R.E.; Al-Sabagh, A.M.; Moustafa, Y.M.; ElKholy, S.G.; Sayed, M.S. Polythiophene modified chitosan/magnetite nanocomposites for heavy metals and selective mercury removal. *Egypt. J. Pet.* **2018**, *27*, 1077–1085. [CrossRef]
37. Tetteh, E.K.; Amo-Duodu, G.; Rathilal, S. Biogas production from wastewater: Comparing biostimulation impact of magnetised-chitosan and -titania chitosan. *Mater. Today: Proc.* **2022**, *62*, S85–S90.
38. Nechita, P. Applications of chitosan in wastewater treatment. *Biol. Act. Appl. Mar. Polysacch.* **2017**, *1*, 209–228.
39. Petcharoen, K.; Sirivat, A. Synthesis and characterization of magnetite nanoparticles via the chemical co-precipitation method. *Mater. Sci. Eng. B* **2012**, *177*, 421–427. [CrossRef]
40. Deravi, L.F. *Piezoelectric Inkjet Printing of Multicomposite Biomaterials*; Vanderbilt University: Nashville, TN, USA, 2009.
41. Rao, C.N.R.; Nath, M. Inorganic nanotubes. In *Advances in Chemistry: A Selection of CNR Rao's Publications (1994–2003)*; World Scientific: Singapore, 2003; pp. 310–333.
42. Taspika, M.; Desiati, R.D.; Mahardika, M.; Sugiarti, E.; Abral, H. Influence of TiO_2/Ag particles on the properties of chitosan film. *Adv. Nat. Sci. Nanosci. Nanotechnol.* **2020**, *11*, 015017. [CrossRef]
43. Mohanasrinivasan, V.; Paliwal, J.S.; Selvarajan, E.; Mishra, M.; Singh, S.K.; Suganthi, V.; Devi, C.S. Studies on heavy metal removal efficiency and antibacterial activity of chitosan prepared from shrimp shell waste. *Biotechnology* **2013**, *4*, 167–175. [CrossRef]
44. Han, X.; Zhang, M.; Zhang, R.; Huang, L.; Jia, X.; Huang, F.; Liu, L. Physicochemical interactions between rice starch and different polyphenols and structural characterization of their complexes. *LWT Food Sci. Technol.* **2020**, *125*, 109227. [CrossRef]
45. Karp, S.G.; Pagnoncelli, M.G.B.; Prado, F.; de Oliveira Penha, R.; Junior, A.I.M.; Kumlehn, G.S.; Soccol, C.R. Chapter5-Starch. *Biomass Biofuels Biochem.* **2021**, 75–100.
46. Azanza, M.P.V.; Alejandro, C.B.S.; Jim, J.T. Impact of Processing Stages and Additives on the Structural Quality of Cornstarch Bihon-Type Noodles. *Philipp. J. Sci.* **2021**, *150*, 1451–1460.

47. Rashid, H.; Mansoor, M.A.; Haider, B.; Nasir, R.; Hamid, S.B.A.; Abdulrahman, A. Synthesis and characterization of magnetite nano particles with high selectivity using in-situ precipitation method. *Sep. Sci. Technol.* **2020**, *55*, 207–1215. [CrossRef]
48. Baumgartner, J.; Dey, A.; Bomans, P.H.; Le Coadou, C.; Fratzl, P.; Sommerdijk, N.A.; Faivre, D. Nucleation and growth of magnetite from solution. *Nat. Mater.* **2013**, *12*, 310–314. [CrossRef]
49. Dhar, P.K.; Shaha, P.; Hasan, M.K; Amin, M. K.; Haque, M.R. Green synthesis of magnetite nanoparticles using *Lathyrus sativus* peel extract and evaluation of their catalytic activity. *Clean. Eng. Technol.* **2021**, *3*, 100117. [CrossRef]
50. Jana, S.; Trivedi, M.K.; Tallapragada, R.M.; Branton, A.; Trivedi, D.; Nayak, G.; Mishra, R.K. Characterization of Physicochemical and Thermal Properties of Chitosan and Sodium Alginate after Biofield Treatment. *Pharm. Anal. Acta* **2015**, *6*, 10. [CrossRef]
51. Bourtoom, T.; Chinnan, M.S. Preparation and properties of rice starchechitosan blend biodegradable film. *Sci. Drirect* **2008**, *41*, 1633–1641.
52. Dey, S.C.; Al-Amin, M.; Rashid, T.U.; Sultan, M.Z.; Ashaduzzaman, M.; Sarker, M.; Shamsuddin, S.M. Preparation, characterization and performance evaluation of chitosan as an adsorbent for re. *Int. J. Latest Res. Eng. Technol.* **2016**, *2*, 52–62.
53. Billah, R.E.K.; Abdellaoui, Y.; AnfarZakaria; Giácoman-Vallejos, G.; Agunaou, M.; Soufi, A. Synthesis and Characterization of Chitosan/Fluorapatite Composites for the Removal of Cr (VI) from Aqueous Solutions and Optimized Parameters. *Water Air Soil Pollut.* **2020**, *231*, 163. [CrossRef]
54. Kumar, S.; Koh, J. Physiochemical, Optical and Biological Activity of Chitosan-Chromone Derivative for Biomedical Applications. *Int. J. Mol. Sci.* **2012**, *13*, 6102–6116. [CrossRef] [PubMed]
55. Bule'on, A.; Colonna, P.; Planchot, V.; Ball, S. Starch granules: Structure and biosynthesis. *Int. J. Biol. Macromol.* **1998**, *23*, 85–112. [CrossRef]
56. Yang, H.; Tang, M.; Wu, W.; Ding, W.; Ding, B.; Wang, X. Study on inhibition effects and mechanism of wheat starch retrogradation by polyols. *Food Hydrocoll.* **2021**, *121*, 106996. [CrossRef]
57. Shih, F.; King, J.; Daigle, K.; An, H.-J.; Ali, R. Physicochemical Properties of Rice Starch Modified by Hydrothermal Treatments. *Cereal Chem.* **2007**, *84*, 527–531. [CrossRef]
58. Matmin, J.; Affendi, I.; Ibrahim, S.I.; Endud, S. Additive-free rice starch-assisted synthesis of spherical nanostructured hematite for degradation of dye contaminant. *Nanomaterials* **2018**, *8*, 702. [CrossRef] [PubMed]
59. Rao, J.S.; Parimalavalli, R.; Jagannadham, K. Impact of Cross-Linking on Physico-Chemical and Functional Properties of Cassava Starch. *Int. J. Adv. Res.* **2014**, *2*, 284–289.
60. Mosaddegh, E.; Hassankhani, A. Application and characterization of eggshell as a new biodegradable and heterogeneous catalyst in green synthesis of 7,8-dihydro-4H-chromen-5(6H)-ones. *Catal. Commun.* **2013**, *33*, 70–75. [CrossRef]
61. Zhang, L.; He, R.; Gu, H.-C. Oleic acid coating on the monodisperse magnetite nanoparticles. *Appl. Surf. Sci.* **2006**, *253*, 2611–2617. [CrossRef]
62. Yew, Y.P.; Shameli, K.; Miyake, M.; Kuwano, N.; Khairudin, N.B.B.A.; Mohamad, S.E.B.; Lee, K.X. Green Synthesis of Magnetite (Fe_3O_4) Nanoparticles Using Seaweed (*Kappaphycus alvarezii*) Extract. *Nanoscale Res. Lett.* **2016**, *11*, 276. [CrossRef] [PubMed]
63. Mohammadi, M.-R.; Ansari, S.; Bahmaninia, H.; Ostadhassan, M.; Norouzi-Apourvari, S.; Hemmati-Sarapardeh, A.; Schaffie, M.; Ranjbar, M. Experimental Measurement and Equilibrium Modeling of Adsorption of Asphaltenes from Various Origins onto the Magnetite Surface under Static and Dynamic Conditions. *ACS Omega* **2021**, *6*, 24256–24268. [CrossRef] [PubMed]
64. Yuvakkumar, R.; Hong, S. Green synthesis of spinel magnetite iron oxide nanoparticles. *Adv. Mater. Res.* **2014**, *1051*, 39–42. [CrossRef]
65. El Knidri, H.; El Khalfaouy, R.; Laajeb, A.; Addaou, A.; Lahsini, A. Eco-friendly extraction and characterization of chitin and chitosan from the shrimp shell waste via microwave irradiation. *Process Saf. Environ. Prot.* **2016**, *104*, 395–405. [CrossRef]
66. Wang, F.; Li, Y.; Gough, C.R.; Liu, Q.; Hu, X. Dual-Crystallizable Silk Fibroin/Poly (L-lactic Acid) Biocomposite Films: Effect of Polymer Phases on Protein Structures in Protein-Polymer Blends. *Int. J. Mol. Sci.* **2021**, *22*, 1871. [CrossRef] [PubMed]
67. Danko, M.; Kronekova, Z.; Krupa, I.; Tkac, J.; Matúš, P.; Kasak, P. Exchange Counterion in Polycationic Hydrogels: Tunability of Hydrophobicity, Water State, and Floating Capability for a Floating pH Device. *Gels* **2021**, *7*, 109. [CrossRef] [PubMed]
68. Ferreira-Villadiego, J.; Garcia-Echeverri, J.; Mejia, M.V.; Pasqualino, J.; Meza-Catellar, P.; Lambis, H. Chemical modification and characterization of starch derived from plantain (*Musa paradisiaca*) Peel waste, as a source of biodegradable material. *Chem. Eng. Trans.* **2018**, *65*, 763–768.
69. Talekar, S.; Pandharbale, A.; Ladole, M.; Nadar, S.; Mulla, M.; Japhalekar, K.; Pattankude, K.; Arage, D. Carrier free co-immobilization of alpha amylase, glucoamylase and pullulanase as combined cross-linked enzyme aggregates (combi-CLEAs): A tri-enzyme biocatalyst with one pot starch hydrolytic activity. *Bioresour. Technol.* **2013**, *147*, 269–275. [CrossRef]
70. Fan, D.; Ma, W.; Wang, L.-J.; Huang, J.-C.H.; Zhao, J.; Zhang, H.; Chen, W. Determination of structural changes in microwaved rice starch using Fourier transform infrared and Raman spectroscopy. *Starch* **2012**, *64*, 598–606. [CrossRef]
71. Hatamie, A.; Parham, H.; Zargar, B.; Heidari, Z. Evaluating magnetic nano-ferrofluid as a novel coagulant for surface water treatment. *J. Mol. Liq.* **2016**, *219*, 694–702. [CrossRef]
72. Choya, S.Y.; Prasada, K.M.N.; Wua, T.Y.; Raghunandan, M.E.; Phang, S.-M.; Juane, J.C.; Ramanan, R.N. Separation of Chlorella biomass from culture medium by flocculation with rice starch. *Algal Res.* **2018**, 162–172. [CrossRef]
73. Huzir, N.M.; Aziz, M.M.A.; Ismail, S.B.; Mahmood, N.A.N.; Umor, N.A.; Muhammad, S.A.F.S. Optimization of coagulation-flocculationprocess for the palm oil milleffluent treatment by using rice husk ash. *Ind. Crops Prod.* **2019**, *139*, 111482. [CrossRef]

74. Asharuddin, S.M.; Othman, N.; Zin, N.S.M.; Tajarudin, H.A.; Din, M.F.; Kumar, V. Performance assessment of cassava peel starch and alum as dual coagulant for turbidity removal in dam water. *Int. J. Integr. Eng.* **2018**, *10*, 185–192. [CrossRef]
75. Sun, Y.; Zhou, S.; Chiang, P.-C.; Shah, K.J. Evaluation and optimization of enhanced coagulation process: Water and energy nexus. *Water Energy Nexus* **2019**, *2*, 25–36. Available online: https://www.researchgate.net/profile/Xiaopeng-Ge/publication/6620675_Optimized_Coagulation_of_High_Alkalinity_Low_Temperature_and_Particle_Water_pH_Adjustment_and_Polyelectrolytes_as_Coagulant_Aids/links/5e65b9ca4585153fb3ce0c56/Optimized-Coagulation-of-High-Alkalinity-Low-Temperature-and-Particle-Water-pH-Adjustment-and-Polyelectrolytes-as-Coagulant-Aids.pdf (accessed on 26 February 2021). [CrossRef]
76. Capodaglio, A.G.; Olsson, G. Energy issues in sustainable urban wastewater management: Use, demand reduction and recovery in the urban water cycle. *Sustainability* **2019**, *12*, 266. [CrossRef]
77. Gomez-Gallegos, M.; Reyes-Mazzoco, R.; Flores-Cervantes, D.; Jarayathne, A.; Goonetilleke, A.; Bandala, E.R.; Sanchez-Salas, J. Role of organic matter, nitrogen and phosphorous on granulation and settling velocity in wastewater treatment. *J. Water Process Eng.* **2021**, *40*, 101967. [CrossRef]
78. Momeni, M.M.; Kahforoushan, D.; Abbasi, F.; Ghanbarian, S. Using chitosan/CHPATC as coagulant to remove color and turbidity of industrial wastewater: Optimization through RSM design. *J. Environ. Manag.* **2018**, *211*, 347–355. [CrossRef]
79. Kakoi, B.; Kaluli, J.W.; Ndiba, P.; Thiong'o, G. Optimization of Maerua Decumbent bio-coagulant in paint industry wastewater treatment with response surface methodology. *J. Clean. Prod.* **2017**, *164*, 1124–1134. [CrossRef]
80. Freitas, T.; Oliveira, V.; De Souza, M.; Geraldino, H.; Almeida, V.; Fávaro, S.; Garcia, J. Optimization of coagulation-flocculation process for treatment of industrial textile wastewater using okra (*A. esculentus*) mucilage as natural coagulant. *Ind. Crops Prod.* **2015**, *76*, 538–544. [CrossRef]
81. Choy, S.Y.; Prasad, K.M.N.; Wu, T.Y.; Raghunandan, M.E.; Yang, B.; Phang, S.-M.; Ramanan, R.N. Isolation, characterization and the potential use of starch from jackfruit seed wastes as a coagulant aid for treatment of turbid water. *Environ. Sci. Pollut. Res. Int.* **2017**, *24*, 2876–2889. [CrossRef] [PubMed]
82. Sibiya, N.; Amo-Duodu, G.; Tetteh, E.K.; Rathilal, S. Response surface optimisation of a magnetic coagulation process for wastewater treatment via Box-Behnken. *Mater. Today Proc.* **2022**, *42*, S121–S126. [CrossRef]
83. Mateus, G.A.P.; Paludo, M.P.; dos Santos, T.R.T.; Silva, M.F.; Nishi, L.; Fagundes-Klen, M.R.; Gomes, R.G.; Bergamasco, R. Obtaining drinking water using a magnetic coagulant composed of magnetite nanoparticles functionalized with *Moringa oleifera* seed extract. *J. Environ. Chem. Eng.* **2018**, *6*, 4084–4092. [CrossRef]

polymers

MDPI

Article

Cellulose Fibre Degradation in Cellulose/Steel Hybrid Geotextiles under Outdoor Weathering Conditions

Avinash Pradip Manian [1], Barbara Paul [1], Helene Lanter [2], Thomas Bechtold [1,*] and Tung Pham [1]

[1] Research Institute of Textile Chemistry and Textile Physics, University of Innsbruck, Hoechsterstrasse 73, 6850 Dornbirn, Austria
[2] Geobrugg AG, CH-8590 Romanshorn, Switzerland
* Correspondence: thomas.bechtold@uibk.ac.at

Abstract: Risks from rockfall and land sliding can be controlled by high-tensile steel nets and meshes which stabilise critical areas. In many cases, a recultivation of the land is also desired. However, high-tensile steel meshes alone are not always sufficient, depending on the location and the inclination of the stabilised slope, to achieve rapid greening. Cellulose fibres exhibit high water binding capacity which supports plant growth. In this work, a hybrid structure consisting of a nonwoven cellulose fibre web and a steel mesh was produced and tested under outdoor conditions over a period of 61 weeks. The cellulose fibres are intended to support plant growth and soil fixation, and thus the biodegradation of the structure is highly relevant, as these fibres will become part of the soil and must be biodegradable. The biodegradation of the cellulose fibres over the period of outdoor testing was monitored by microscopy and analytical methods. The enzymatic degradation of the cellulose fibres led to a reduction in the average degree of polymerisation and also a reduction in the moisture content, as polymer chain hydrolysis occurs more rapidly in the amorphous regions of the fibres. FTIR analysis and determination of carboxylic group content did not indicate substantial changes in the remaining parts of the cellulose fibre. Plant growth covered geotextiles almost completely during the period of testing, which demonstrated their good compatibility with the greening process. Over the total period of 61 weeks, the residual parts of the biodegradable cellulose web merged with the soil beneath and growing plants. This indicates the potential of such hybrid concepts to contribute a positive effect in greening barren and stony land, in addition to the stabilising function of the steel net.

Keywords: geotextiles; cellulose; biodegradation; moisture sorption; cellulase

Citation: Manian, A.P.; Paul, B.; Lanter, H.; Bechtold, T.; Pham, T. Cellulose Fibre Degradation in Cellulose/Steel Hybrid Geotextiles under Outdoor Weathering Conditions. *Polymers* **2022**, *14*, 4179. https://doi.org/10.3390/polym14194179

Academic Editors: Alexandre Vetcher and Alexey Iordanskii

Received: 26 August 2022
Accepted: 30 September 2022
Published: 5 October 2022

1. Introduction

Serious concerns about micro-plastics in the environment have added credence to the constantly growing need to substitute petrol-based polymers with nature-based materials. Because of its abundant availability in nature, cellulose has been studied for many applications and its biodegradability is claimed as a positive argument in numerous publications [1].

Polymer degradation can be initiated by a chemical or physical attack to cellulose-based material and leads to significant losses in physical and mechanical properties. The chemical degradation of cellulose has been studied extensively to understand ageing of paper, where hydrolytic and oxidative reactions represent the main pathways in degradation [2]. The end-wise degradation (peeling) of cellulose also has been studied intensively under the conditions of regenerated cellulose fibre production [3–5]. Biodegradation occurs under the action of fungi, bacteria and algae via extracellular and intracellular enzymes and leads to complete mineralisation of the polymer [6]. In the case of the aerobic biodegradation of cellulose, the products are then CO_2, water and biomass [7]. Important groups of extracellular enzymes are endo-β-1,4-glucanases and cellobiohydrolases (exo-β-1,4-glucanases), which lead to cleavage of the cellulose chain, preferably in amorphous regions, and release cellobiose [8,9]. The disaccharide is then hydrolysed into glucose by the action of β-1,4-glucosidase [8,10]. Additionally, in anaerobic cellulose biodegradation, the first step is the enzymatic hydrolysis to glucose and oligosaccharides, which are then digested by the anaerobic microbial community [11].

The biodegradation of cellulose-based textile products under simulated conditions has been demonstrated in the literature [12]. Lignin components were shown to have an inhibitory effect on the cellulolytic activity of cellulases, which is attributed to protein precipitation effects rather than to inhibitory effects on the cellulose-enzyme complex [13]. Additionally, the presence of finishing agents e.g., cationic softeners, can lead to retarded biodegradation of treated cellulose fibre-based textiles [14].

The biodegradability of a polymer depends on the physical and chemical properties of the material, as well as on the conditions under which biodegradation is considered. Thus, the conditions of biodegradation vary between composting, soil biodegradation, marine biodegradation, sewage sludge biodegradation, anaerobic biodegradation, landfill biodegradation and others [15]. As an example, lower biodegradation of cellulose fibres was observed under seawater conditions, in particular at 7 °C water temperature, when compared to conditions present during soil biodegradation [16].

The presence of cellulosic fibres in polyhydroxyalkanoates (PHA) leads to improved degradation behaviour under various conditions of degradation, thus indicating the positive influence of a readily degradable fibre on the biodegradation of the PHA matrix [15].

In soil burial tests, the degradation of cellulose fibres occurs via disintegration of the fibre structure and mineralisation of the fragments. In such wet soil, cellulose-based test fabrics were broken down completely within 4 weeks of biodegradation [17].

In forest soil, the degradation of cellulose is dependent on the type of soil, as well as on the climatic conditions, as a lower cellulose degradation rate was observed during dry seasons [18]. Additionally, the type of cellulose fibre determines the observed rate of fibre biodegradation, e.g., under warm, moist, aerobic soil conditions, more rapid degradation of viscose fibres was observed when compared to cotton and lyocell fibres [19]. Compared to the rapid biodegradation of the cellulose fibres in soil tests, no significant biodegradation was observed in the same tests for synthetic fibres, e.g., produced from polylactic acid (PLA), polyethylene terephthalate (PET) or polyacrylonitrile (PAN) [20].

Numerous standards to assess the biodegradability of a substance have been recommended; however, the specific conditions of the application of a product also determine the final observed biodegradation. Well-defined laboratory experiments and standardised biodegradation experiments permit access to reproducible results; however; the analysis of the specific behaviour of a polymer during biodegradation and the complex interaction with the environment requires the set-up of adequate field tests [21].

Depending on the test protocols used, full mineralisation of cellulose fibres into CO_2 and H_2O can reach values of 60–70%. This can value be understood as full mineralisation, as part of the fibre mass is also transformed with an increase in biomass [14].

The use of cellulose-based fibrous material as strengthening fibres in geopolymers has been extensively studied by researchers with a focus on strength, toughness of fibres against matrix cracking, degradation in the presence of chlorides, degradation by temperature and repeated wetting–drying cycles [22].

Non-woven structures, e.g., from polypropylene, are frequently used as the ground layer in the construction of roads to prevent road stones from sinking into the ground. Coarse jute-based fabrics are mounted as biodegradable geotextiles to stabilise steeper sites and to support their revegetation when only limited mechanical stress is expected to occur. Polypropylene, polyethylene and polyethyleneterephthalate-based products have been recognised as sources for microplastics, while the biobased cellulose fibres viscose, modal and lyocell exhibit biodegradability. Differences in the polymer packing and order between viscose, modal and lyocell, as well as cotton modulate the rate of biodegradation, however, do not affect the biodegradability of the cellulose fibres (Röder et al., 2019 [23]).

High-tensile steel nets and meshes are frequently used to secure landslide prone slopes above roads and railways, for example. Such structures provide mechanically excellent long-term stabilisation of critical sites. Depending on the inclination and orientation of said slope, revegetation occurs naturally.

However, when rapid vegetation of an area is at risk because of the steepness of the slope, the orientation is unfavourable because it dries out the soil, or when there is a risk of soil erosion through water percolation, a combination of the steel nets with geotextiles is necessary to achieve revegetation. There are two types of geotextiles that can be used depending on the site. When revegetation might take several years to settle definitively, a long-term geotextile is requested which does almost not degrade in the first years. On the other hand, many sites are easily revegetated within a season's cycle if the right support of initial plant growth is given, and the lifetime of the geotextile layer does not need to exceed this period. This can be achieved, for example, with water sorbing cellulose fibre webs.

While the steel net provides mechanical stability and safety over many years, the cellulose fibre web should contribute to the formation of humus and greening. Such geotextiles are frequently mounted in an ecologically sensitive environment; thus, knowledge about interaction with plant growth and biodegradability of the fibre web under conditions of simulated application are required to understand the behaviour of such fibres during biodegradation. In this application, the cellulose fibres form the first surface layer on the ground, thus a specific setup to monitor biodegradation under real conditions is required to assess the behaviour of the cellulose fibres in the environment.

In this particular application, the cellulose part of the hybrid geotextiles should stabilise the ground during the period of plant growth followed by a phase of biodegradation and integration into the soil. At present, no scientific results about the biodegradation of cellulose fibres when positioned at the surface of the ground have been reported in the literature. In this paper, a long-term study to monitor the biodegradation of cellulose in cellulose/steel hybrid geotextiles under field test conditions is presented. Larger dimension prototypes (1.3×1.3 m^2) were prepared, mounted on the ground and monitored over a period of 427 days. The prototypes were prepared in cooperation with a viscose fibre producer, thus viscose fibres were used for this study. Samples of the cellulose fibres were taken in regular intervals and characterised by the analytical techniques: Fourier transform infrared spectroscopy (FTIR), moisture sorption, viscometry, laser scanning microscopy and carboxyl group content to monitor fibre degradation during the period of observation. The set of elaborated data contributes significant information to the complex behaviour of cellulose fibres in the environment and also describes the behaviour of the new cellulose/steel hybrids under real growth conditions.

2. Experimental

2.1. Materials and Chemicals

Throughout this work, 100% viscose fibres (CV) with a linear density of 1.7 dtex, kindly supplied from Kelheim Fibres GmbH (Kelheim, Germany), were used as received. Norafin Industries GmbH (Mildenau, Germany) manufactured the nonwoven material made from the viscose fibres. For the preparation of the geo textiles, the steel mesh DELTAX® G80/2 from Geobrugg AG (Romanshorn, Switzerland) was used and two starch products were used as fibre binding agents: Maltodextrin starch (S1) and corn starch (S2) (Maizena, Unilever Austria GmbH, Wien, Austria).

Methylene blue (MB), borate, HCl, $FeCl_3 \cdot 6H_2O$, tartaric acid and NaOH were analytical grade chemicals.

2.2. Production of Geo Textile Prototypes/Wet-Laying Process

The prototypes were manufactured as a combination of steel mesh, viscose fibre and viscose hydroentangled nonwoven in a size of 130×130 cm^2. The starch solution was added to strengthen the bonding between open CV fibres, the CV nonwoven and supported the anchoring of the cellulose web to the steel fabric.

In a first step, the steel mats were cut to size and, depending on the type of construction (Table 1), a selected nonwoven fabric of the same size was placed underneath. Un-crimped fibres with an approximate length of 100 mm were opened by hand and placed loosely on the steel mesh.

Table 1. Compositions of various geotextile prototypes for the degradation tests (binder: S1 maltodextrin, S2 corn starch).

Prototype	Mass per Area Viscose Fibres $g\ m^{-2}$	Mass per Area Nonwoven Fabric $g\ m^{-2}$	Ratio of Binding Agent per Fibre, Binder $g\ m^{-2}$
P 1	200	200	0.25 S1
P 2	200	200	0.25 S1
P 3	200	200	0.85 S2
P 4	200	200	0.25 S2
P 5	200	-	0.25 S2
P 6	130	200	0.25 S2

A wet-laying system was built (Figure 1) that allows water to continuously drain through the floor and filter out the fibres to form a matted web, which is then dried to fix the fabric. The starch was dissolved in hot water and evenly distributed on the layer. The prototypes were removed from the frame after a drying time of about 48 h.

(a) (b)

Figure 1. Preparation of a prototype, (**a**) Viscose fibres are distributed randomly on the steel mesh, underneath there is a viscose fleece, (**b**) prototype after wet-laying process.

2.3. Biodegradation Tests

For the degradation tests and the investigation of the biological degradation of the cellulose fibre structure, the geotextiles were mounted outdoors on a small slope with full sun exposure near the location of the producer Geobrugg in Romanshorn in Switzerland (Elevation: 406 m (1332 ft) in summer 2019. Accurate location of the experimental site was defined by GPS coordinates (47°33′24.4″ N, 9°21′59.9″ E). Fibre samples were taken at regular intervals and brought to the laboratory for analysis.

2.4. Determination of Moisture Content and Carboxyl Group Content

The collected cellulose samples were conditioned at 21 ± 2 °C at 65 ± 2% RH for at least 24 h. To determine the exact dry weight, the samples were weighed and dried at 105 °C overnight. They were then cooled in a desiccator with P_2O_5 and weighed in dry state to determine the moisture content from the weight loss. The values are given as the mean of a double determination [24].

The methylene blue sorption measurements were carried out with 0.5 g of dry sample [25,26]. First, 0.30 g methylene blue was dissolved in 750 mL deionised water. Then, the pH was adjusted to 8.5 by the addition of NaOH and the solution was filled to 1 L. Two samples were analysed for each specimen. The dried sample was chopped and mixed with

10 mL of 0.30 g L^{-1} MB solution and 25 mL borate buffer pH 8.5 and then made up to 50 mL with deionised water. After 1 h of stirring at room temperature, the suspensions were filtered through a glass filter. Then, 10 mL of the filtrate was diluted in 10 mL 0.1 M HCl and made up to 50 mL with deionised water. The diluted solution was analysed by UV–Vis spectrophotometry at a wavelength of 664.5 nm. (Spectrophotometer HITACHI U-2000, Inula, Vienna, Austria, 10 mm cuvette) Borate buffer served as reference. The absorbance of the solution was used to determine the total content of acid sites in the geotextiles. The carboxyl group content was then calculated as mmol kg^{-1} cellulose material. To ensure the repeatability of the methylene blue results, the absorbance was additionally measured using 5 mL of the filtered suspension. The results were given as mean value and the standard deviation of six independent experiments.

2.5. Determination of Viscosity-Average Degree of Polymerisation

A viscometric determination of the average degree of polymerization (average DP_v) was performed using an Ubbelohde capillary viscometer (Type No 501, Capillary 12/Ic) (SI Analytics GmbH, Mainz, Germany) and fresh ferric sodium tartrate (FeTNa) solvent (0.3 M $FeCl_3 \cdot 6H_2O$, 0.98 M tartaric acid, and 5.36 M NaOH). Cellulose samples (20 mg) were dissolved in 50 mL FeTNa solution. Argon gas was purged into the bottle to remove air and prevent oxidation during dissolution. The bottle was closed and shaken for more than 16 h to make sure that cellulose was completely dissolved. The viscosity η of a cellulose solution was measured from the efflux time at 20 ± 1 °C through the viscometer and the intrinsic viscosity was calculated by equation (DIN 54270-3). The results are given as mean and standard deviation of a double determination.

$$\eta = \frac{(\eta_{rel} - 1)/c}{1 + k \cdot (\eta_{rel} - 1)} \tag{1}$$

η = intrinsic viscosity (mL g^{-1})
c = concentration of viscose sample in the solution (g mL^{-1})
k = 0.339
$\eta_{rel} = \frac{t}{t_0}$ with t (in seconds) and t0 being the efflux time of the solution and the solvent respectively.

The viscosity average DPv is calculated from the intrinsic viscosity η using Equation (2) [27].

$$\eta = 4.85 \times DP_v^{0.61} \tag{2}$$

2.6. Confocal 3D Laser-Scanning Microscopy and FTIR Spectroscopy

The nonwoven fabric samples were examined with a confocal 3D laser-scanning microscope (Keyence VK-X150, Keyence, Neu-Isenburg, Germany) at 20×, 50× and 100× magnification, using the on board software.

The Fourier transform infrared spectra (FTIR) of the fibres were recorded in the spectral range 4000 cm^{-1} to 500 cm^{-1} using an attenuated total reflectance (ATR) unit equipped with a diamond crystal (Bruker Vector 22, Karlsruhe, Germany; resolution 2 cm^{-1}, and 128 scans).

3. Results and Discussion

3.1. Exposure to Weathering and Visual Inspection

Biodegradation of the cellulose fibres under real weathering conditions depends on the actual climate conditions during the exposure. The prototypes were installed directly on the ground on a slope in Romanshorn (CH) (Elevation: 406 m, 1,332 ft, GPS coordinates 47°33′24.4″ N, 9°21′59.9″ E). Bacterial growth and plant growth in the structure depend on the actually prevalent temperature, precipitation and season, respectively, thus the climate data given in Figure 2 provide an overview about most important conditions during the 61 weeks degradation test in the moderate climate of the region. The simulated raw data were provided by meteoblue (meteoblue.com, accessed on 2 August 2022). Red triangles in the diagram show when the samples were taken for laboratory analysis.

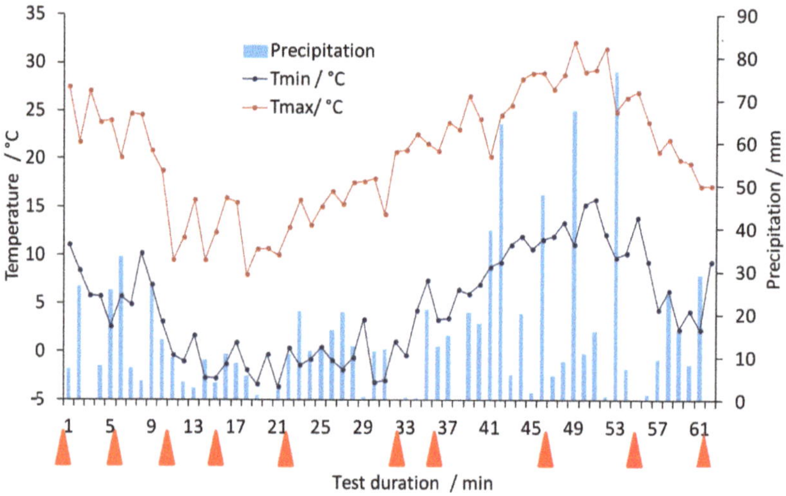

Figure 2. Weather data simulated for the site of the out-door prototype testing (Romanshorn, CH) during the period of outdoor degradation tests (data provided by meteoblue.com).

Table 2 shows the detailed sampling time and the overall duration of the exposure to outdoor weathering conditions.

Table 2. Period of the degradation tests and sampling dates.

		Month	Weeks	Days
	Start of weather trials	August	0	1
1	sampling	October	6	42
2	sampling	November	11	77
3	sampling	December	15	108
4	sampling	January	22	157
5	sampling	April	32	228
6	sampling	May	35	248
7	sampling	July	47	335
8	sampling	September	55	365
9	End of weather trials	October	61	427

Over the period of 61 weeks, an accumulated total amount of precipitation of 900 mm water was calculated. Precipitation occurred in quite regular intervals without a long dry period. The temperature mainly ranged from 5 °C to 20 °C, with a minimal temperature of −4 °C in winter and a maximum temperature of 30 °C during summer.

The biological degradation of the cellulose web was also monitored by visual and microscopic inspection and with chemical analytical methods.

The biological decomposition of the materials can be easily assessed visually. Photographs of the prototypes were taken at the same time when material samples were collected for laboratory analysis. The appearance of the geotextiles during the weathering is shown in Figures 3 and 4, using photographs of prototype 1. The integration of the geotextiles into nature began quite quickly: after 11 weeks, about half of the geotextile was already covered with vegetation and firmly anchored to the soil. The exposure to sun and rain during the first period of the installation (August–November), along with the open structures in the geotextile, ensured the rapid development of the vegetation and overgrowth with plant material. After 35 weeks, the prototypes were almost completely overgrown with vegetation; however, the cellulose web can still be recognised in a close-up view (Figure 3f).

Figure 3. Photographs of the installed prototypes, (**a**) installed prototype 1–6, (**b**) day of installation, (**c**) after 11 weeks, (**d**) after 22 weeks, (**e**) after 35 weeks (**f**) close-up view after 35 weeks.

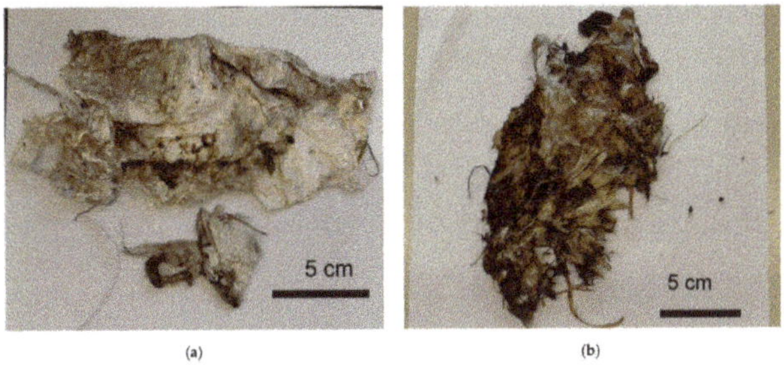

Figure 4. Photographs of a cellulose fibre sample from prototype 1, (**a**) 15 weeks, (**b**) 47 weeks of outdoor installation.

The process of degradation was also indicated by the dark discolouration of the sample pieces in Figure 4. Rapid progress of plant growth and integration of the chosen geotextile structure also accelerates the access of microorganisms. The results show that the degradation is also co-initiated by soil microorganisms, as with longer degradation time, the brown discolouration intensifies [20]. The weather trials started in the summer with higher temperatures, followed by the wet autumn season; both factors also accelerate the growth of microorganisms and, thus, fibre degradation.

The biological attack and degradation also led to changes in the flexibility of the material. The nonwoven structures became more brittle and thinner over the weeks and completely fragmented at the end of the trials (Figure 5). As a result, sampling of pure cellulose fibre containing specimen became impossible and the presence of soil and plant residues made determining the degree of polymerisation impossible. A photodocumentation of the degradation of the 6 prototypes is given in the supplementary information (Figure S1).

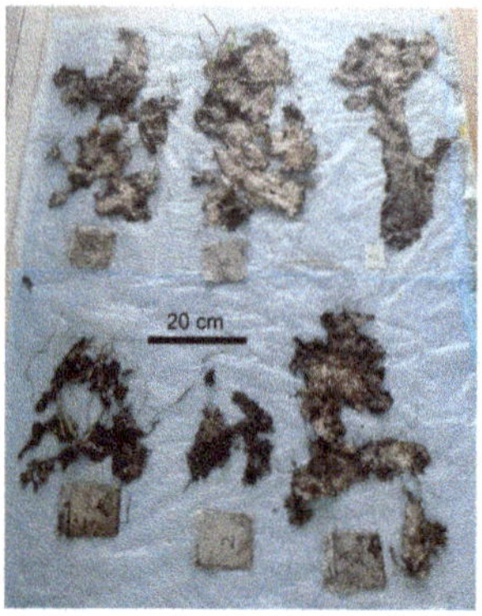

Figure 5. Photographs of samples from prototype 1–6 after 61 weeks of outdoor weathering.

These observations already indicate the rapid decomposition of the chemical and morphological structure of the fabric samples within the test period.

Figure 6 shows the photomicrographs of samples after different times of weathering. It can be observed that microorganisms settled on the surface of the fibres. After more than a year of weathering tests, the geotextiles were noticeably degraded, and the integrity and structure of the fabric almost collapsed. Remarkably, the remaining material still exhibited the shape of a fibrous structure and very little fibrillation or fibre fragments was detected in the photomicrographs. Most probably, the fibres biodegraded from their more accessible fibre ends, where enzymatic hydrolysis then leads to the release of cellobiose and glucose oligomers. This is in agreement with the proposed mechanism for enzymatic cellulose degradation, where amorphous zones of the cellulose fibre are hydrolysed more rapidly. Free chain ends are formed by the endo-glucanases and are then hydrolysed to cellobiose by exo-glucanases [6].

Figure 6. Photomicrographs of viscose fibres as function of time in outdoor weathering. (**a**) pristine CV fibres, (**b**,**c**) with use of higher magnification; (**d**) fibres after 15 weeks of outdoor weathering, (**e**,**f**) with use of higher magnification; (**g**) fibres after 47 weeks of outdoor weathering, (**h**,**i**) with use of higher magnification; (**j**) fibres after 61 weeks of outdoor weathering, (**k**,**l**) with use of higher magnification.

3.2. Monitoring of Fibre Degradation

While visual inspection and microscopic analysis of the prototypes during the period of outdoor testing provide information about physical damage, structure loss and plant growth, analytical methods have to be applied to monitor cellulose biodegradation on a molecular level. An important measure indicating cellulose degradation is obtained from measurements of the dynamic viscosity of the cellulose after dissolution in ferric sodium tartrate (FeTNa) solvent.

In case biodegradation of the cellulose polymer occurs in the fibre structure, a reduction in the dynamic viscosity of the solution should be detected when compared to pristine viscose fibres. In Figure 7, the change in the DP_v values as a function of weathering time are given. DP_v changes from about 400 for after 60 weeks to 100 for the cellulose fibres after a degradation period of 22 weeks. After longer weathering time (32 weeks or more), the fibres could no longer be dissolved completely by ferric sodium tartrate (FeTNa) solvent, which also could contribute to an apparently lower DP_v value. The DP_v value of the pristine viscose fibres at the beginning ranges at 400 ± 53 glucose units per cellulose molecule. Up to 11 weeks of weathering, the reduction in DP_v value is not significant. After 22 weeks of outdoor weathering, a reduction to DP_v of 200 is indicated by the measurement of the dynamic viscosity. The remarkably low reduction in DP_v can be explained with the mechanism of biodegradation. Where small fragments of cellobiose and glucose oligomers are removed from the fibre surface as products from the enzymatic hydrolysis, the DP_v of the cellulose in the remaining fibre bulk will not be affected substantially.

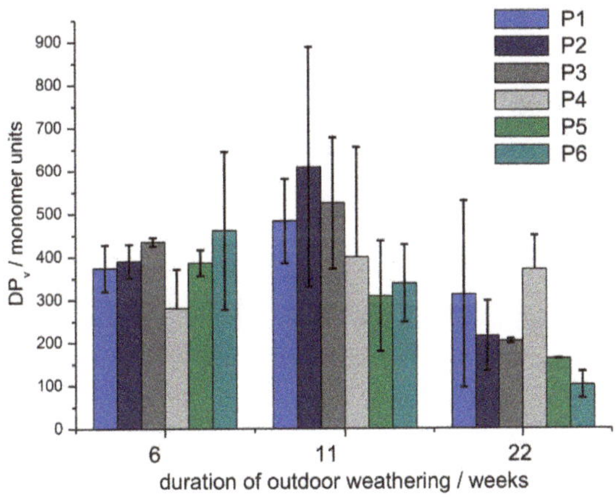

Figure 7. Degree of polymerization of the cellulose as function of time under outdoor conditions.

The results for DP_v measured at the samples taken after 11 weeks of outdoor degradation indicate a slight increase in DP_v, which, however, must be interpreted under consideration of the relatively high standard deviation. To bring the result to a statistically significant level, a very high number of additional repetitions would be required.

3.3. Determination of Moisture Content and Carboxyl Group Content

The collected samples were characterised by determining their moisture content and by analysing the carboxylic group content.

The initial moisture content of the viscose fibre was determined with 12 %wt, which is the expected value for regenerated cellulose fibres produced by the xanthogenate route. Following a model of structural weakening and opening of the fibre structure, an increase in moisture sorption during biodegradation would be expected initially. Remarkably, the moisture content decreased during the phase of biodegradation. The decrease from 12.5 ± 0.5 %wt at week 6 to 11.1 ± 0.5 %wt after week 32 is statistically significant at a two-sided confidence level of 99.9%. This finding can be understood on the basis of the enzymatic processes, which lead to more rapid degradation of the amorphous parts of the fibre. As the amorphous parts of a cellulose fibre are responsible for moisture sorption biodegradation processes, which reduce the share of the amorphous regions preferentially, it will lead to a reduced sorption capacity for water of the remaining fibre structure. The overall change in moisture sorption of the cellulose fibres as a function of time is shown in Figure 8. A few samples exhibit a rather high standard deviation, which is indicated by the comparable high error bars. This can be explained with the presence of other impurities, e.g., soil, plant material, which could not be removed completely from the specimen.

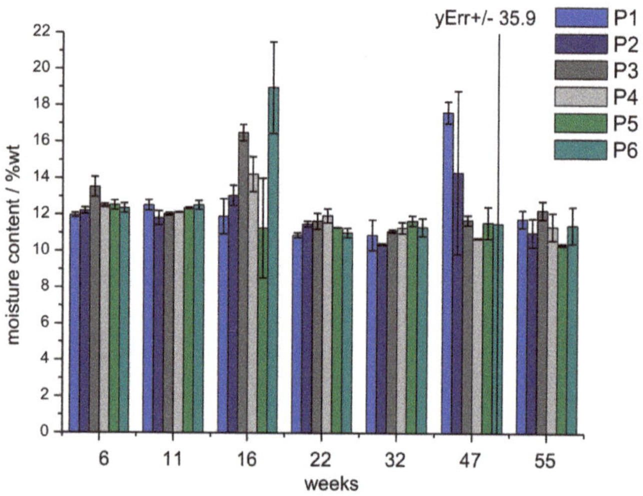

Figure 8. Moisture content of the viscose fibres of the prototypes during outdoor weathering as function of time (mean and standard deviation as error bar).

During viscose fibre manufacturing, the reducing end of the cellulose chain is oxidised and, instead of the aldehyde group, carboxylic groups are formed. Representative values for the carboxylic group content of viscose fibres are near 20 mmol kg^{-1} fibre material [28]. The carboxylic group content in viscose fibres used for these experiments was determined with 17.3 ± 0.1 mmol kg^{-1}.

During the biodegradation of the cellulose fibres, weathering hydrolysis of the glycosidic bonds in the cellulose chains and oxidation of the formed aldehyde groups could occur. As a consequence, an increase in carboxylic group content should be observed. The carboxylic group content in the collected samples as a function of weathering time is given in Figure 9.

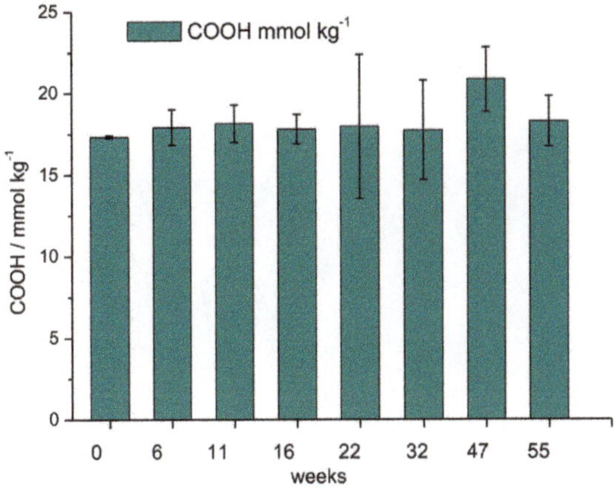

Figure 9. Carboxyl group content of weathered viscose samples as function of time.

The analysed content of the carboxylic groups remained constant over the period of outdoor weathering, which indicates that light-induced cellulose chain breakage followed by photo-oxidation of the end groups formed to carboxylic groups is a minor pathway, compared to the enzymatic degradation processes. The higher scattering of the results with increasing time outdoors can be explained with the sensitivity of the methylene blue sorption method to other impurities present in the specimen. The presence of plant residues and microorganisms on the fibre specimen (Figure 6), as well as soil residues, can lead to higher variations in methylene blue sorption and, thus, increase the variation in experimental results.

FTIR-ATR analysis of the samples was performed to identify the appearance of new functional groups, e.g., carbonyl groups or carboxyl groups as the results of the weathering tests. Representative spectra of samples after different time in outdoor weathering are shown in Figure 10. A zoomed-in portion of the fingerprint region is shown in Figure 11.

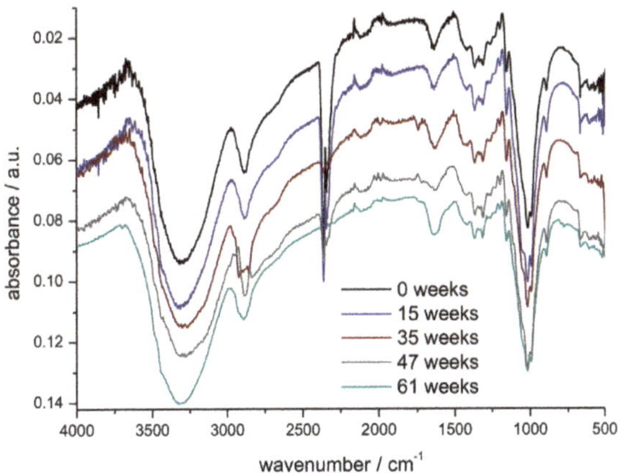

Figure 10. FTIR-ATR spectra of prototype 4 as function of exposure time in outdoor weathering (week 0, 15, 35, 47, 61).

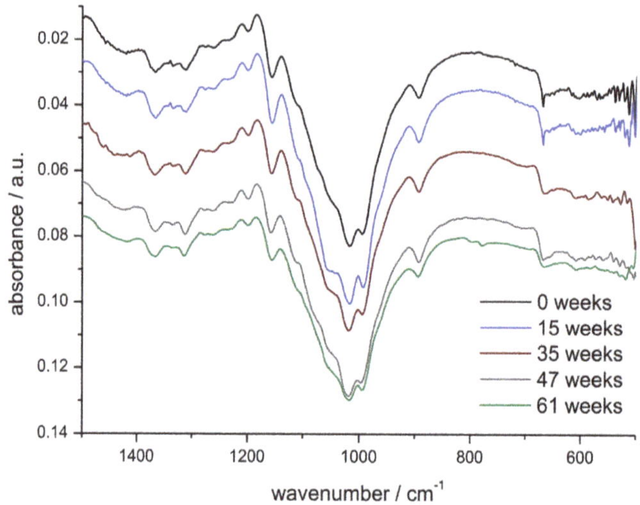

Figure 11. Zoomed-in section of the fingerprint region of theFTIR-ATR spectra of prototype 4 as function of exposure time in outdoor weathering (week 0, 15, 35, 47, 61).

The spectra shown in Figure 10 exhibit the typical absorbance pattern for cellulose. A detailed table of characteristic peak values is shown in Table 3.

Table 3. Characteristic absorption peaks in the infrared spectrum of the viscose fibres and literature reference values for cellulose [29–31].

Wavenumber cm^{-1}	Literature cm^{-1}	Type of Vibration
3000–3500	3000–3500	O-H hydrogen bonded stretching
2926–2886	2892	C-H stretching
1417–1420	1430	H-C-H and O-C-H in plane bending
1365–1366	1375	C-O-C, C-C-O, C-C-H deformation and stretching
1310–1313	1312	O-H bending
1156–1157	1157	C-O asym. valence
1018	1026	C-O-C pyranose ring skeletal
893	892	C-O-C valence
666–668	668	C-OH out of plane bending
1636, 1641	1635, 1638	Adsorbed water

The broad absorbance between 3500 cm^{-1} and 3000 cm^{-1} can be attributed to the stretching vibration of the O-H groups and the absorbance near 2900 cm^{-1} can be attributed to the C-H stretching vibration. C-H and CH_2 bending vibrations are observed from 1500 cm^{-1} to 1100 cm^{-1}, and the C-O stretching vibration leads to the strong absorbance at 1100 cm^{-1} [29]. Only minor differences are seen in the FTIR spectra, with longer exposure time in outdoor weathering. Smaller variations in the wavenumber region 2750 cm^{-1}–3000 cm^{-1} (C-H stretching vibrations) may be due to the presence of higher amounts of microorganisms, as well as adhering soil residues.

In the FTIR of the sample collected after 35 weeks, a smaller signal is detected around 1750 cm^{-1}, which most probably is due to the presence of other material (plant material, microbial growth) at the particular site where the FTIR spectrum was measured. At this wavenumber, ester groups, e.g., present in membrane lipids and cell wall pectines, are observed [32].

4. Conclusions

Hybrid structures made from cellulose fibre nonwovens and steel wire provide two complementary functions when applied as a combination of geotextiles and slope stabilisation solutions. The steel mesh contributes with the required physical strength to prevent rock fall and land sliding. The hydrophilic viscose fibres serve as a biodegradable water reservoir, which supports greening in a stabilised area. During the period of greening, biodegradation of the cellulose fibres may occur.

A long-term outdoor study with prototypes of the geotextiles was executed with a total duration of 61 weeks of weathering to monitor the biodegradation of the cellulose fibres and plant growth. Samples of the cellulose nonwovens were collected regularly over the test period and the biodegradation was monitored by microscopy and instrumental analytical methods. The carboxylic group content of the residual material did not change during the period of weathering and the formation of additional functional groups, e.g., carbonyl groups, was not indicated by FTIR analysis. However, a significant decrease in the moisture content was observed with the duration of the outdoor weathering. These findings are in agreement with the theory of enzymatic degradation of cellulose fibres [6]. The biodegradation of cellulose occurs via the action of cellulases, which hydrolyse the cellulose chains and leads to the release of cellobiose and anhydroglucose oligomers. This reaction preferentially occurs at the amorphous domains of the cellulose fibres. The amorphous domains are responsible for the moisture sorption in the fibres; thus, degradation of these domains will also lead to reduced water sorption behaviour. A reduction in the average chain length of cellulose was also detected by determining the degree of polymerisation. During the phase of biodegradation, the nonwoven cellulose structure partially disintegrated and plant growth through the nonwoven structure occurred.

The results demonstrate that the desired functionality of a short-term biodegradable flexible structure with a durable steel backbone, which supports soil fixation, contributes to water storage and enables plant growth successfully, could be achieved with the cellulose/steel hybrid structure.

Future research could address the preparation of regenerated cellulose fibres from cellulose containing wastes, e.g., from flax production as well as from recycled cellulose material, to reduce the ecological impact of fibre production [32,33]. Such a comparison should also include the different processes for fibre production, e.g., the lyocell process and the use of ionic liquids.

Supplementary Materials: The following supporting information can be downloaded at: https://www.mdpi.com/article/10.3390/polym14194179/s1, Figure S1: Photodocumentation of the biodegradation of the test prototypes P1–P6 as function of test time.

Author Contributions: Data curation, A.P.M. and B.P.; Formal analysis, B.P.; Funding acquisition, T.P.; Investigation, B.P.; Methodology, A.P.M.; Resources, H.L. and T.P.; Supervision, T.P.; Validation, H.L.; Writing–original draft, T.B.; Writing–review & editing, T.B. All authors have read and agreed to the published version of the manuscript.

Funding: Financial support is gratefully acknowledged to the COMET Project "Textile Competence Center Vorarlberg 2–FFG 882502", funded within COMET–Competence Centers for Excellent Technologies–by BMK, BMDW as well as co-financing federal province Vorarlberg. The COMET-Funding Program is managed by the Austrian Research Promotion Agency FFG.

Institutional Review Board Statement: Not applicable.

Data Availability Statement: The data presented in this study are available on request from the corresponding author. The data are not publicly available due to confidentiality reasons.

Acknowledgments: Authors want to thank meteoblue.com for providing detailed weather data for the testing period.

Conflicts of Interest: The authors declare no conflict of interest.

References

1. Eriksson, K.-E.L.; Blanchette, R.A.; Ander, P. Biodegradation of Cellulose. In *Microbial and Enzymatic Degradation of Wood and Wood Components*; Springer: Berlin/Heidelberg, Germany, 1990; pp. 89–180. [CrossRef]
2. Potthast, A.; Ahn, K.; Becker, M.; Eichinger, T.; Kostic, M.; Böhmdorfer, S.; Jeong, M.J.; Rosenau, T. Acetylation of Cellulose—Another Pathway of Natural Cellulose Aging during Library Storage of Books and Papers. *Carbohydr. Polym.* **2022**, *287*, 119323. [CrossRef]
3. Mayr, G.; Zeppetzauer, F.; Zweckmair, T.; Bauer, D.; Hild, S.; Potthast, A.; Rosenau, T.; Roeder, T. The Reactions of Cellulose and Hemicellulose Degradation Products in the Viscose Fibre Spin Bath. *Lenzing. Ber.* **2015**, *92*, 53–58.
4. Roeder, T.; Kraft, G.; Borgards, A.; Zuckerstaetter, G.; Rosenau, T. Cellulose Degradation during Viscose Processing. In *Abstracts of Papers, 241st ACS National Meeting & Exposition, Anaheim, CA, USA, 27–31 March 2011*; American Chemical Society: Washington, DC, USA, 2011; p. CELL CO-69NZUR.
5. Roeder, T.; Kliba, G.; Milacher, W.; Kraft, G.; Potthast, A.; Rosenau, T. Cellulose Degradation during Deformation Processing and Analytics. In *Abstracts of Papers, 249th ACS National Meeting & Exposition, Denver, CO, USA, 22–26 March 2015*; American Chemical Society: Washington, DC, USA, 2015; p. CELL CO-69TQWW.
6. Bisaria, V.S.; Ghose, T.K. Biodegradation of Cellulosic Materials: Substrates, Microorganisms, Enzymes and Products. *Enzyme Microb. Technol.* **1981**, *3*, 90–104. [CrossRef]
7. Yu, Z.; Ji, Y.; Bourg, V.; Bilgen, M.; Meredith, J.C. Chitin- and Cellulose-Based Sustainable Barrier Materials: A Review. *Emergent Mater.* **2020**, *3*, 919–936. [CrossRef]
8. Beguin, P.; Aubert, J.-P. The Biological Degradation of Cellulose. *FEMS Microbiol. Rev.* **1994**, *13*, 25–58. [CrossRef] [PubMed]
9. Polman, E.M.N.; Gruter, G.J.M.; Parsons, J.R.; Tietema, A. Comparison of the Aerobic Biodegradation of Biopolymers and the Corresponding Bioplastics: A Review. *Sci. Total Environ.* **2021**, *753*, 141953. [CrossRef] [PubMed]
10. Berdugo-Clavijo, C.; Scheffer, G.; Sen, A.; Gieg, L.M. Biodegradation of Polymers Used in Oil and Gas Operations: Towards Enzyme Biotechnology Development and Field Application. *Polymers* **2022**, *14*, 1871. [CrossRef]
11. Tsavkelova, E.A.; Netrusov, A.I. Biogas Production from Cellulose-Containing Substrates: A Review. *Appl. Biochem. Microbiol.* **2012**, *48*, 421–433. [CrossRef]
12. Raschle, P. Microbial Influence on Cellulosic Textiles and Microbiological Testing. *Int. Biodeterior.* **1989**, *25*, 237–244. [CrossRef]
13. Sewalt, V.J.H.; Glasser, W.G.; Beauchemin, K.A. Lignin Impact on Fiber Degradation. 3. Reversal of Inhibition of Enzymatic Hydrolysis by Chemical Modification of Lignin and by Additives. *J. Agric. Food Chem.* **1997**, *45*, 1823–1828. [CrossRef]
14. Lykaki, M.; Zhang, Y.Q.; Markiewicz, M.; Brandt, S.; Kolbe, S.; Schrick, J.; Rabe, M.; Stolte, S. The Influence of Textile Finishing Agents on the Biodegradability of Shed Fibres. *Green Chem.* **2021**, *23*, 5212–5221. [CrossRef]
15. Meereboer, K.W.; Misra, M.; Mohanty, A.K. Review of Recent Advances in the Biodegradability of Polyhydroxyalkanoate (PHA) Bioplastics and Their Composites. *Green Chem.* **2020**, *22*, 5519–5558. [CrossRef]
16. Kim, S.; Cho, Y.; Park, C.H. Effect of Cotton Fabric Properties on Fiber Release and Marine Biodegradation. *Text. Res. J.* **2022**, *92*, 2121–2137. [CrossRef]
17. Arshad, K.; Mujahid, M. Biodegradation of Textile Materials. Master's Thesis, University of Boras, Boras, Sweden, 2011.
18. Hayakawa, C.; Funakawa, S.; Fujii, K.; Kadono, A.; Kosaki, T. Effects of Climatic and Soil Properties on Cellulose Decomposition Rates in Temperate and Tropical Forests. *Biol. Fertil. Soils* **2014**, *50*, 633–643. [CrossRef]
19. Warnock, M.; Davis, K.; Wolf, D.; Gbur, E. Biodegradation of Three Cellulosic Fabrics in Soil. In *Summaries of Arkansas Cotton Research 2009*; Oosterhuis, D.M., Ed.; Arkansas Agricultural Experiment Station, University of Arkansas: Favetteville, NC, USA, 2010.
20. Sülar, V.; Devrim, G. Biodegradation Behaviour of Different Textile Fibres: Visual, Morphological, Structural Properties and Soil Analyses. *Fibres Text. East. Eur.* **2019**, *27*, 100–111. [CrossRef]
21. Haider, T.P.; Völker, C.; Kramm, J.; Landfester, K.; Wurm, F.R. Plastics of the Future? The Impact of Biodegradable Polymers on the Environment and on Society. *Angew. Chemie—Int. Ed.* **2019**, *58*, 50–62. [CrossRef]
22. Liu, J.; Lv, C. Durability of Cellulosic-Fiber-Reinforced Geopolymers: A Review. *Molecules* **2022**, *27*, 796. [CrossRef]
23. Röder, T.; Kogler, M.; Innerlohinger, J.; Schuster, K.C. Regenerated Cellulose—Developments and Biodegradability. In Proceedings of the 6th EPNOE International Polysaccharide Conference, Aveiro, Portugal, 21–25 October 2019; p. 93.
24. Jaturapiree, A.; Manian, A.P.; Bechtold, T. Sorption Studies on Regenerated Cellulosic Fibers in Salt-Alkali Mixtures. *Cellulose* **2006**, *13*, 647–654. [CrossRef]
25. Philipp, B.; Rehder, W.; Lang, H. Determination of the Carboxyl Content of Dissolving Pulp. *Papier* **1965**, *19*, 1–9.
26. Fitz-Binder, C.; Bechtold, T. One-Sided Surface Modification of Cellulose Fabric by Printing a Modified TEMPO-Mediated Oxidant. *Carbohydr. Polym.* **2014**, *106*, 142–147. [CrossRef]
27. Klemm, D.; Philipp, B.; Heinze, T.; Heinze, U.; Wagenknecht, W. *Comprehensive Cellulose Chemistry*; Volume I: Fundamentals and Analytical Methods; Wiley-VCH Verlag: Weinheim, Germany, 1998.
28. Fitz-Binder, C.; Bechtold, T. Ca2+ Sorption on Regenerated Cellulose Fibres. *Carbohydr. Polym. Technol. Appl.* **2012**, *90*, 937–942. [CrossRef] [PubMed]

29. Široký, J.; Blackburn, R.S.; Bechtold, T.; Taylor, J.; White, P. Attenuated Total Reflectance Fourier-Transform Infrared Spectroscopy Analysis of Crystallinity Changes in Lyocell Following Continuous Treatment with Sodium Hydroxide. *Cellulose* **2010**, *17*, 103–115. [CrossRef]
30. Reddy, K.O.; Maheswari, C.U.; Shukla, M. Physico-Chemical Characterization of Cellulose Extracted from Ficus Leaves. *J. Biobased Mater. Bioenergy* **2013**, *7*, 496–499. [CrossRef]
31. Oh, S.Y.; Il Yoo, D.; Shin, Y.; Seo, G. FTIR Analysis of Cellulose Treated with Sodium Hydroxide and Carbon Dioxide. *Carbohydr. Res.* **2005**, *340*, 417–428. [CrossRef] [PubMed]
32. Makarov, I.S.; Golova, L.K.; Smyslov, A.G.; Vinogradov, M.I.; Palchikova, E.E.; Legkov, S.A. Flax Noils as a Source of Cellulose for the Production of Lyocell Fibers. *Fibers* **2022**, *10*, 45. [CrossRef]
33. Munasinghe, P.; Druckman, A.; Dissanayake, D.G.K. A Systematic Review of the Life Cycle Inventory of Clothing. *J. Clean. Prod.* **2021**, *320*, 128852. [CrossRef]

polymers

MDPI

Article

Mechanism of Multi-Stage Degradation in Hot Bitumen of Micronized Powder Elastomeric Modifiers from Worn-Out Tire's Rubber

Vadim Nikol'skii [1], Tatiana Dudareva [1], Irina Krasotkina [1], Irina Gordeeva [1,*], Viktoriya Gorbatova [1], Alexandre A. Vetcher [2,3,*] and Alexander Botin [2,4,*]

[1] N.N. Semenov Federal Research Center of Chemical Physics, Russian Academy of Sciences, 4, Kosygin Str., 119991 Moscow, Russia
[2] Institute of Biochemical Technology and Nanotechnology (IBTN) of the Peoples' Friendship University of Russia (RUDN), 6 Miklukho-Maklaya Str., 117198 Moscow, Russia
[3] Complimentary and Integrative Health Clinic of Dr. Shishonin, 5 Yasnogorskaya Str., 117588 Moscow, Russia
[4] N.V. Sklifosovsky Institute of Emergency Medicine, 129090 Moscow, Russia
* Correspondence: ivgordeeva@bk.ru (I.G.); avetcher@gmail.com (A.A.V.); botin_as@pfur.ru (A.B.)

Abstract: For the first time, by atomic force microscopy (AFM) methods, micro- and nanofragments of micronized powder elastomeric modifier (PEM) formed at the short-term (3 min at 160 °C) interaction of PEM with hot bitumen have been demonstrated. It is the technology of high-temperature shear-induced grinding of a worn-out tire's crumb rubber or its co-grinding with styrene–butadiene–styrene (SBS) block copolymer which provides the creation of the PEM structure inclined to rapid degradation in hot bitumen. The formation just after the preparation process of a new structure of a modified binder, more resistant to external effects, is supported by the data of rheological tests. Performance tests for a modified binder using Superpave standard adopted by the road industry for bituminous binders showed an extended temperature range, resistance to rutting, and low-temperature and fatigue cracking. The better resistance to low-temperature and fatigue cracking is certainly related to energy absorption and crack growth stopping in the presence of micron and submicron resilient PEM fragments in accordance with the mechanism of increasing impact toughness in plastics.

Keywords: powder elastomeric modifiers; hybrid powder; high-temperature shear-induced grinding; bitumen; AFM; rheology; modification mechanism

Citation: Nikol'skii, V.; Dudareva, T.; Krasotkina, I.; Gordeeva, I.; Gorbatova, V.; Vetcher, A.A.; Botin, A. Mechanism of Multi-Stage Degradation in Hot Bitumen of Micronized Powder Elastomeric Modifiers from Worn-Out Tire's Rubber. *Polymers* **2022**, *14*, 4112. https://doi.org/10.3390/polym14194112

Academic Editor: Yung-Sheng Yen

Received: 9 September 2022
Accepted: 29 September 2022
Published: 30 September 2022

1. Introduction

Crumb rubber, obtained using the technology of cryogenic grinding or mechanical crushing at an ambient temperature of worn-out tires, is used in the road industry as a modifier of bitumen and/or asphalt mixtures [1].

One of the main aims of bitumen modification is to extend the temperature range of the bitumen and to increase the resistance to pavement defects (rutting, temperature, and fatigue cracking), to the formation of which bitumen contributes significantly [2]. When using a modifier, it is important to understand whether it can increase the bitumen resistance to permanent deformation (rutting) at high in-service pavement temperatures and whether it has an enhancing or weakening effect on the structuring and cracking processes of bitumen at intermediate and low temperatures. For example, there are AFM data showing that when bitumen cools, rigid "bee-like" structures form, which can be cracking concentrators, especially under cyclic loading [3]. Most modifiers increase the resistance of bitumen to rutting at high in-service temperatures. The improvement in pavement defect resistance at lower temperatures by modification is not so obvious [4].

When using crumb rubber (CR) as a modifier, the main features of the modification process are: the number of technological stages, apparatus design, temperature, duration of the process, and the efficiency of modification, which are mainly determined by the method of obtaining CR, its size, and morphology [5–8]. Two methods of modification are used in the application of CR. The first method is the introduction of CR directly into the mixer during asphalt mixture preparation. Most often in this case, CR replaces some fractions of the mineral components of the asphalt mixture and does not give a noticeable result in terms of improving the properties of bitumen, because the mixing time at high temperature (up to 1 min) is too short [5,8–10]. The second method aims to improve the rheological properties of bitumen and is a separate technological step—the preparation of rubber–bitumen binder, preceding the preparation of asphalt mixture. In this case, CR with an initial size of up to 1 mm, and in recent years up to 0.6 mm, is subjected to a sufficiently long mixing process (at least 1 hr) with bitumen at relatively high temperatures (170÷200 °C). In long-term contact with hot bitumen, some of the CR particles, mostly smaller than 200–300 μm in size, noticeably swell and break down along the cracks formed in the crushing of the worn-out tire. Nevertheless, many particles often do not undergo visible changes even after prolonged agitation. Thus, micromechanical models are developed for the complex shear modulus of rubber-modified binder, and swelled particles of the same size as the original are considered [11]. However, there are available data showing that a significant increase in the fatigue cracking resistance of modified bitumen is observed only when the amount of CR particles smaller than 75 μm is increased by at least 5% compared to the original amount. [9]. To increase the compatibility of CR with bitumen, various methods of chemical or physical–mechanical treatment are used to change the surface structure of the particles. In this case, a layer of partially regenerated rubber is formed on the surface of the particles. However, the process of bitumen modification, even in this case, requires high temperature and long mixing time [12–15].

The goal of this study was to investigate the degradation mechanism of micronized powder elastomeric modifiers (PEM) particles during interaction with hot bitumen. Micronized PEMs are produced by the high-temperature shear-induced grinding (HTSG) of CR or the co-grinding of CR and styrene–butadiene–styrene block copolymer (SBS). Such modifiers are active powders of discretely devulcanized rubber (APDDR) and hybrid (APDDR-SBS) powder. PEM is injected into the asphalt mixture during its preparation. The process of PEM production is realized in specialized equipment: a rotary dispergator. HTSG is intermediate in temperature of the grinding process (160–170 °C) among tire-rubber-recycling methods between its mechanical crushing and obtaining the regenerated rubber by devulcanization [16]. In the grinding zone of the rotary dispergator, due to high temperature and significant shear forces applied to the compressed layer of material in the modulated mode, along with a decrease in the size of the processed particles, discrete devulcanization of the rubber occurs in the locations where optimal conditions are created. It is assumed that the intermolecular bonds are disrupted and rearranged, and, in the case of co-grinding, the components are combined on the micro- and nano level. It is also noted that there is almost no change in molecular weight distribution of rubber molecules, since at such temperatures, macromolecules usually have time to relax without breaking the molecular chain. According to electron microscopy data, a PEM particle is an agglomerate of micron- and submicron-size blocks, connected by strands of different thickness (from several nanometers to several microns). According to atomic force microscopy data, the structure of PEM particles can be defined as having a strong phase heterogeneity (at the level of 0.1–0.2 μm or even less) [16,17].

Previously, the method of electron scanning microscopy (SEM) showed that after the solvent washing of binders prepared by mixing bitumen with PEM for 1–40 min in the temperature range 120–180 °C, fragments of PEM in the form of particles up to 100–200 nm or in the form of micron-sized thin films were observed [17]. However, it is important to test the presence of such fragments in the modified bitumen. In this study, comparative research by atomic force microscopy (AFM) of the surface structure of the original bitumen and PEM-modified binder prepared under temperature and time conditions close to the preparation conditions of asphalt mixtures (3 min at 160 °C) was carried out. To estimate the effectiveness of modification after a short-term interaction of PEM with hot bitumen for the original (no-aged) samples of bitumen and freshly prepared PEM-modified binders (which were not subjected to additional temperature effects), comparative performance tests in a wide temperature range by using Superpave standards adopted by the road industry for bitumen binder were conducted [13,18,19].

2. Materials and Methods

2.1. Materials

The following materials were investigated in this work:

- Micronized powder elastomeric modifiers (PEM): APDDR and hybrid (APDDR-SBS) powder, obtained by high-temperature shear-induced grinding of crumb rubber in a rotary dispergator. APDDR was produced by grinding of CR of worn-out tires and hybrid powder by co-grinding of CR and SBS L 30-01 (content of bound styrene, wt%—29-31; molecular weight—around 78 KD). The APDDR and hybrid powder were homogeneous black powders. The properties of PEM are listed in Table 1. The specific surface was determined by the BET method at T = 77 K using the adsorption analyzer of surface area "NOVA 1200e" (Quantachrome Instruments, Ltd.; Boynton Beach, FL, USA). Particle size distribution parameters were determined using wet laser diffraction using "ANALYSETTE 22 NanoTec plus" (Fritsch GmbH – Mahlen und Messen, Idar-Oberstein, Germany).

Table 1. The PEM properties.

PEM	Particle Size Distribution Parameters, μm			Specific Surface, m^2/g
	D_{10}	D_{50}	D_{90}	
APDDR	40	140	300	0.49
Hybrid (APDDR+SBS)	50	160	340	0.45

- Blown bitumen grade BND 60/90 was used as a basis for preparation of modified binders for AFM research and performance testing.
- Blown bitumen grade BND 100/130 was used for preparation of modified binders and performance testing.

The coding and properties of bitumen samples are shown in Table 2.

Table 2. Properties of bitumen.

Penetration Grade	BND 60/90	BND 100/130
Bitumen code	Bit-A	Bit-B
Penetration@ 25 °C, dmm	60	110
Softening Point, R&B, °C	48	44
Fraass Breaking Point, °C	−18	−25

Modified binders were prepared:

- For AFM research by mixing, 10÷20 wt.% APDDR or hybrid 80/20 powder (by co-grinding of 80 wt.% CR and 20 wt.% SBS L 30-01) with 90÷80 wt.% bitumen Bit-A heated to 160 °C with 180 rpm of paddle stirrer (IKA HB10 DIGITAL) for 1 min;
- For performance testing by mixing, APDDR or hybrid 80/20 or hybrid 95/5 powder (cogrinding of 95 wt.% CR and 5 wt.% SBS L 30-01) with bitumen heated to 160 °C with 600 rpm of paddle stirrer for 3 min. The coding and composition of modified binder samples are presented in Table 3.

Table 3. The coding and composition of modified binder samples.

Modified Binder Code	Modified Binder Composition,%				
	Bit-A	Bit-B	APDDR	Hybrid80/20	Hybrid95/5
MB-A(0)	87.5		12.5		
MB-A(20)	87.5			12.5	
MB-B(0)		87.5	12.5		
MB-B(5)		87.5			12.5

2.2. Research Methods

2.2.1. AFM Study

Research of structural and morphological features of the bitumen and MB sample's surface was carried out on an NtegraPrima AFM (NT-MDT Spectrum Instruments, LLC; Moscow, Russia) in a semicontact mode. AFM topography and phase imaging were obtained at room temperature at a scanning speed of 0.6÷1 Hz. Gold-plated silicon cantilevers were used (rounding radius 10 nm; resonance frequency 150÷240 kHz).

2.2.2. Rheological Study

Rheological tests were carried out on a dynamic shear rheometer MCR 702e (Anton Paar, GmbH, Gratz, Austria) using a parallel geometry measuring system:

- 25 mm for a multiple stress creep recovery (MSCR) test [20] (10 cycles, creep phase 1 s, recovery phase 9 s at different stress levels—0.1, 3.2, 6.5, 10, and 12 kPa at 64 °C) and for a dynamic (oscillatory) shear test at an angular frequency of 10 rad/s and shear strain 12% according to [21];

- 8 mm diameter for a linear amplitude sweep (LAS) test [22] (at 0.1 Hz in oscillatory mode in increasing strain: 0.1%, 1% and further up to 30% in steps of 1% (total 3100 cycles) at 16 °C).

The measuring gap during the tests was 1 mm for the MSCR test and for the dynamic shear test and 2 mm for the LAS test. The specimens were poured into silicone molds of 25 or 8 mm diameter. Rheological tests of the samples were carried out not earlier than 15 min and not later than 2 h after placing them into the molds. The samples were deposited in a rheometer at 58 °C.

Low-temperature-cracking parameters were determined on an Asphalt Binder Cracking Device (ABCD) (Infratest, LLC; Moscow, Russia) under static conditions according to the ABCD test [23] while cooling four parallel samples in a climatic chamber at a rate of 18÷20 °C/h.

3. Results and Discussion

3.1. AFM Study

Atomic force microscopy has been used for more than 25 years to study the microstructure of bitumen. Particular attention of the researchers was focused on the so-called "bee-like" structure [24]. Creep measurements demonstrated that the microstructure of the "bee-like" phase has 40–50% higher stiffness than the surrounding matrix phase [25]. Studies of the binders' resistance to low-temperature cracking after applied load showed the appearance of cracks at the "bee"-matrix phase interface [3], which allowed the visualization of theoretical ideas justifying the possibility of binder cracking under climatic factors and transport loading due to its inherent heterogeneity [26].

Figure 1a shows the surface topography of bitumen Bit-A, which illustrates the structural heterogeneity of bitumen and the presence of a pronounced microheterogeneity in it. The surface of the bitumen sample examined 2 days after preparation showed the presence of a "periphase" surrounded by "bee-like" structures with a maximum size of 5–7 μm, with a tendency to form "star bee-like" structures. Storage of this bitumen at ambient temperature for 1.5 months led to an increase in the maximum size of the "bee-like" structures by 1.5–2 times—up to 15–20 μm.

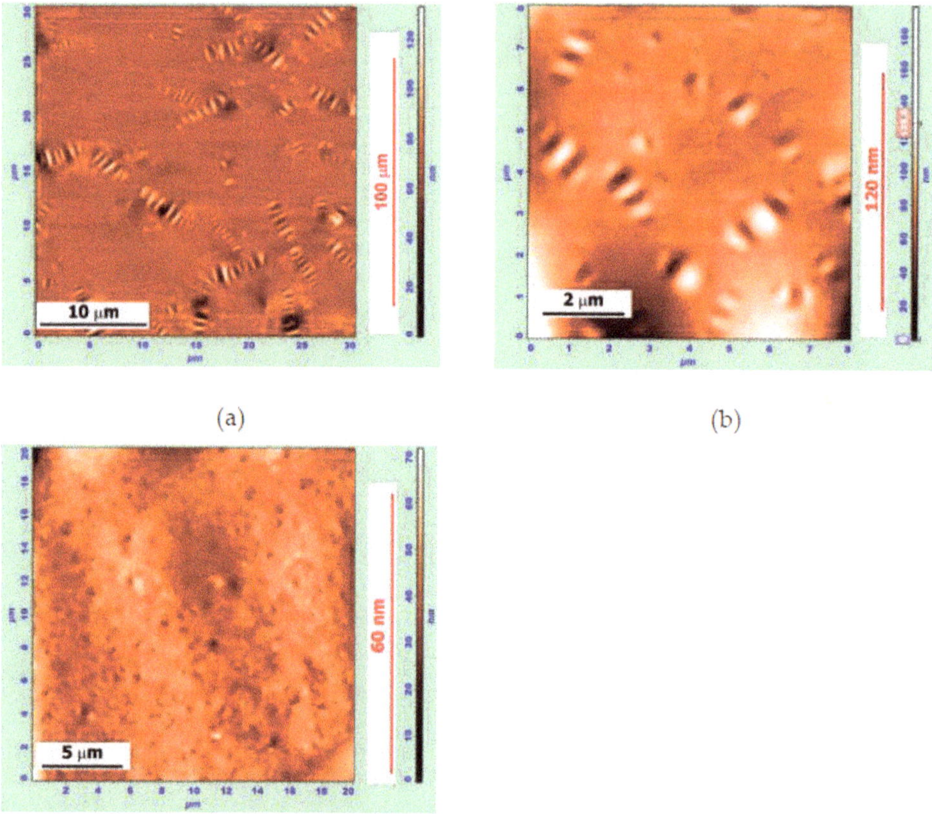

Figure 1. AFM images of the surface topography of (**a**) the sample of Bit-A, obtained 2 days after sample preparation (30 × 30 μm, Z-coordinate—130 nm); (**b**) the sample of MB of 90 wt.% Bit-A+ 10 wt.% APDDR, obtained 2 days after sample preparation (8 × 8 μm, Z-coordinate—190 nm); (**c**) the sample of MB of 80 wt.% Bit-A+ 20 wt.% APDDR obtained 1.5 months after sample preparation (20 × 20 μm, Z-coordinate—70 nm).

The ©ntroduction of APDDR into bitumen primarily had an effect on the formation of a "bee-like" structure.

For MB with 10 wt.% APDDR content, a significant decrease in the length of "bees" to the maximum 2–3 μm was observed (Figure 1b). When the APDDR content in the MB was increased to 15 wt.% and higher, no "bees" were observed even after 1.5 months of sample storage (Figure 1c).

The structure of the fragmented PEM particles observed using AFM was compared with the SEM data for the modified binder, where the APDDR content was 15 wt.%. The measurements were carried out 2 days after MB preparation (3 minutes' mixing at 160 °C).

In Figure 2a,b, one can see an elongated APDDR particle of the size of about 10 μm. When the image is zoomed in (Figure 2c,d), it is clearly seen that the particle consists of two different-sized fragments: the upper one is about 4÷5 μm long, and the lower one is about 2 μm. These fragments are connected by strands up to 2μm-long and a few tenths of a micron thick. We can also assume that the upper fragment of this APDDR particle consists of at least three parts. It can be hypothesized that this particle is the decay product of a larger particle, whose fragments were interconnected by strands.

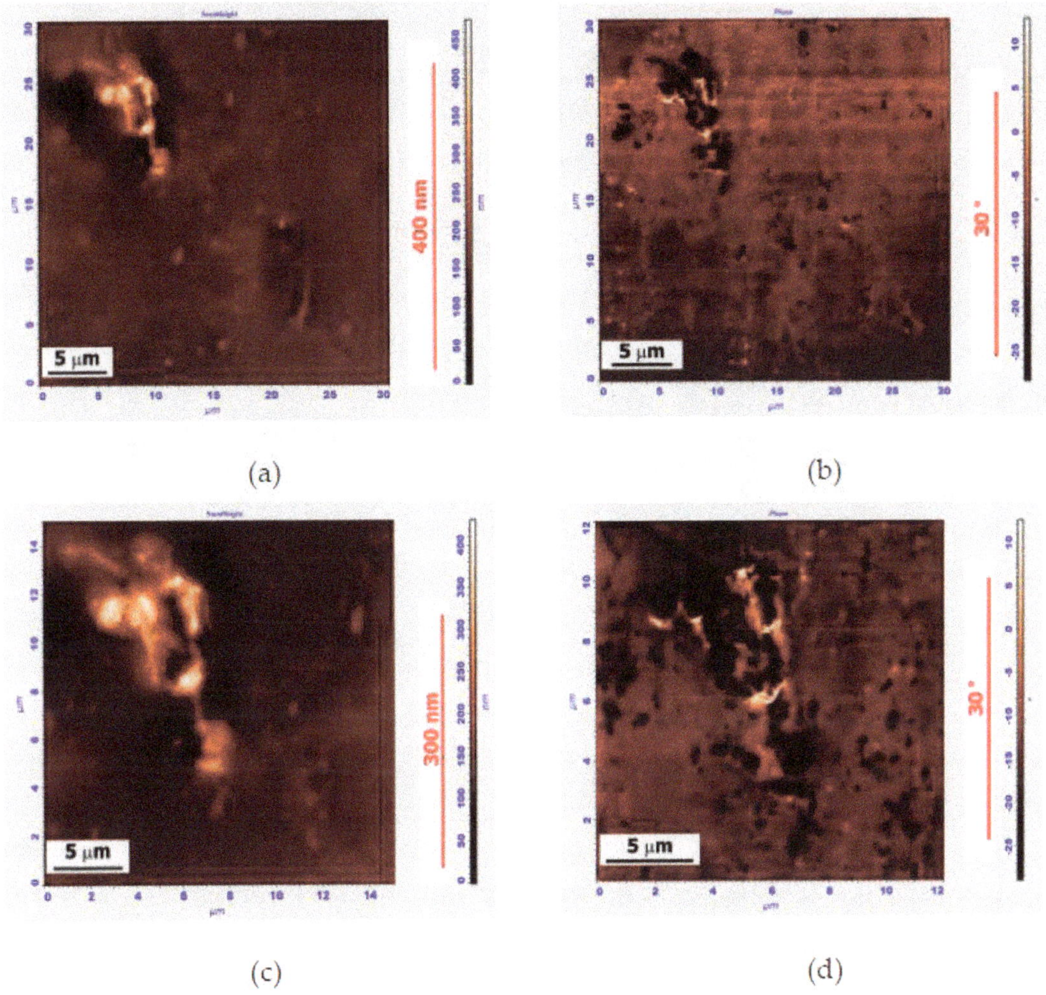

Figure 2. AFM surface images of the MB sample of 85 wt.% Bit-A + 15 wt.% APDDR, obtained 2 days after sample preparation. Topography (left), phase image (right): (**a**,**b**)—30 × 30 μm, (**a**) Z-coordinate—470 nm; (**c**,**d**)—12 × 12 μm, (**c**) Z-coordinate—470 nm.

Smaller fragments of the original APDDR particles of 100÷1000 nm are clearly visible in Figure 3. Similar particles of the same size and similar (self-similar) agglomerative structure were observed in SEM images after washing the MB with solvent [17].

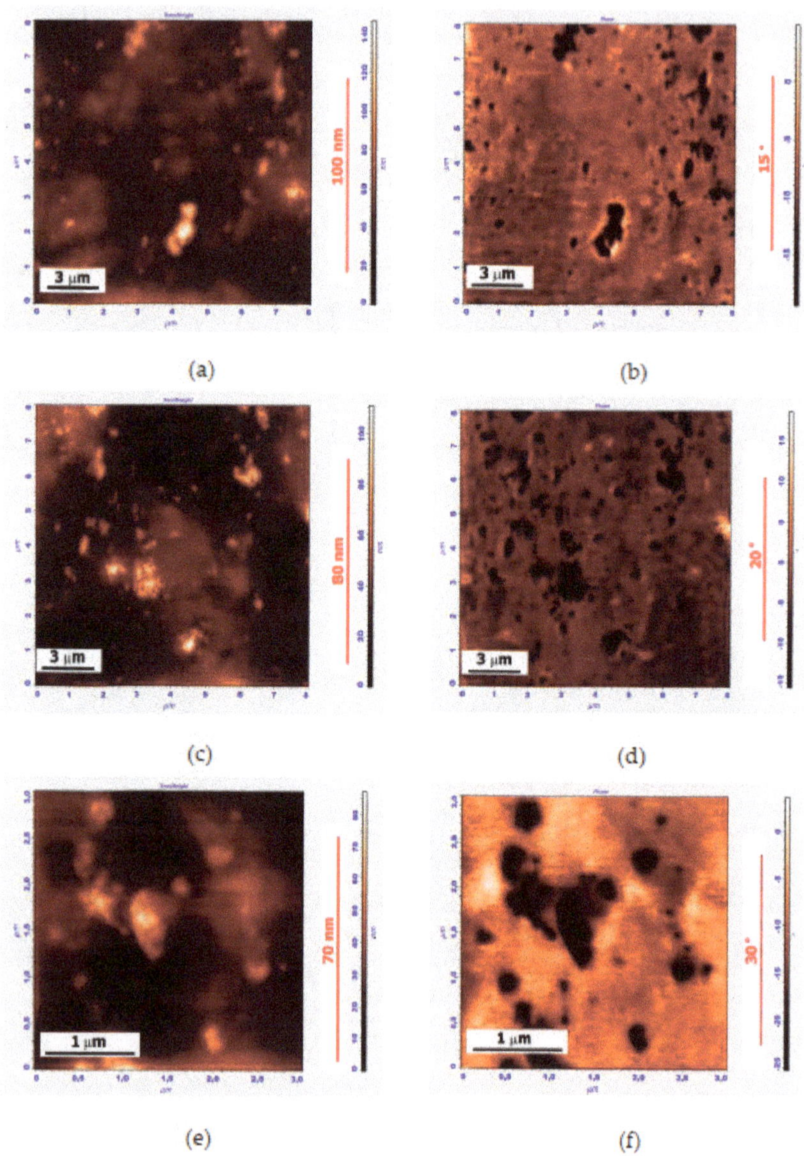

Figure 3. AFM surface images of the MB sample of 85 wt.% Bit-A + 15 wt.% APDDR, obtained 2 days after sample preparation. Topography (left), phase image (right): (**a**,**b**)—8 × 8 µm, (**a**) Z-coordinate—140 nm; (**c**,**d**)—8 × 8 µm, (**c**) Z-coordinate—110 nm; (**e**,**f**)—3 × 3 µm, © Z-coordinate—90 nm.

The images in Figure 4 illustrate the stage preceding the separation of the 50÷200 nm APDDR fragments from a larger particle (Figure 4a,b) and the formation of the spatial structure (Figure 4c,d). The data presented in Figures 3 and 4 indicate the formation of a physical spatial network of rubber fragments. Since the distances between the fragments are comparable to their fragment sizes, the percolation threshold's conditions are satisfied.

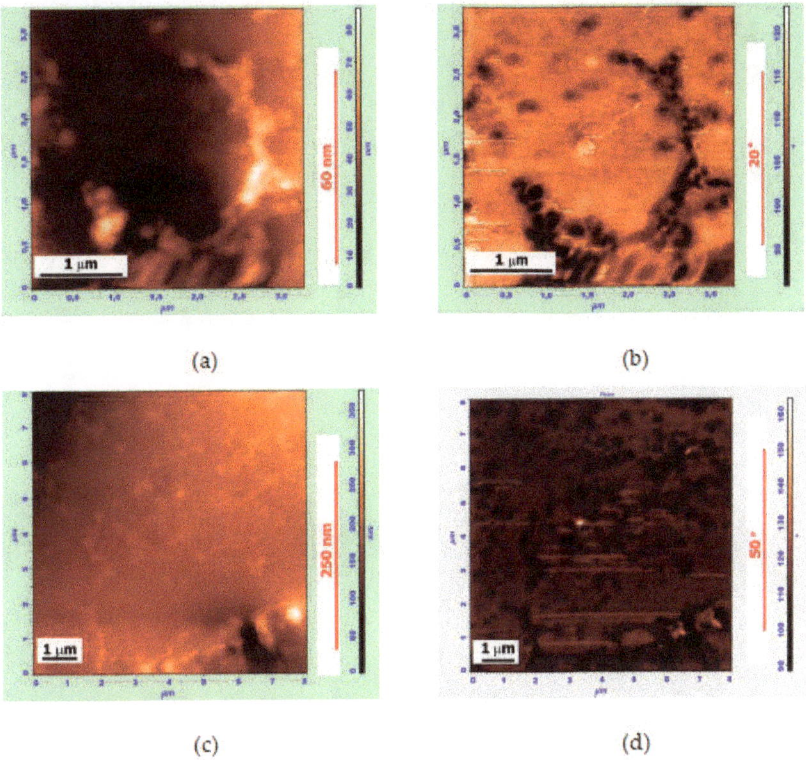

Figure 4. AFM images of the MB surface with the gel structure area. Topography (left), phase image (right): (**a**,**b**)—3 × 3 µm, (**a**) Z-coordinate—85 nm; (**c**,**d**)—8 × 8 µm, (**c**) Z-coordinate—375 nm.

Figure 5 shows AFM images of the MB surface (the MB is based on Bit-A and hybrid powder; the Bit-A/hybrid powder is 85/15 wt.%). The measurements were carried out two days after sample preparation.

In a series of zoomed images, the different stages and mechanisms of hybrid particle disintegration can be observed. In Figure 5a,b at the top left, a round-shaped hybrid particle of 2–3 µm in size can be observed connected by strands with smaller fragments up to 0.2 µm in size. One such fragment is shown in Figure 5c,d. The appearance of these fragments correlates very well with SEM images of the APDDR fragments [17]. Elements of an agglomerative structure are also traced.

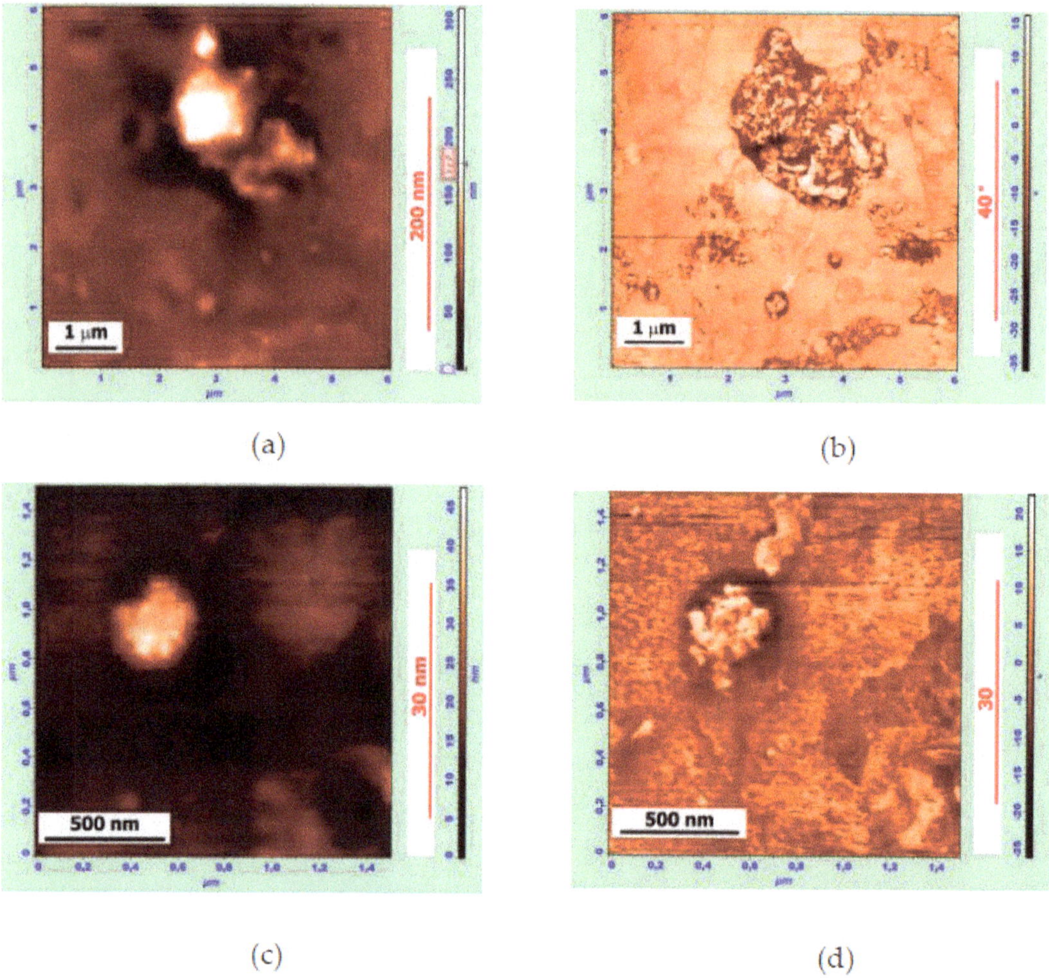

Figure 5. AFM images of the surface of the sample of the MB with composition of 85 wt.% Bit-A +15 wt.% hybrid powder, obtained 2 days after sample preparation. Topography (left), phase image (right): (**a**,**b**)—6 × 6 µm, (**a**) Z-coordinate—310 nm; (**c**,**d**)—1.4 × 1.4 µm, (**c**) Z-coordinate—50 nm.

The AFM images of thin films observed on the surface of the MB with a 90 wt.% Bit-A+10 wt.% APDDR composition are shown in Figure 6. In our opinion, the separation of such films occurs as a result of multidirectional swelling forces from the surface of microblocks with a denser structure than others. In our opinion, these images have similarities with the fragments of APDDR particles in the form of films observed earlier using electron scanning microscopy [17]. Figure 6b shows a three-dimensional image of the MB surface, from which we can see that the film is partially located on the bee structure. Thus, it is verified that the films were formed by the rapid decay of PEM particles, because the formation of the "bee" takes longer.

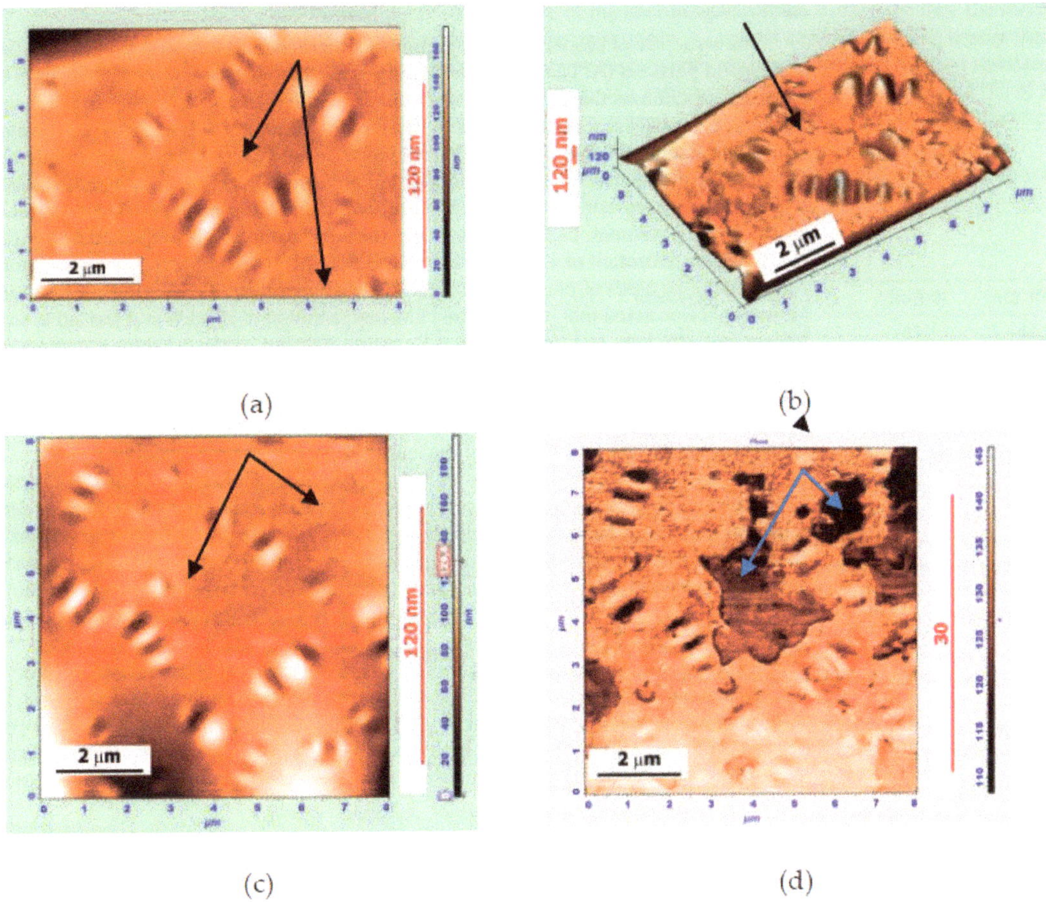

Figure 6. AFM surface images of the MB sample of composition 90 wt.% Bit-A + 10 wt.% APDDR, obtained after 2 days after sample preparation. (**a**,**c**)—topography, (**b**)—AFM-3D image; (**d**)—phase image. (**a**,**b**)—8 × 6 μm, (**a**) Z-coordinate—175 nm; (**c**,**d**)—8 × 8 μm, (**c**) Z-coordinate—195 nm. Thin films are indicated by arrows.

The results of AFM studies confirm the main conclusions drawn based on SEM studies [17]: already at the early stage of interaction, PEM particles break down into micro- and nanofragments with an agglomerative structure, similar (self-similar) to that of the original modifier particles given in [17], as well as present in the form of thin films. Additional information obtained on the basis of AFM images concerns the formation of gel structures on the basis of broken PEM particles. Additional information obtained on the basis of AFM images concerns the formation of gel structures on the basis of the broken PEM particles. Such a spatial network of nano- and microfragments should hinder the processes of diffusion and crystallization of the waxes present in bitumen. The result is the disappearance or a significant decrease in the size of "bee-like" formations, which is observed at PEM concentrations above 10 wt.%. A more homogeneous structure of the modified binder is created.

Figure 8 shows the dependence tangent of phase angle (tg¦Ä) versus shear strain (γ) recorded during testing at 16 and 7 °C. As can be seen, the modified binders in all cases show less sensitivity to cyclic deformation.

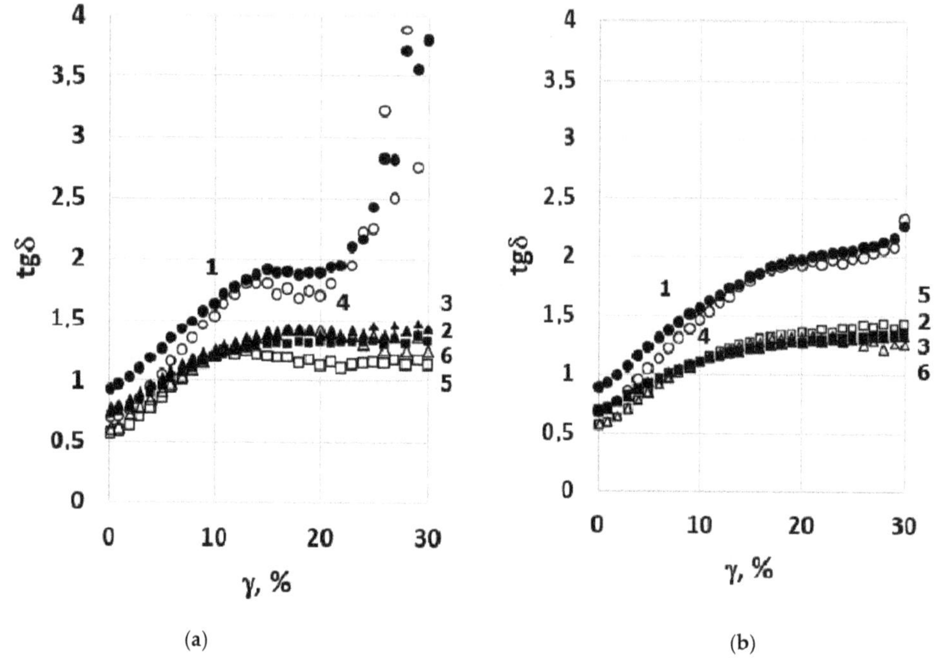

Figure 8. Plots of the tgδ versus shear strain (γ) in the LAS test at temperature 16 (1, 2, and 3) and 7 °C (4, 5, and 6) for Bit-A (**a**—1 and 4); MB-A(0) (**a**—2 and 5); MB-A(20) (**a**—3 and 6); Bit-B (**b**—1 and 4); MB-B(0) (**b**—2 and 5); MB-B(5) (**b**—3 and 6).

Table 6 shows the number of cycles to failure (N_f) at strains of 2.5 and 5%, calculated according to [22] based on LAS test data. As can be seen, the modification significantly improved the resistance to cyclic strain. Recall that Table 6 shows the data obtained for no-aged specimens of the modified binders. After RTFO and after PAV aging, the best results were observed for the modified binder based on hybrid PEM particles.

Table 6. LAS test results at temperatures 16 and 7 °C: number of cycles to failure (N_f) (VECD analysis) at 2.5 and 5% strain.

Binder		Bit-A	MB-A(0)	MB-A(20)	Bit-B	MB-B(0)	MB-B(5)
$N_{f2,5\%}$	16 °C	54,000	353,500	257,500	68,600	3,480,000	1,840,000
	7 °C	7350	47,500	28,700	7830	189,000	63,400
$N_{f5\%}$	16 °C	4500	17,000	10,600	4600	110,900	56,500
	7 °C	325	1200	780	290	3700	1300

The comparative rheological tests of bitumen samples and PEM-modified binder produced in conditions close to the temperature–time conditions of road mixture production (3 min mixing at T = 160 °C) and not subjected to additional temperature influence showed that even such a short time of interaction of the hot bitumen and PEM provides an expansion of the performance temperature range of modified bitumen and its increased resistance to cyclic loads.

The improvement in rheological indicators confirms that the rapid degradation of micronized powders of elastic modifiers into micro- and nanofragments leads simultaneously to the formation of a new structure of the modified binder that is more resistant to external influences. It can also be assumed that the improvement in low-temperature and fatigue-cracking resistance may be due to energy absorption and crack growth stopping in the presence of micron and submicron PEM elastic fragments in accordance with the mechanism of increasing impact toughness in plastics [30–32].

The results obtained confirm the effective recycling of worn-out tire rubber by the HTSG method in obtaining PEM for further use to modify bitumen directly in the production of road asphalt mixtures or to reduce the time and energy costs for the preparation of modified bitumen binder for road construction.

4. Conclusions

In this paper, the structure and performance properties of bitumen and modified binders obtained as a result of the short-term interaction of hot bitumen and micronized powder elastomeric modifiers (PEM) were investigated. PEM was produced by the high-temperature shear-induced grinding of a worn-out tire's crumb rubber or its co-grinding with butadiene styrene thermoplastic elastomer.

Atomic force microscopy (AFM) studies have shown that a short-term interaction between hot bitumen with PEM particles having a highly developed surface leads to their degradation into numerous micro- and nanofragments. It is hypothesized that the rapid degradation of PEM particles occurs under the action of multidirectional swelling forces in locations of stressed-bond concentration or locations of rubber discrete devulcanization. An agglomerative structure of fragments similar (self-similar) to the structure of the original particles of the modifier (given in [17]) forms. The fragments in the form of thin films apparently form when the swollen rubber peels off the surface of the particle.

AFM studies also showed the formation of a more homogeneous binder structure at PEM concentrations greater than 10 wt.% by significantly reducing the length of the "bees" up to the disappearance of "bee-like" units.

The results of rheological tests of bitumen and freshly prepared (not subjected to additional temperature treatment) samples of modified binder showed an improvement in binder performance at a wide range of temperatures and load levels, including:

- Expansion of the operating temperature range;
- Increased resistance to permanent deformation (rutting) at high temperature, including high creep resistance, sensitivity to load increase, and elastic recovery;
- Improved resistance to low-temperature and fatigue cracking.

The improvement in rheological performance confirms that the rapid degradation of micronized powders of elastic modifiers into micro- and nanofragments leads simultaneously to the formation of a new binder structure that is more resistant to external influences. It can also be assumed that the improvement in cracking resistance in modified binders is due to energy absorption and crack growth being stopped by the spatial network of micron and submicron resilient PEM fragments.

The results obtained clarify the effectiveness of PEM in the most cost-effective way: the injection of PEM into the asphalt mixture just at the time of its preparation.

Further work will be aimed at studying the structure and properties of mastics containing PEM and assessing the PEM effect on the interfacial interaction between modified binder and mineral filler.

Author Contributions: Conceptualization, V.N., T.D., I.K. and I.G.; methodology, T.D. and I.K.; validation, V.N., T.D. and I.K.; investigation, I.G., V.G., A.B. and A.A.V.; data curation, I.K.; writing—original draft preparation, A.A.V. and T.D.; Writing—Review and Editing, A.A.V., V.N., I.K., I.G. and V.G.; visualization, I.K., I.G., A.A.V. and A.B.; supervision, V.N. and I.K.; project administration, T.D. All authors have read and agreed to the published version of the manuscript.

Funding: Work was carried out at the expense of the subsidy allocated by N.N. Semenov Federal Research Center of Chemical Physics, Russian Academy of Sciences to fulfill the state assignment. This paper has been also supported by the RUDN University Strategic Academic Leadership Program (recipient A.A.V.)

Institutional Review Board Statement: Not applicable.

Informed Consent Statement: Not applicable.

Data Availability Statement: The data presented in this study are available on request from the corresponding author(s).

Conflicts of Interest: The authors declare no conflict of interest.

References

1. *UNEP/CHW.10/6/Add.1/Rev. 1*; Basel Convention Technical Guidelines for the Environmentally Sound Management of Used And Waste Pneumatic Tvres. Global Mercury Partnership: Basel, Switzerland, 11 November 2011.
2. Partl, M.N.; Bahia, H.U.; Canestrari, F.; de la Roche, C.; di Benedetto, H.; Piber, H.; Sybilski, D. Advances in interlaboratory testing and evaluation of bituminous materials. In *The International Union of Laboratories and Experts in Construction Materials, Systems and Structures (RILEM)*; Springer: Cham, Switzerland, 2013. [CrossRef]
3. Das, P.K.; Jelagin, D.; Birgisson, B.; Kringos, N. Atomic force microscopy to characterize the healing potential of asphaltic materials. In *Atomic Force Microscopy—Imaging, Measuring and Manipulating Surfaces at the Atomic Scale*; Bellitto, V., Ed.; Intechopen: London, UK, 2012; pp. 200–230. [CrossRef]
4. Bahia, H.U.; Hanson, D.I.; Zeng, M.; Zhai, H.; Khatri, M.A.; Anderson, M.R. NCHRP report 459, Characterization of modified asphalt binders in Superpave mix design. In *Prepared for National Cooperative Highway Research Program*; Transportation Research Board, National Research Council: Washington, DC, USA, 2001.
5. Bressi, S.; Fiorentini, N.; Huang, J.; Losa, M. Crumb Rubber Modifier in Road Asphalt Pavements: State of the Art and Statistics. *Coatings* **2019**, *9*, 384. [CrossRef]
6. Han, L.; Zheng, M.; Wang, C. Current status and development of terminal blend tyre rubber modified asphalt. *Constr. Build. Mater.* **2016**, *128*, 399–409. [CrossRef]
7. Hassan, N.A.; Airey, G.; Jaya, R.P.; Mashros, N.; Aziz, M.A. A Review of Crumb Rubber Modification in Dry Mixed Rubberised Asphalt Mixtures. *J. Teknol.* **2014**, *70*, 127–134. [CrossRef]
8. Porto, M.; Caputo, P.; Loise, V.; Eskandarsefat, S.; Teltayev, B.; Oliviero Rossi, C. Bitumen and Bitumen Modification: A Review on Latest Advances. *Appl. Sci.* **2019**, *9*, 742. [CrossRef]
9. Kutay, E.; Ozturk, H. An Analysis Using 3D Microtomography Imaging. In Proceedings of the Asphalt Rubber 2012 Conference, Munich, Germany, 23–26 October 2012; pp. 503–512.
10. Nguyen, H.T.; Tran, T.N. Effects of crumb rubber content and curing time on the properties of asphalt concrete and stone mastic asphalt using dry process. *Int. J. Pavement Res. Technol.* **2018**, *11*, 236–244. [CrossRef]
11. Wang, H.; Liu, X.; Zhang, H.; Apostolidis, P.; Erkens, S.; Skarpas, A. Micromechanical modelling of complex shear modulus of crumb rubber modified bitumen. *Mater. Des.* **2020**, *188*, 108467. [CrossRef]
12. Li, B.; Li, H.; Wei, Y.; Zhang, X.; Wei, D.; Li, J. Microscopic Properties of Hydrogen Peroxide Activated Crumb Rubber and Its Influence on the Rheological Properties of Crumb Rubber Modified Asphalt. *Materials* **2019**, *12*, 1434. [CrossRef]
13. Xu, M.; Liu, J.; Li, W.; Duan, W. Novel Method to Prepare Activated Crumb Rubber Used for Synthesis of Activated Crumb Rubber Modified Asphalt. *J. Mater. Civ. Eng.* **2015**, *27*, 5. [CrossRef]
14. Sheng, Y.; Li, H.; Geng, J.; Tian, Y.; Li, Z.; Xiong, R. Production and performance of desulfurized rubber asphalt binder. *Int. J. Pavement Res. Technol.* **2017**, *10*, 262–273. [CrossRef]
15. Chen, Z.; Wang, T.; Pei, J.; Amirkhanian, S.; Xiao, F.; Ye, Q.; Fan, Z. Low temperature and fatigue characteristics of treated crumb rubber modified asphalt after a long term aging procedure. *J. Clean. Prod.* **2019**, *234*, 1262–1274. [CrossRef]
16. Berlin, A.A.; Dudareva, T.V.; Krasotkina, I.A.; Nikol'Skii, V.G. Tire-Rubber-Waste Recycling and Active Powder of Discretely Devulcanized Rubber. *Polym. Sci. Ser. D* **2018**, *11*, 323–329. [CrossRef]
17. Nikol'skii, V.; Dudareva, T.; Krasotkina, I.; Gordeeva, I.; Vetcher, A.A. Ultra-dispersed powders produced by high-temperature shear-induced grinding of worn-out tire and products of their interaction with hot bitumen. *Polymers* **2022**, *14*, 3627. [CrossRef] [PubMed]
18. Chen, H.; Zhang, Y.; Bahia, H.U. The role of binders in mixture cracking resistance measured by ideal-CT test. *Int. J. Fatigue* **2020**, *142*, 105947. [CrossRef]
19. Kim, S.-S. The Asphalt Binder Cracking Device Test. *TR News* **2013**, *284*, 51–53.
20. *AASHTO T 350*; Standard Method of Test for Multiple Stress Creep Recovery (MSCR) Test of Asphalt Binder Using a Dynamic Shear Rheometer (DSR). AASHTO: Washington, DC, USA, 2019.
21. *AASHTO T315-20*; Standard Method of Test for Determining the Rheological Properties of Asphalt Binder Using a Dynamic Shear Rheometer (DSR). AASHTO: Washington, DC, USA, 2021.

22. *AASHTO Designation T 391-20*; Standard Method of Test for Estimating Fatigue Resistance of Asphalt Binders Using the Linear Amplitude Sweep. AASHTO: Washington, DC, USA, 2021.
23. *AASHTO TP 92-14*; Determining the Cracking Temperature of Asphalt Binder Using the Asphalt Binder Cracking Device (ABCD). AASHTO: Washington, DC, USA, 2018.
24. Loeber, L.; Sutton, O.; Morel, J.; Valleton, J.-M.; Muller, G. New direct observations of asphalts and asphalt binders by scanning electron microscopy and atomic force microscopy. *J. Microsc.* **1996**, *182*, 32–39. [CrossRef]
25. Allen, R.G.; Little, D.N.; Bhasin, A. Structural Characterization of Micromechanical Properties in Asphalt Using Atomic Force Microscopy. *J. Mater. Civ. Eng.* **2012**, *24*, 1317–1327. [CrossRef]
26. Jahangir, R.; Little, D.; Bhasin, A. Evolution of asphalt binder microstructure due to tensile loading determined using AFM and image analysis techniques. *Int. J. Pavement Eng.* **2014**, *16*, 337–349. [CrossRef]
27. *AASHTO T 240 and ASTM D 2872*; Effect of Heat and Air on a Moving Film of Asphalt (Rolling Thin-Film Oven Test). AASHTO: Washington, DC, USA, 2021.
28. *AASHTO R 28*; Accelerated Aging of Asphalt Binder Using a Pressurized Aging Vessel (PAV). AASHTO: Washington, DC, USA, 2021.
29. Berlin, A.A.; Nikol'Skii, V.G.; Krasotkina, I.A.; Dudareva, T.V.; Gorbatova, V.N.; Gordeeva, I.V.; Sorokin, A.V.; Lobachev, V.A.; Dubina, S.I.; Sinkevich, M.Y. Rubber and Rubber–Polymer Modifiers of Asphalt–Concrete Mixes Prepared by High-Temperature Shear-Induced Grinding. Part 3. Evaluation of Effectiveness of Modification. *Polym. Sci. Ser. D* **2022**, *15*, 71–78. [CrossRef]
30. Newman, S. Modification of plastics with rubbers. In *Polymer Mixtures*; Paul, D.R., Newman, S., Eds.; Academic Press: Cambridge, MA, USA, 1978; Volume 2, pp. 63–89.
31. Kaplan, A.M.; Chekunaev, N.I.; Nikol'skii, V.G. High temperature failure of stressed polymers by the mechanism of avalanche-like cracking. *Chem. Phys. Rep.* **2003**, *2*, 247.
32. Michler, G.H.; Godehardt, R.; Adhikari, R.; Kim, G.-M.; Henning, S.; Seydewitz, V.; Lebek, W. *Electron Microscopy of Polymers*; Springer: Berlin/Heidelberg, Germany, 2008.

polymers MDPI

Article

Acceleration of Polybutylene Succinate Biodegradation by *Terribacillus* sp. JY49 Isolated from a Marine Environment

Su Hyun Kim [1], Jang Yeon Cho [1], Do Hyun Cho [1], Hee Ju Jung [1], Byung Chan Kim [1], Shashi Kant Bhatia [1,2], See-Hyoung Park [3], Kyungmoon Park [3] and Yung-Hun Yang [1,2,*]

1 Department of Biological Engineering, College of Engineering, Konkuk University, Seoul 05029, Korea
2 Institute for Ubiquitous Information Technology and Applications, Konkuk University, Seoul 05029, Korea
3 Department of Biological and Chemical Engineering, Hongik University, Sejong 30016, Korea
* Correspondence: seokor@konkuk.ac.kr; Tel.: +82-2-450-2-3936

Abstract: Polybutylene succinate (PBS) is a bioplastic substitute for synthetic plastics that are made from petroleum-based products such as polyethylene and polypropylene. However, the biodegradation rate of PBS is still low and similar to that of polylactic acid (PLA). Moreover, our knowledge about degrader species is limited to a few fungi and mixed consortia. Here, to identify a bacterial degrader to accelerate PBS degradation, we screened and isolated *Terribacillus* sp. JY49, which showed significant degradability. In order to optimize solid and liquid culture conditions, the effect of factors such as temperature, additional carbon sources, and salt concentrations on degradation was confirmed. We observed a degradation yield of 22.3% after 7 days when adding 1% of glucose. Additionally, NaCl was added to liquid media, and degradation yield was decreased but PBS films were broken into pieces. Comparing the degree of PBS degradation during 10 days, the degradation yield was 31.4% after 10 days at 30 °C. Alteration of physical properties of films was analyzed by using scanning electron microscopy (SEM), gel permeation chromatography (GPC), and Fourier transform infrared (FT-IR). In addition, *Terribacillus* sp. JY49 showed clear zones on poly(butylene adipate-co-terephthalate) (PBAT), polycaprolactone (PCL), and copolymers such as P(3HB-*co*-3HV) and P(3HV-*co*-4HB), exhibiting a broad spectrum of degradation activities on bioplastics. However, there was no significant difference in absorbance when esterase activity was examined for different types of bioplastics. Overall, *Terribacillus* sp. JY49 is a potential bacterial strain that can degrade PBS and other bioplastics, and this is the first report of *Terribacillus* sp. as a bioplastic degrader.

Keywords: polybutylene succinate (PBS); *Terribacillus goriensis*; biodegradation

Citation: Kim, S.H.; Cho, J.Y.; Cho, D.H.; Jung, H.J.; Kim, B.C.; Bhatia, S.K.; Park, S.-H.; Park, K.; Yang, Y.-H. Acceleration of Polybutylene Succinate Biodegradation by *Terribacillus* sp. JY49 Isolated from a Marine Environment. *Polymers* **2022**, *14*, 3978. https://doi.org/10.3390/polym14193978

Academic Editors: Alexey Iordanskii and Alexandre Vetcher

Received: 23 August 2022
Accepted: 19 September 2022
Published: 23 September 2022

1. Introduction

Pollution in the oceans is a serious issue that is rapidly increasing due to the waste discharged into water bodies. Increasing marine waste, water pollution, and the formation of garbage islands in the ocean threaten the habitats of marine organisms [1,2]. Synthetic plastics greatly contribute to pollution because of their high usage and improper recycling. Synthetic plastics are artificially produced from petroleum oil with long chains of monomers [3,4]. They are inexpensive and easy to process. Therefore, they are found in manufacturing and packaging goods and pharmaceutical industries [5–8]. Because of their low cost and lightweight nature, we use them frequently in our day-to-day lives [9,10]. Moreover, their chemical structure allows their easy manipulation into shapes, makes them non-reactive, and increases their durability. However, their inert nature and durability make them hard to decompose in nature [11,12], rendering them difficult to dispose of, thus leading to serious environmental problems.

Therefore, the use of biodegradable bioplastics made from bio-based materials is increasing. Bio-based bioplastics are produced from biomass such as sugarcane, cellulose, or corn instead of fossil fuel. They can even be produced from microbes under various culture conditions [13]. Biodegradable bioplastics are converted to natural substances such as water and carbon dioxide by microorganisms [14]. Therefore, they are gaining attraction as substitutes for petroleum-based plastics. There are various kinds of bioplastics: bio-based, such as polyhydroxyalkanoate (PHA) and polylactic acid (PLA), and fossil-based, such as polybutylene succinate (PBS), poly(butylene adipate-*co*-terephthalate) (PBAT), and polycaprolactone (PCL). These bioplastics can be degraded by microbial enzymes or directly by microorganisms, helping with waste management and decreasing environmental pollution [15–17].

Polybutylene succinate (PBS) is a bioplastic synthesized by polycondensation between 1,4-butanediol and succinic acid [18]. Succinic acid can be obtained from renewable feedstock such as glucose, starch, or sucrose via fermentation by microorganisms [19]. To synthesize 1,4-butanediol, petroleum-based feedstocks are used [20]. It is an aliphatic polyester that has thermal and chemical resistance, flexibility, and melt processability. It has polyethylene (PE)- and polypropylene (PP)-like properties [21] and is widely used and discarded in everyday life. It can also be used for mulching films, compostable bags, and packaging [22], or can be mixed with other bioplastics to improve its properties. This makes PBS a promising material to replace synthetic plastics, preventing the accumulation of plastic waste. PBS can be degraded into water and carbon dioxide by hydrolytic or enzymatic degradation. In a previous study, when the hydrolytic method was applied, the weight loss of PBS was 31% [23]. Lipase produced by *Pseudomonas cepacia* can degrade PBS, resulting in 3.5% weight loss in 12 days [24]. However, the speed of PBS degradation is relatively slow, and using enzymes is not economically efficient. Therefore, we need to identify or engineer microorganisms that can degrade PBS directly. However, except for some fungi and mixed consortia species, we know little about other degrading microorganisms [25]. In this study, we screened for degrading strains and confirmed their biodegradation ability to expand the range of strains known and available to the scientific community.

To identify a species of degrading bacteria, in this study, we screened PBS-degrading strains from marine samples and selected one with the best degradation activity using solid and liquid cultures at 30 °C, the temperature at which they were isolated. Additionally, we characterized the selected strain, *Terribacillus* sp. JY49, optimized the degradation conditions and observed the physical properties of the films (Table 1). Lastly, the degradability of other plastics was confirmed using clear-zone tests and esterase assay.

Table 1. List of PBS-degrading strains reported in previous studies.

Strain	Type	Temp	Period (Days)	Weight Loss (%)	Biodegradation Percentage (%)	Condition	Reference
Fusarium sp. FS1301	Fungi	30 °C	21	80%	-	Liquid	[26]
Bionectria ochroleuca BFM-X1	Fungi	30 °C	30	60%	-	Soil	[27]
Aspergillus fumigatus	Fungi	30 °C	30	80%	-	Soil	[28]
Fusarium solani	Fungi	-	14	-	2.8%	Soil	[29]
Aspergillus versicolor, Penicillium, Bacillus, Thermopolyarpora	Consortia	-	90	-	71.9% (powder) 60.7% (film) 14.1% (granule)	Soil	[30]
Terribacillus goriensis	Bacteria	30 °C	10	-	31.4%	Liquid	This paper

2. Materials and Methods

2.1. Chemicals

All chemicals used in this study were of analytical grade. Chloroform, acetonitrile, poly-3-hydroxybutyrate-co-3-hydroxy-valerate(P(3HB-*co*-3HV)) pellets, *p*-nitrophenyl acetate, butyrate, octanoate, decanoate, and dodecanoate were obtained (Sigma-Aldrich, St. Louis, MO, USA). We obtained *p*-nitrophenyl hexanoate (Tokyo Chemical Industry, Tokyo, Japan). PHB pellets were obtained(Goodfellow Cambridge Ltd., Huntingdon, UK). PBAT and PBS pellets were obtained (Gio Soltech Co., Ltd. Wonju, Korea). Poly-3-hydroxybutyrate-co-4-hydroxybutyrate (P(3HB-co-4HB)) pellets were obtained (CJ, Suwon, Korea). Dichloromethane (DCM), fructose, and sucrose were obtained (Junsei Chemical

Co., Tokyo, Japan). Glucose and xylose were obtained (Duksan Pure Chemicals, Ansan, Korea). Galactose was obtained (Daejung Chemicals, Siheung, Korea).

2.2. Preparing Solid Media Containing Plastic

For the preparation of media plates containing different plastics, 1 g of plastic pellets were dissolved in 40 mL DCM in a water bath at 60 °C. After adding 100 mL of distilled water, 2 mL of 2% Sarkosyl NL was added to the boundary of water and DCM [31,32]. The mixture was sonicated for 10 min with a 15 s pulse using Vibra Cell VCX500 (Sonics & Materials, Inc., Newtown, CT, USA). The amplitude was set at 30% to mix the contents uniformly. After sonication, the solvent was completely evaporated using a stirrer. Next, 1 g/L of plastic emulsion, uniformly dissolved in the aqueous phase of the solvent, was added to the marine broth (MB; Difco Laboratories, Detroit, MI, USA) containing peptone (5.0 g/L), yeast extract (1.0 g/L), ferric citrate (0.1 g/L), sodium chloride (19.45 g/L), magnesium chloride (5.9 g/L), magnesium sulfate (3.24 g/L), calcium chloride (1.8 g/L), potassium chloride (0.55 g/L), sodium bicarbonate (0.16 g/L), potassium bromide (0.08 g/L), strontium chloride (34.0 mg/L), boric acid (22.0 mg/L), sodium silicate (4.0 mg/L), sodium fluoride (2.4 mg/L), ammonium nitrate (1.6 mg/L), disodium phosphate (8.0 mg/L), and 2% of agarose. All the mixtures were then autoclaved for 15 min at 121 °C [33,34].

2.3. Screening Microorganisms for Their Ability to Degrade PBS

Soil samples were collected from different shores in Korea. These samples were diluted using autoclaved distilled water and spread on the MB agar plate containing 1 g/L of PBS. After incubation for 3–5 days at 30 °C, colonies that showed clear zones were isolated from MB-PBS agar plates. Each colony was inoculated in liquid MB media for 1 day, and stocks containing 20% (*w*/*v*) glycerol were stored at −80 °C for further use. Isolated colonies that produced clear zones were identified at the species level using 16s rRNA sequencing performed by Bionics (Seoul, Korea). Sequencing was conducted using the universal primer 27F. Partial sequences were aligned in the NCBI GeneBank database using BLASTn.

2.4. Characterizing the PBS-Degrading Strain

Antibiotic resistance (ampicillin, spectinomycin, gentamicin, kanamycin, and chloramphenicol) and hydrolase activity tests (amylase, protease, chitinase, and lipase) were conducted for the PBS-degrading strains. For the antibiotic resistance test, the strains were cultured on an MB agar plate containing 100 μg/mL ampicillin, 100 g/mL spectinomycin, 35 μg/mL chloramphenicol, 50 μg/mL kanamycin, and 25 μg/mL gentamicin. The appearance of colonies was considered a positive result [35].

Amylase activity was confirmed on starch agar plates containing 10 g/L soluble starch, 1 g/L yeast extract, and 15 g/L agar. After incubation, residual starch sources were stained using a 0.1 N iodine solution for 2 min. Protease activity was screened on skim milk agar plates containing 28 g/L skim milk, 5 g/L tryptone, 2.5 g/L yeast extract, 1 g/L glucose, and 15 g/L agar. The chitin agar plate was prepared by adding M9 medium, 1% soluble chitin, and 15 g/L agar and adjusting pH to 7. For lipase activity test, an agar plate containing 1% olive oil, 0.01% phenol red, 10 mM $CaCl_2$, and 20 g/L agar was prepared, and the pH was adjusted to 7.3–7.4. All agar plates contained 2% salt to maintain salinity similar to that in the marine broth medium. In the case of lipase activity, only the formation of yellow zones was considered a positive result [35].

2.5. Solid and Liquid Culture to Monitor PBS Degradation

To characterize and optimize each microorganism, clear zone tests with various conditions were performed. To determine the optimal temperature to degrade PBS, each microbe was precultured in an optimal liquid medium for 24 h at 30 °C. Next, paper discs (Toyo Roshi Kaisha, Tokyo, Japan) were placed on the plate [36,37] and 10 μL of precultured cells were inoculated on the paper disc and incubated at 20 °C, 30 °C, 37 °C, and 42 °C for 7 days. We also inoculated the precultured cells on plates with 1% carbon source and 1%, 2%, 3%, and 4% of NaCl concentration and incubated them at 30 °C. The radius of clear zones was confirmed by measuring the distance between the paper disc and the endpoint of the clear zone. All experiments were performed in duplicates.

For liquid culture, PBS films were prepared using the conventional solvent-cast method [38]. We dissolved 0.2 g of PBS pellets in 100 mL of chloroform and heated the solution in a water bath at 60 °C until the pellets completely dissolved. The solvent containing dissolved PBS was kept in a fume hood until the solvent was completely evaporated and plastic films were formed. These films were cut into 40 mg pieces and sterilized with 70% ethanol and UV radiation on a clean bench. Prepared films were cultured in 100 mL flasks with 40 mL MB liquid medium. The liquid culture was performed under the same conditions as the solid culture. The time-dependent degradation rate was measured after 3, 5, 7, and 10 days of cultivation. We inoculated 2% JY49 and cultured it in a rotary shaker at 200 rpm. For further analysis, residual films were recovered, washed with distilled water several times, and freeze-dried for preparation of GC-MS samples. All experiments were performed in duplicate.

2.6. GC-MS Analysis

The amount of residual PBS and degradation yield were confirmed using GC-MS. To prepare GC-MS samples, the culture medium was centrifuged at 10,000× *g* for 10 min, and the residual PBS films were collected and washed several times with distilled water to remove residual medium components. Collected samples were lyophilized in Teflon-stoppered glass vials to completely remove water. For methanolysis of PBS, 1 mL of methanol/sulfuric acid (85:15 v/v) and 1 mL of chloroform were added to the vials and heated for 2 h at 100 °C. After 2 h, these vials were cooled down at room temperature, 1 mL of HPLC grade water was added to the vials, and the samples were vortexed for 1 min. The organic phase was extracted using a pipette and transferred to an e-tube containing anhydrous sodium sulfate to remove water. Samples were filtered (pore size, 0.2 µm) before injecting them into a GC-MS (Perkin Elmer, Waltham, MA, USA) equipped with a fused silica capillary column (Elite-5 ms, 30 m × 0.25 mm i.d. × 0.25 µm) and subjected to a linear temperature gradient for analysis (50 °C for 1 min, increased at 15 °C/min to 120 °C for 2 min and then increased at 10 °C/min to 300 °C for 10 min). The injector port temperature was 250 °C. Mass spectra were obtained using electron impact ionization at 70 eV, and scan spectra were obtained within the range of 45–450 m/z [39]. Selected ion monitoring was used for the detection and fragmentation analysis of the major products. To quantify the amount of residual PBS films, a calibration curve was obtained and degradation yield was calculated to indicate the degree of PBS degradation compared with the initial amount of PBS.

2.7. Analyzing the Physical Properties of PBS Films

To compare the change in the films' surface after degradation, scanning electron microscopy (SEM) was used. To prepare the sample for SEM, degradation by *Terribacillus* sp. JY49 was performed for 0, 3, 5, 7, and 10 days, and residual PBS films were collected. Recovered films were washed with distilled water to remove the medium components and lyophilized. The films were coated with gold dust at 5 mA for 300 s, and back-scatter electron images were acquired using a TM4000Plus SEM instrument (Hitachi, Tokyo, Japan) at 5 kV [38]. To observe the films at high magnification, SU-8010 FE-SEM (Hitachi, Tokyo, Japan) was used at 5 kV.

To determine the molecular weight changes of the PBS films during degradation, we used gel permeation chromatography (YOUNG IN Chromass, Anyang, Republic of Korea). To prepare the samples, PBS films were recovered after degradation, dissolved in 1 mL of chloroform, and heated on a heat block at 60 °C. This solution was filtered through a syringe filter (pore size, 0.2 µm; Chromdisc, Daegu, Korea). A high-performance liquid chromatography (HPLC) apparatus was used for the analysis, consisting of a loop injector (Rheodyne 7725i), an isocratic pump with dual heads (YL9112), column oven (YL9131), columns (Shodex, K-805, 8.0 mm I.D. × 300 mm; Shodex, K-804, 8.0 mm I.D. × 300 mm), and an RI detector (YL9170). We injected 60 μL of the sample without air bubbles. Chloroform was used as the mobile phase, and the flow rate was maintained at 1.0 mL/min at 35 °C. The data were analyzed using YL-Clarity software for a single YL HPLC instrument (YOUNG IN Chromass). The molecular weight was analyzed in relation to polystyrene standards ranging from 5000 to 2,000,000 g/mol [32].

The changes in functional groups in the PBS films were observed using Fourier-transform infrared spectroscopy (Nicolet 6700, Thermo Fisher Scientific, Waltham, MA, USA). We recorded 32 scans for each spectrum using an auto base in the scanning range of 4000 to 600 cm^{-1} and a resolution of 4 cm^{-1}.

2.8. Confirming the PBS Monomer Effect

Succinic acid and 1,4-butanediol were added to MB medium and cultured with *Terribacillus* sp. JY49 to confirm the effect of PBS monomers. This culture was incubated at 30 °C and 200 rpm by adding 10 mM of each monomer. The pH range was adjusted to 7.6 ± 0.2. We sampled 300 μL of culture medium after 24 h and 48 h to measure the optical density. After checking the optical density of each sample, the remaining culture medium was centrifuged at 13,000 rpm for 10 min to obtain the supernatant to check the monomer consumption using high-performance liquid chromatography (HPLC). Supernatants were diluted 10-fold with HPLC water and filtered (pore size, 0.2 μm). The HPLC (Perkin Elmer, Waltham, MA, USA) was equipped with a refractive index detector and a UV-vis detector. PBS monomers were separated on an Aminex HPX-87H column (300 × 7.8 mm internal diameter) (Bio-Rad, Hercules, CA, USA). The flow rate of the mobile phase was maintained at 0.6 mL/min using 0.004 mol/L sulfuric acid (H_2SO_4). The oven temperature was set to 60 °C during the operation [40].

We also confirmed the minimal inhibitory concentration (MIC) to check the monomers' effect on the growth of *Terribacillus* sp. JY49. Monomer solution at a concentration of 128 mM was prepared. In a 96-well plate, monomer concentration ranging from 0.25–64 mM was prepared, and 2% of strain was inoculated. After incubating for 24 h, optical density was measured to check the minimal inhibitory concentration of PBS monomers.

2.9. Esterase Activity Assay with p-Nitrophenyl Esters

To check the esterase activity of the degrading strains, six *p*-nitrophenyl esters were used as substrates: *p*-nitrophenyl acetate, *p*-nitrophenyl butyrate, *p*-nitrophenyl hexanoate, *p*-nitrophenyl octanoate, *p*-nitrophenyl decanoate, and *p*-nitrophenyl dodecanoate. The enzyme reactions comprised 180 μL of 50 mM phosphate buffer (pH 7.4), 10 μL of supernatant (collected from the liquid culture centrifuged at 4 °C and 13,000 rpm for 10 min), 5 μL of substrates dissolved in acetonitrile, and 5 μL of ethanol. These mixtures were reacted in an incubator at 37 °C for 10 min. Absorbance was measured at 405 nm in a 96-well plate to confirm esterase activity [41].

3. Results

3.1. Screening Polybutylene Succinate (PBS)-Degrading Strains from Marine Samples

To screen the strains with the ability to degrade PBS, we collected various marine samples. These samples were spread on marine broth-containing agar plates containing 1% PBS emulsion, and strains forming clear zones were selected. Five different microbes were screened from these samples (Table S1), and all strains were isolated at 30 °C. To select one strain with the best degradation activity towards PBS, the radius of clear zones was compared, and the residual amount of PBS was analyzed using GC-MS. In the case of liquid culture, 20mg of PBS films were added to the media. These experiments were conducted at 30 °C, the temperature at which the strains were isolated. In clear-zone tests, JY52 had the largest radius until 11 days, but JY49 showed larger and more transparent clear zones after 21 days (Figure 1a). Comparing GC-MS results of the two, the degradation yield of JY49 was higher than that of other strains (Figure 1b), and it was selected for subsequent experiments.

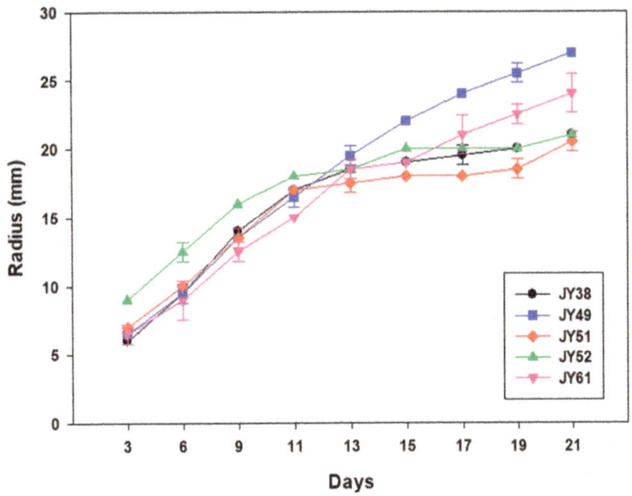

(a)

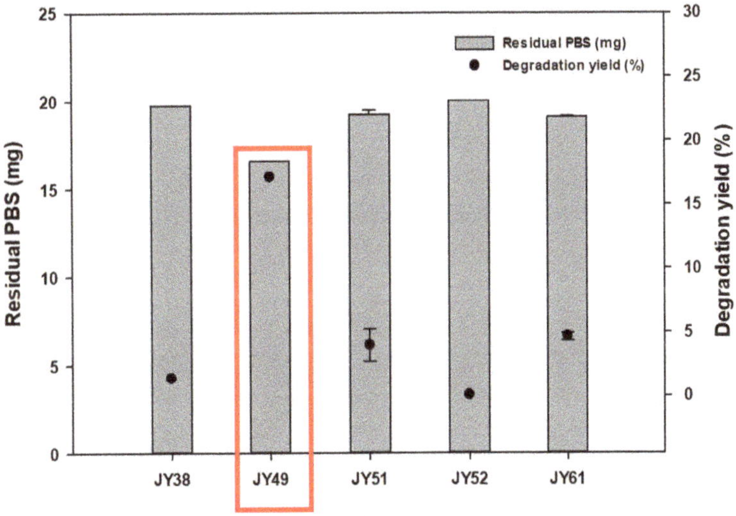

(b)

Figure 1. Comparing PBS-degrading strains using solid and liquid cultures and identifying the selected strains. (**a**) The changes in radius size of clear zones for 21 days at 30 °C. (**b**) Amount of residual PBS and degradation yield of five PBS-degrading strains using GC-MS analysis.

We found JY49 to have the highest similarity with *Terribacillus gorinesis* (99.51%) using 16S rRNA sequencing (Figure 2). *Terribacillus* sp. are gram-positive strains that can live in aerobic and halophilic conditions. To characterize this strain, hydrolysis activity and antibiotic resistance tests were performed. Except for protease activity, activity was confirmed in other hydrolysis tests, and the strain was not resistant to any antibiotics (Table S2). Several types of *Terribacillus* sp. strains have been published in previous studies (Table 2). However, there have been few reports of producers or degraders of substances. Therefore, this study is the first report of PBS-degrading *Terribacillus* species.

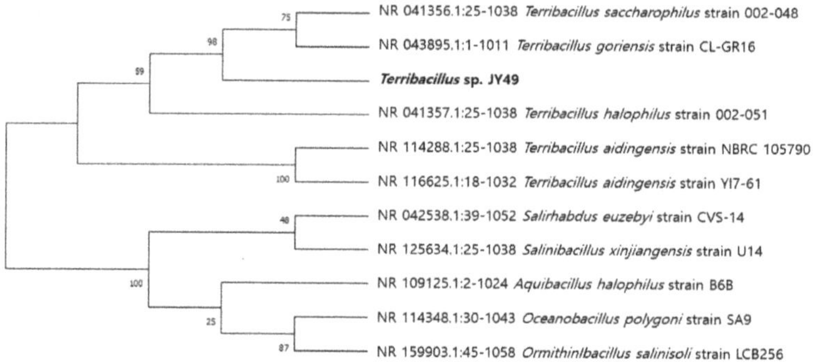

Figure 2. Phylogenetic tree of *Terribacillus* sp. JY49 according to 16S rRNA sequencing.

Table 2. List of previously reported *Terribacillus* sp.

Strains	Strain No.	Remark	Isolated Site	Reference
Terribacillus saccharophilus	KCTC 13936, DSM 21620	Characterization *	Field soil	[42]
Terribacillus halophilus	KCTC 13936, DSM 21620	Characterization *, Antimicrobial behavior **	Field soil	[42,43]
Terribacillus aidingensis	DSM 28352, CGMCC 1.8913	Characterization *	Soil from lake	[44]
Terribacillus goriensis	KCCM 42329, DSM 18252	Characterization *	Surface of sea water	[45]

* Characterization: report of screened and novel strains with basic experiments; ** Antimicrobial behavior: report of antimicrobial activity of strains.

3.2. Optimizing Temperature Conditions for PBS Degradation by JY49

Terribacillus sp. JY49 was isolated from MB medium at 30 °C. Reportedly, the optimal growth temperature of this strain is 30 °C [46]. To test the best temperature to degrade PBS, clear-zone tests and liquid culture were conducted at various temperatures (20 °C, 30 °C, 37 °C, and 42 °C) for 7 days. At 37 °C, clear zones formed rapidly in the beginning, but the largest and most transparent clear zone appeared at 30 °C after 7 days. Moreover, *Terribacillus* sp. JY49 had the ability to degrade PBS at 42 °C, but the size of the clear zone was small and did not increase rapidly compared with those at 30 °C and 37 °C. However, a clear zone was not visible at 20 °C (Figure 3a). After liquid culture, a residual amount of PBS films was measured using GC-MS. *Terribacillus* sp. JY49 can degrade PBS at all temperatures. At 30 °C, degradation yield was the highest (Figure 3b). We expect that the optimal growth temperature affects PBS degradation.

3.3. Effect of Carbon and Salt Concentration on PBS Degradation

To confirm the effect of carbon sources on PBS degradation, five carbon sources (glucose, fructose, galactose, xylose, and sucrose) were selected. We compared the size of clear zones on the MB agar media containing 1% of carbon sources and PBS emulsion at 30 °C. After 7 days, clear zones were observed for all carbon sources except sucrose. The radius size was similar to that in the control in the case of xylose and glucose. Plates containing fructose also showed clear zones, but with poor transparency compared with other carbon sources. GC-MS results showed that xylose-containing media had a degradation yield of 22.3%, which was higher than the 18.8% in the control (Figure 3c). The degradation yield was 20.8% for glucose. However, degradation yield decreased compared with that in the control when fructose, galactose, and sucrose were added to the medium. These carbon sources are expected to affect the growth of strain and enzyme production involved in degradation, and therefore, it appears that different degradation yield depending on carbon sources are shown in the above results.

Clear zone size and residual amount of PBS were evaluated by adding NaCl to the solid and liquid medium. When 1% of NaCl was added, this strain showed a larger clear zone radius compared with the control on solid medium. At other salt concentrations, clear zones were observed, but the transparency and size were weaker than that for 1% NaCl. In liquid culture, degradation yield decreased after adding NaCl to the medium (Figure 3d). Compared with the control, when NaCl was added to 1% and 2% to liquid medium, the degradation yield was 6.2% and 7.0%. In the presence of NaCl, PBS could be fragmented by JY49, but the amount of residual film did not change significantly. *Terribacillus* sp. is moderately halotolerant [46]. Therefore, in the presence of excess NaCl, degradation can occur even though degradation yield was not high compared with control conditions.

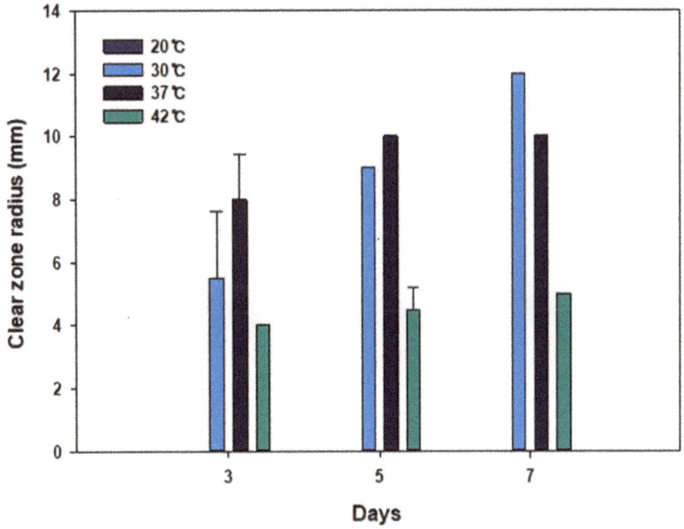

(a)

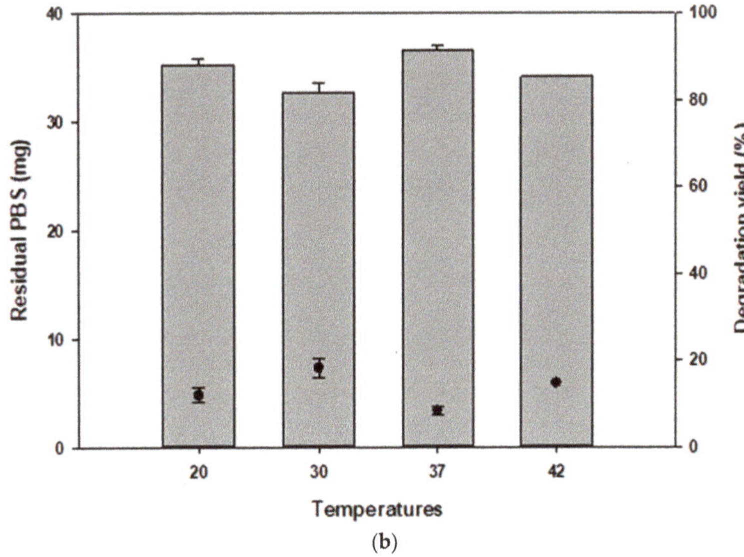

(b)

Figure 3. *Cont.*

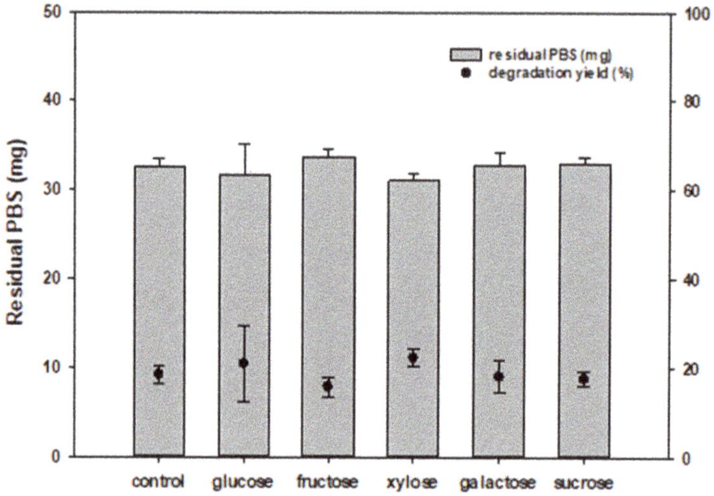

(c)

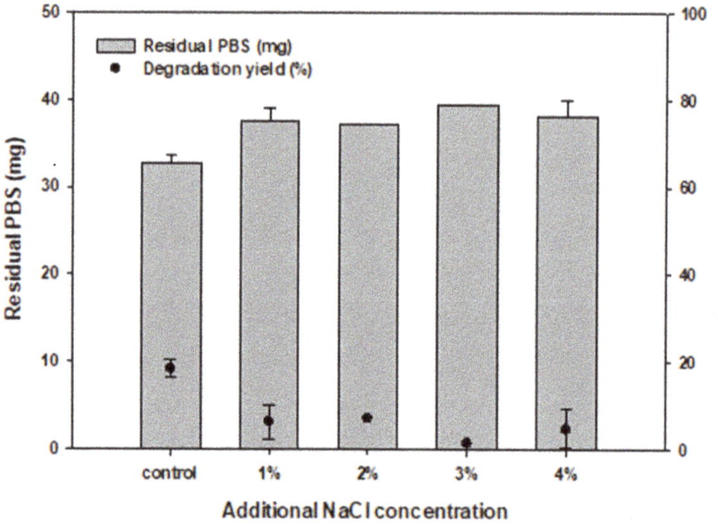

(d)

Figure 3. Optimizing the conditions for degrading PBS. (**a**) The changes in clear zone size at different temperatures (20 °C, 30 °C, 37 °C, 42 °C). (**b**) The results of GC-MS analysis to compare the amounts of residual PBS and degradation yield according to temperatures. (**c**) Five different carbon sources (glucose, fructose, xylose, galactose, and sucrose) were used to compare their effect on degradation at 30 °C using GC-MS. (**d**) Comparison of degradation yield according to salt concentration at 30 °C.

3.4. Time-Dependent Monitoring of PBS Degradation

We analyzed the degradation pattern in liquid culture at different time intervals. This experiment was conducted in MB liquid medium at 30 °C with 40 mg of PBS films. The results showed that the film was gradually degraded, and after 10 days, the residual amount of PBS was 27.5 mg and the degradation yield was 31.4% (Figure 4a). Compared with control films, the shape of the films changed, and they gradually fragmented (Figure 4b). However, after 10 days, PBS film degradation yield did not increase (data not shown). This might be due to the metabolites produced from degrading PBS (such as PBS monomers) that might inhibit the growth of *Terribacillus* sp. JY49, or the production of enzymes associated with degradation. In previous studies, several systems were screened for degrading PBS, and they showed high weight loss in liquid and soil conditions. However, it was confirmed that most of the degradation strains were fungi and a mixed consortium of bacteria. Therefore, *Terribacillus* sp. JY49 is an important PBS-degrading species.

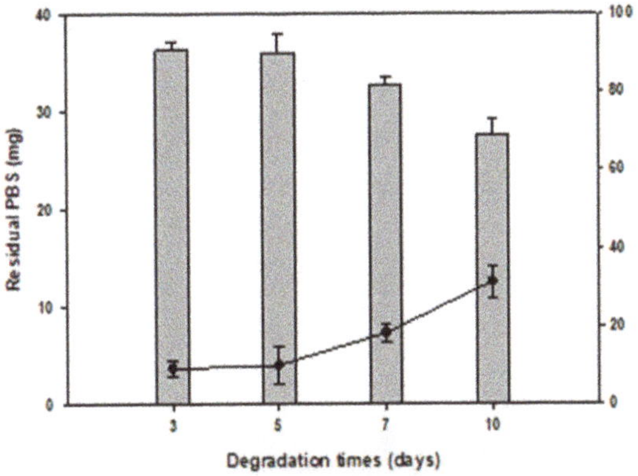

(a)

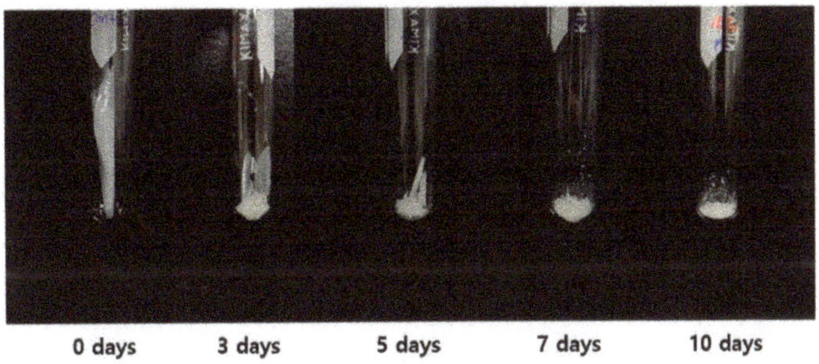

(b)

Figure 4. Confirming the degree of PBS degradation based on cultivation period (**a**) Residual amount of PBS and degradation yield was confirmed using GC-MS. (**b**) Recovered PBS films after lyophilization. The amount of PBS films decreased, and the films fragmented with time.

3.5. Changes in Physical Properties of PBS Films

As PBS degrades, the physical properties of films may be affected, such as the surface, molecular weight, and functional groups. Changes in the surface of PBS films were observed using SEM and FE-SEM with high magnification PBS film surface before degradation was smooth without any cracks and scratches (Figure 5a). At the beginning of the degradation, cracks began to appear, and as the degradation progressed, the number and depth of the cracks increased. After 10 days, PBS films with large cracks and rough surface were observed. FE-SEM showed similar results. Surface changes in the films demonstrate that *Terribacillus* sp. JY49 can degrade PBS (Figure 5b).

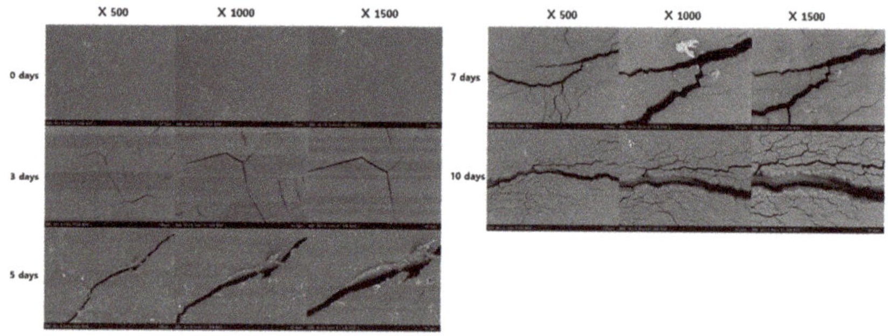

(**a**)

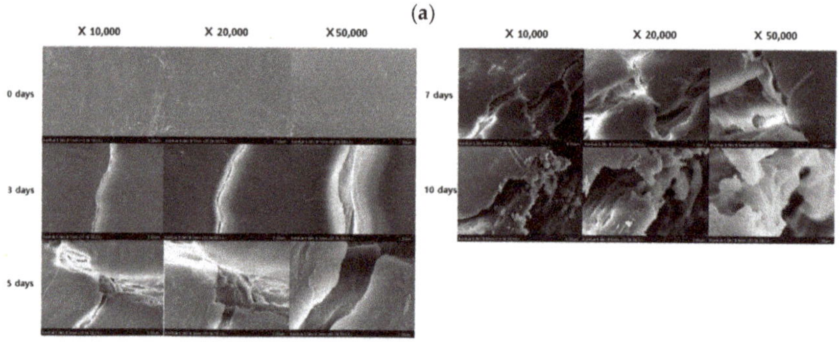

(**b**)

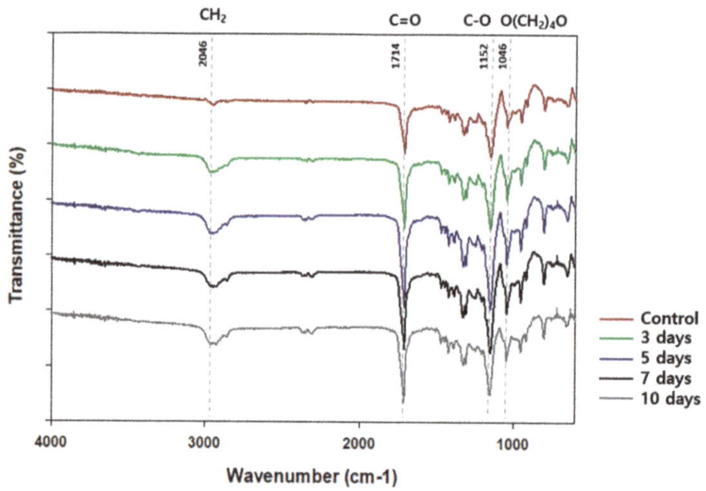

(**c**)

Figure 5. Changes in the surface and functional group of PBS films after degradation. (**a**) The surface of PBS was observed using SEM after 3, 5, 7, and 10 days of cultivation. (**b**) Cracks and rough surfaces formed after degradation were also observed using FE-SEM with high magnification. (**c**) Changes in functional groups were confirmed using FI-IR analysis. After degradation by *Terribacillus* sp. JY49, peak intensities changed with time.

Additionally, molecular weight was measured using gel permeation chromatography during the degradation (Table 3). Number average molecular weight (M_n), weight average molecular weight (M_w), and polydispersity index (PDI) of PBS films were determined. Before degradation, the values of M_w and M_n were 3.87×10^4 and 9.49×10^4, respectively. As the degradation progressed, the M_n and M_w values decreased because the molecular chain was degraded by *Terribacillus* sp. JY49. After 10 days, films that were recovered from culture medium had M_n of 2.88×10^4 and M_w of 6.82×10^4. However, PDI values did not show significant changes like in the case of PHA and PBAT degradation [32,40]. Usually, PBS is degraded because enzymes produced from microorganisms attack the ester bond in the main chain of PBS. Lipase and cutinase, which are esterases, can attack and degrade PBS. The esterase-based enzyme produced by *Terribacillus* sp. JY49 might cut the end of polymers. However, it did not cut the inside of the polymer and hence, did not affect the molecular weight distribution.

Table 3. Change in molecular weight of PBS films, analyzed using gel permeation chromatography.

Day	$M_n \times 10^4$	$M_w \times 10^4$	PDI
0	3.87	9.49	2.45
3	3.57	8.56	2.40
5	3.38	8.31	2.46
7	3.11	7.63	2.45
10	2.88	6.82	2.37

PBS films were analyzed after degradation using Fourier transform infrared spectroscopy to check the change in functional groups (Figure 5c). *Terribacillus* sp. JY49 was cultured with PBS films for 10 days, and the residual PBS films were recovered. After PBS degradation, the peak intensity at 2945 cm^{-1} had changed. This peak indicates a CH_2 stretching bond. The change of this peak intensity indicates a variation in alkane groups. A change of peak intensity at 1714 cm^{-1} and 1152 cm^{-1}, corresponding to a C=O stretching bond and C–O bond, could be confirmed. These bonds are a part of the ester bond. Because PBS has ester bonds, the changes in the two peaks indicate that PBS was degraded by this strain. In addition, the peak in the range of 1046 cm^{-1} indicates the $O(CH_2)_4O$ vibration. This functional group exists in PBS structures, therefore, degradation by *Terribacillus* sp. JY49 can change the chemical structure of PBS.

3.6. Effect of PBS Monomers such as Succinic Acid and 1,4-Butanediol on the Growth of JY49

PBS is composed of succinate and 1,4-butanediol monomers. Hence, when JY49 biodegrades PBS, these monomers are released and can affect the growth of this strain. Therefore, we confirmed monomer consumption and growth of JY49 through LC analysis and O.D. measurement, respectively, in a pH-controlled environment. Compared with control, the residual amount of succinic acid was decreased and completely consumed after 48 h (Figure S1a). 1,4-butanediol consumption was confirmed by its decreasing concentration with time. Increased growth was also observed after adding these two monomers (Figure S1b). This means that monomers of PBS were not significantly inhibitory factors in an environment with adjusted pH.

We also checked minimal inhibitory concentration (MIC) of the monomers. When the concentration of succinic acid reached ≥8 mM, the O.D value decreased sharply, indicating that high concentration of succinic acid inhibited the growth of JY49 (Figure S2a). However, the concentration of 1,4-butanediol was not significantly related to the growth of JY49 (Figure S2b). Therefore, as degradation continues, the released monomers may affect the growth of JY49, which is also expected to affect the degradation yield. This may be one of the reasons for the stalled degradation of PBS.

3.7. Degradability of Other Bioplastics by Terribacillus sp. JY49

To test the degradability of other plastics by *Terribacillus* sp. JY49, a clear zone test was performed. Experiments were performed using an MB agar medium containing different plastics at 30 °C. This strain showed clear zones on PBAT, PCL, P(3HB-*co*-3HV), and P(3HV-

co-4HB)-containing plates (Figure 6a), which indicates that *Terribacillus* sp. JY49 can degrade other bioplastics. Among them, the plate containing P(3HV-*co*-4HB) showed the largest and most transparent clear zone. However, there were no formations of clear zones on PLA, PHB, P(3HB-*co*-3HV-*co*-3HHx). The possibility of different plastic degradability of *Terribacillus* sp. JY49 was confirmed through clear zone testing.

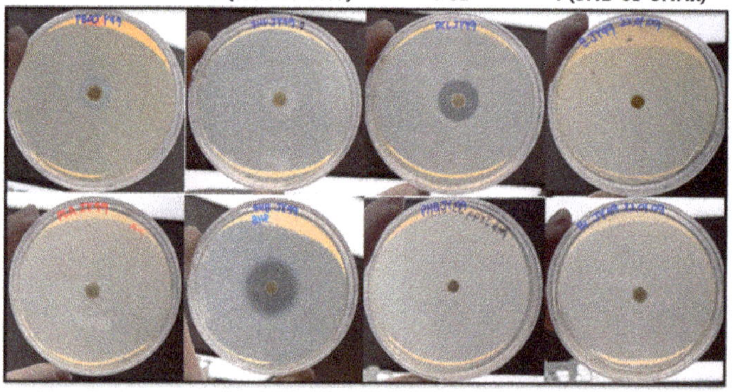

(**a**)

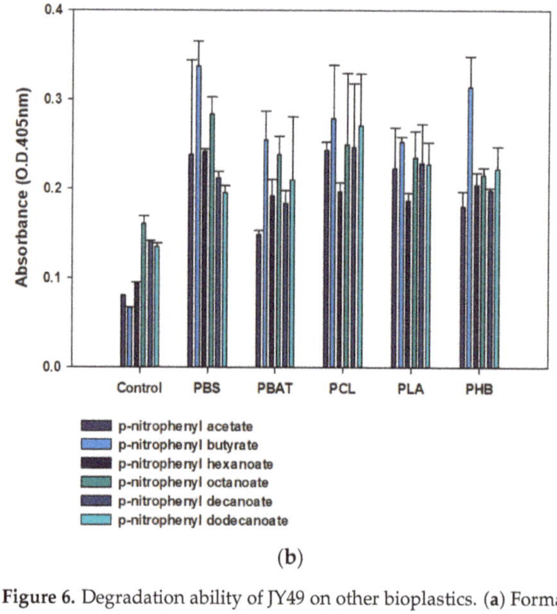

(**b**)

Figure 6. Degradation ability of JY49 on other bioplastics. (**a**) Formation of clear zone was tested on plates containing other plastics at 30 °C for 7 days. (**b**) Esterase assay was conducted with various p-nitrophenyl esters, and absorbance was measured at 405 nm.

To check the esterase activity that could degrade other bioplastics and monitor whether each enzyme produced with different bioplastics has a different activity towards substrates of different lengths, different kinds of substrates were selected (*p*-nitrophenyl acetate, *p*-nitrophenyl-butyrate, *p*-nitrophenyl hexanoate, *p*-nitrophenyl octanoate, *p*-nitrophenyl decanoate, and *p*-nitrophenyl dodecanoate) [41,47]. Enzyme reaction was performed on 96-well plates for 10 min, and the absorbance was measured at 405 nm. After reaction, the color changes to yellow in the wells due to the extracellular esterase enzyme activity from

Terribacillus sp. JY49. This strain showed little difference in esterase activity with respect to different bioplastics (Figure 6b), and any small change could be due to the many different enzymes such as protease and cutinase degrading the different bioplastics.

4. Conclusions

The use of bioplastics is drawing attention as a solution to alleviate environmental problems caused by plastic waste. Among them, PBS is a promising replacement for synthetic plastics derived from petroleum. Because the mechanism of PBS degradation is still unknown and the number of degrading strains is small, we aim to find PBS-degrading bacterial strains and confirmed the biodegradation activity. In this study, we screened *Terribacillus* sp. JY49, which has the highest similarity with *Terribacillus goriensis*, and which is the best PBS degrader. It has optimum degradation activity at 30 °C in solid and liquid cultures, and the degradation yield changes with carbon sources and salt concentration. We analyzed the roughness and erosions on the surface of the films through SEM, the reduction in molecular weight using gel permeation chromatography, and change in functional groups via FT-IR. This strain also has esterase activity—confirmed by measuring the absorbance—which can aid the degradation of PBS. Additionally, it showed degradability for other bioplastics such as PBAT, PCL P(3HB-*co*-3HV), and P(3HV-*co*-4HB). Therefore, *Terribacillus* sp. JY49 can be a useful and promising bacterial strain that can be applied for PBS degradation.

Supplementary Materials: The following supporting information can be downloaded at: https://www.mdpi.com/article/10.3390/polym14193978/s1. Table S1: PBS-degrading strains isolated from marine samples and identified via 16s rRNA sequencing, Table S2: Characterization of *Terribacillus* sp. JY49 using hydrolysis and antibiotic resistance test, Figure S1. Confirmation of monomer consumption and growth of *Terribacillus* sp. JY49, Figure S2: Minimal inhibitory concentration (MIC) of the monomers were checked.

Author Contributions: S.H.K., S.-H.P., K.P. and Y.-H.Y. conceived and designed the study; S.H.K., J.Y.C., D.H.C., H.J.J. and B.C.K. performed the experiments; S.H.K. drafted the manuscript; S.K.B. and Y.-H.Y. interpreted the experimental results; and S.H.K., J.Y.C., S.-H.P., K.P. and Y.-H.Y. revised the manuscript. All authors have read and agreed to the published version of the manuscript.

Funding: This study was supported by the National Research Foundation of Korea (NRF) (NRF-2022R1A2C2003138, NRF-2022M3I3A1082545) and the R&D Program of MOTIE/KEIT (grant number 20009508, 20018337 and 20018132).

Institutional Review Board Statement: Not applicable. No new data were created or analyzed in this study. Data sharing is not applicable to this article.

Informed Consent Statement: Not applicable.

Data Availability Statement: Not applicable. No new data were created or analyzed in this study. Data sharing is not applicable to this article.

Conflicts of Interest: The authors declare no conflict or interest.

References

1. Landrigan, P.J.; Stegeman, J.J.; Fleming, L.E.; Allemand, D.; Anderson, D.M.; Backer, L.C.; Brucker-Davis, F.; Chevalier, N.; Corra, L.; Czerucka, D.; et al. Human Health and Ocean Pollution. *Ann. Glob. Health.* **2020**, *86*, 151. [CrossRef] [PubMed]
2. Bahl, S.; Dolma, J.; Singh, J.J.; Sehgal, S. Biodegradation of Plastics: A State of the Art Review. *Mater. Today Proc.* **2021**, *39*, 31–34. [CrossRef]
3. Evode, N.; Qamar, S.A.; Bilal, M.; Barceló, D.; Iqbal, H.M.N. Plastic Waste and Its Management Strategies for Environmental Sustainability. *Case Stud. Chem. Environ. Eng.* **2021**, *4*, 100142. [CrossRef]
4. Chia, W.Y.; Tang, D.Y.Y.; Khoo, K.S.; Lup, A.N.K.; Chew, K.W. Nature's Fight against Plastic Pollution: Algae for Plastic Biodegradation and Bioplastics Production. *Environ. Sci. Ecotechnol.* **2020**, *4*, 100065. [CrossRef]
5. Sivan, A. New Perspectives in Plastic Biodegradation. *Curr. Opin. Biotechnol.* **2011**, *22*, 422–426. [CrossRef] [PubMed]
6. DiGregorio, B.E. Biobased Performance Bioplastic: Mirel. *Chem. Biol.* **2009**, *16*, 1–2. [CrossRef]

7. Nilani, P.; Raveesha, P.; Kasthuribai, B.R.N.N.; Duraisamy, B.; Dhamodaran, P.; Elango, K. Formulation and Evaluation of Polysaccharide Based Biopolymer—An Ecofriendly Alternative for Synthetic Polymer. *J. Pharm. Sci. Res.* **2010**, *2*, 178–184.
8. Muneer, F.; Rasul, I.; Azeem, F.; Siddique, M.H.; Zubair, M.; Nadeem, H. Microbial Polyhydroxyalkanoates (PHAs): Efficient Replacement of Synthetic Polymers. *J. Polym. Environ.* **2020**, *28*, 2301–2323. [CrossRef]
9. Shah, M.; Rajhans, S.; Pandya, H.A.; Mankad, A.U. Bioplastic for Future: A Review Then and Now. *World J. Adv. Res. Rev.* **2021**, *9*, 56–67. [CrossRef]
10. Shah, A.; Alshehrei, F. Biodegradation of Synthetic and Natural Plastic by Microorganisms. *J. Appl. Environ. Microbiol.* **2017**, *5*, 8–19.
11. Sharma, S.R. Bioremediation of Polythenes and Plastics: A Microbial Approach. In *Nanotechnology in the Life Sciences*; Springer: Cham, Switzerland, 2018; pp. 97–114.
12. Bano, N.; Younas, T.; Shoaib, F.; Rashid, D.; Jaffri, N. Plastic: Reduce, Recycle, and Environment. In *Environmentally-Benign Energy Solutions*; Springer: Cham, Switzerland, 2020; pp. 191–208.
13. Ben, M.; Mato, T.; Lopez, A.; Vila, M.; Kennes, C.; Veiga, M.C. Bioplastic Production Using Wood Mill Effluents as Feedstock. *Water Sci. Technol.* **2011**, *63*, 1196–1202. [CrossRef] [PubMed]
14. Chin, I.-J. Biodegradable Polymers and Plastics. *Chem. Int.-Newsmag. IUPAC* **2004**, *26*, 28–29.
15. Scaffaro, R.; Maio, A.; Sutera, F.; Gulino, E.F.; Morreale, M. Degradation and Recycling of Films Based on Biodegradable Polymers: A Short Review. *Polymers* **2019**, *11*, 651. [CrossRef]
16. Patel, M.; Bastioli, C.; Marini, L.; Würd, D.E. *Biopolymers Online—Environmental Assessment of Bio-Based Polymers and Natural Fibres*; Utrecht University: Utrecht, The Netherlands, 2005.
17. Dilshad, E.; Waheed, H.; Ali, U.; Amin, A.; Ahmed, I. General Structure and Classification of Bioplastics and Biodegradable Plastics. In *Bioplastics for Sustainable Development*; Kuddus, M., Roohi, Eds.; Springer Nature: Singapore, 2021; pp. 61–82.
18. Jiang, L.; Zhang, J. Biodegradable and Biobased Polymers. In *Applied Plastics Engineering Handbook: Processing, Materials, and Applications*, 2nd ed.; William Andrew Publishing: Norwich, NY, USA, 2017; pp. 127–143. [CrossRef]
19. Pinazo, J.M.; Domine, M.E.; Parvulescu, V.; Petru, F. Sustainability Metrics for Succinic Acid Production: A Comparison between Biomass-Based and Petrochemical Routes. *Catal. Today* **2015**, *239*, 17–24. [CrossRef]
20. Datta, J.; Kasprzyk, P.; Błażek, K.; Włoch, M. Synthesis, Structure and Properties of Poly(Ester-Urethane)s Obtained Using Bio-Based and Petrochemical 1,3-Propanediol and 1,4-Butanediol. *J. Therm. Anal. Calorim.* **2017**, *130*, 261–276. [CrossRef]
21. Kumari Pallathadka, P.; Koh, X.Q.; Khatta, A.; Luckachan, G.E.; Mittal, V. Characteristics of Biodegradable Poly(Butylene Succinate) Nanocomposites with Thermally Reduced Graphene Nanosheets. *Polym. Compos.* **2017**, *38*, E42–E48. [CrossRef]
22. Rafiqah, S.A.; Khalina, A.; Harmaen, A.S.; Tawakkal, I.A.; Zaman, K.; Asim, M.; Nurrazi, M.N.; Lee, C.H. A Review on Properties and Application of Bio-based Poly(Butylene Succinate). *Polymers* **2021**, *13*, 1436. [CrossRef] [PubMed]
23. Sheikholeslami, S.N.; Rafizadeh, M.; Taromi, F.A.; Shirali, H.; Jabbari, E. Material Properties of Degradable Poly(Butylene Succinate-Co-Fumarate) Copolymer Networks Synthesized by Polycondensation of Pre-Homopolyesters. *Polymer* **2016**, *98*, 70–79. [CrossRef]
24. Kong, X.; Qi, H.; Curtis, J.M. Synthesis and Characterization of High-Molecular Weight Aliphatic Polyesters from Monomers Derived from Renewable Resources. *J. Appl. Polym. Sci.* **2014**, *131*, 4401–4404. [CrossRef]
25. Adhikari, D.; Mukai, M.; Kubota, K.; Kai, T.; Kaneko, N.; Araki, K.S.; Kubo, M. Degradation of Bioplastics in Soil and Their Degradation Effects on Environmental Microorganisms. *J. Agric. Chem. Environ.* **2016**, *5*, 23. [CrossRef]
26. Mao, H.; Liu, H.; Gao, Z.; Su, T.; Wang, Z. Biodegradation of Poly(Butylene Succinate) by Fusarium Sp. FS1301 and Purification and Characterization of Poly(Butylene Succinate) Depolymerase. *Polym. Degrad. Stab.* **2015**, *114*, 1–7. [CrossRef]
27. Mei, X.; Tian, C.; Dong, Q.; Liang, Y. Influencing Factors and Process on In Situ Degradation of Poly(*Butylene Succinate*) Film by Strain *Bionectria ochroleuca* BFM-X1 in Soil. *J. Environ. Prot.* **2012**, *3*, 523–532. [CrossRef]
28. Ishii, N.; Inoue, Y.; Tagaya, T.; Mitomo, H.; Nagai, D.; Kasuya, K.I. Isolation and Characterization of Poly(Butylene Succinate)-Degrading Fungi. *Polym. Degrad. Stab.* **2008**, *93*, 883–888. [CrossRef]
29. Abe, M.; Kobayashi, K.; Honma, N.; Nakasaki, K. Microbial Degradation of Poly(Butylene Succinate) by Fusarium Solani in Soil Environments. *Polym. Degrad. Stab.* **2010**, *95*, 138–143. [CrossRef]
30. Zhao, J.H.; Wang, X.Q.; Zeng, J.; Yang, G.; Shi, F.H.; Yan, Q. Biodegradation of Poly(Butylene Succinate) in Compost. *J. Appl. Polym. Sci.* **2005**, *97*, 2273–2278. [CrossRef]
31. Uchida, H.; Nakajima-Kambe, T.; Shigeno-Akutsu, Y.; Nomura, N.; Tokiwa, Y.; Nakahara, T. Properties of a Bacterium Which Degrades Solid Poly(Tetramethylene Succinate)-Co-Adipate, a Biodegradable Plastic. *FEMS Microbial Lett.* **2000**, *189*, 25–29. [CrossRef] [PubMed]
32. Park, S.L.; Cho, J.Y.; Kim, S.H.; Bhatia, S.K.; Gurav, R.; Park, S.H.; Park, K.; Yang, Y.H. Isolation of Microbulbifer Sp. Sol66 with High Polyhydroxyalkanoate-Degrading Activity from the Marine Environment. *Polymers* **2021**, *13*, 4257. [CrossRef]
33. Park, S.L.; Cho, J.Y.; Choi, T.R.; Song, H.S.; Bhatia, S.K.; Gurav, R.; Park, S.H.; Park, K.; Joo, J.C.; Hwang, S.Y.; et al. Improvement of Polyhydroxybutyrate (PHB) Plate-Based Screening Method for PHB Degrading Bacteria Using Cell-Grown Amorphous PHB and Recovered by Sodium Dodecyl Sulfate (SDS). *Int. J. Biol. Macromol.* **2021**, *177*, 413–421. [CrossRef]
34. Park, Y.L.; Bhatia, S.K.; Gurav, R.; Choi, T.R.; Kim, H.J.; Song, H.S.; Park, J.Y.; Han, Y.H.; Lee, S.M.; Park, S.L.; et al. Fructose Based Hyper Production of Poly-3-Hydroxybutyrate from Halomonas Sp. YLGW01 and Impact of Carbon Sources on Bacteria Morphologies. *Int. J. Biol. Macromol.* **2020**, *154*, 929–936. [CrossRef]

35. Lee, S.M.; Lee, H.J.; Kim, S.H.; Suh, M.J.; Cho, J.Y.; Ham, S.; Jeon, J.M.; Yoon, J.J.; Bhatia, S.K.; Gurav, R.; et al. Screening of the Strictly Xylose-Utilizing Bacillus Sp. SM01 for Polyhydroxybutyrate and Its Co-Culture with Cupriavidus Necator NCIMB 11599 for Enhanced Production of PHB. *Int. J. Biol. Macromol.* **2021**, *181*, 410–417. [CrossRef]
36. Cho, J.Y.; Park, S.L.; Lee, H.J.; Kim, S.H.; Suh, M.J.; Ham, S.; Bhatia, S.K.; Gurav, R.; Park, S.H.; Park, K.; et al. Polyhydroxyalkanoates (PHAs) Degradation by the Newly Isolated Marine Bacillus Sp. JY14. *Chemosphere* **2021**, *283*, 131172. [CrossRef] [PubMed]
37. Bhatia, S.K.; Yoon, J.J.; Kim, H.J.; Hong, J.W.; Hong, Y.G.; Song, H.S.; Moon, Y.M.; Jeon, J.M.; Kim, Y.G.; Yang, Y.H. Engineering of Artificial Microbial Consortia of Ralstonia Eutropha and Bacillus Subtilis for Poly(3-Hydroxybutyrate-Co-3-Hydroxyvalerate) Copolymer Production from Sugarcane Sugar without Precursor Feeding. *Bioresour. Technol.* **2018**, *257*, 92–101. [CrossRef] [PubMed]
38. Jung, H.R.; Choi, T.R.; Han, Y.H.; Park, Y.L.; Park, J.Y.; Song, H.S.; Yang, S.Y.; Bhatia, S.K.; Gurav, R.; Park, H.A.; et al. Production of Blue-Colored Polyhydroxybutyrate (PHB) by One-Pot Production and Coextraction of Indigo and PHB from Recombinant Escherichia Coli. *Dye. Pigment.* **2020**, *173*, 107889. [CrossRef]
39. Choi, T.-R.; Park, Y.-L.; Song, H.-S.; Lee, S.M.; Park, S.L.; Lee, H.S.; Kim, H.-J.; Bhatia, S.K.; Gurav, R.; Choi, K.-Y.; et al. Fructose-Based Production of Short-Chain-Length and Medium-Chain-Length Polyhydroxyalkanoate Copolymer by Arctic Pseudomonas Sp. B14-6 Tae-Rim. *Polymers* **2021**, *13*, 1398. [CrossRef] [PubMed]
40. Cho, J.Y.; Park, S.L.; Kim, S.H.; Jung, H.J.; Cho, D.H.; Kim, B.C.; Bhatia, S.K.; Gurav, R.; Park, S.H.; Park, K.; et al. Novel Poly(Butylene Adipate-Co-Terephthalate)-Degrading Bacillus Sp. JY35 from Wastewater Sludge and Its Broad Degradation of Various Bioplastics. *Waste Manag.* **2022**, *144*, 1–10. [CrossRef]
41. Yun, H.J.; Lee, Y.J.; Yeo, S.H.; Choi, H.S.; Park, H.Y.; Park, H.D.; Baek, S.Y. The Isolation and Culture Characterization of a Lipolytic Enzyme Producing Strain from Meju. *Korean J. Microbiol. Biotechnol.* **2012**, *40*, 98–103. [CrossRef]
42. An, S.Y.; Asahara, M.; Goto, K.; Kasai, H.; Yokota, A. Terribacillus Saccharophilus Gen. Nov., Sp. Nov. and Terribacillus Halophilus Sp. Nov., Spore-Forming Bacteria Isolated from Field Soil in Japan. *Int. Syst. Evol. Microbiol.* **2007**, *57*, 51–55. [CrossRef]
43. Essghaier, B. Antimicrobial Behavior of Intracellular Proteins from Two Moderately Halophilic Bacteria: Strain J31 of Terribacillus Halophilus and Strain M3-23 of Virgibacillus Marismortui. *J. Plant. Pathol Microbiol* **2014**, *5*, 1. [CrossRef]
44. Liu, W.; Jiang, L.; Guo, C.; Yang, S.S. Terribacillus Aidingensis Sp. Nov., a Moderately Halophilic Bacterium. *Int. J. Syst. Evol. Microbiol.* **2010**, *60*, 2940–2945. [CrossRef]
45. Krishnamurthi, S.; Chakrabarti, T. Proposal for Transfer of Pelagibacillus Goriensis Kim et al. 2007 to the Genus Terribacillus as Terribacillus Goriensis Comb. Nov. *Int. J. Syst. Evol. Microbiol.* **2008**, *58*, 2287–2291. [CrossRef]
46. Kim, Y.G.; Hwang, C.Y.; Yoo, K.W.; Moon, H.T.; Yoon, J.H.; Cho, B.C. Pelagibacillus Goriensis Gen. Nov., Sp. Nov., a Moderately Halotolerant Bacterium Isolated from Coastal Water off the East Coast of Korea. *Int. J. Syst. Evol. Microbiol.* **2007**, *57*, 1554–1560. [CrossRef] [PubMed]
47. Nagaroor, V.; Gummadi, S.N. Biochemical Characterization of an Esterase from Clostridium Acetobutylicum with Novel GYSMG Pentapeptide Motif at the Catalytic Domain. *J. Ind. Microbiol. Biotechnol.* **2020**, *47*, 169–181. [CrossRef] [PubMed]

MDPI
St. Alban-Anlage 66
4052 Basel
Switzerland
www.mdpi.com

Polymers Editorial Office
E-mail: polymers@mdpi.com
www.mdpi.com/journal/polymers

www.ingramcontent.com/pod-product-compliance
Lightning Source LLC
Chambersburg PA
CBHW040837250926
51850CB00007BA/1338